Soil Carbon Management

Economic, Environmental and Societal Benefits

AF766663

Edited by

J. M. Kimble • C. W. Rice • D. Reed
S. Mooney • R. F. Follett • R. Lal

CRC Press
Taylor & Francis Group
Boca Raton London New York

CRC Press is an imprint of the
Taylor & Francis Group, an **informa** business

CRC Press
Taylor & Francis Group
6000 Broken Sound Parkway NW, Suite 300
Boca Raton, FL 33487-2742

First issued in paperback 2019

ISBN-13: 978-1-4200-4407-2 (hbk)
ISBN-13: 978-0-367-38910-9 (pbk)

Library of Congress Cataloging-in-Publication Data

Soil carbon management : economic, environmental and societal benefits /
editors, John M. Kimble ... [et al.].
p. cm.
Includes bibliographical references and index.
ISBN-13: 978-1-4200-4407-2 (alk. paper)
ISBN-10: 1-4200-4407-9 (alk. paper)
1. Soils--Carbon content. 2. Soil management--Economic aspects. 3. Soil management--Environmental aspects. 4. Soil management--Social aspects. I. Kimble, J. M. (John M.)

S592.6.C35S64 2007
631.4'1--dc22 2006102349

Visit the Taylor & Francis Web site at
http://www.taylorandfrancis.com

and the CRC Press Web site at
http://www.crcpress.com

Contents

SECTION III Off-Site Benefits

Foreword

Complex policy and development issues affect the agricultural sector and our natural resources base, and mounting pressures continue to weigh on these natural resources and their functions. Population and development growth, and increased demand for food, fiber, and, increasingly, biofuels all lead to increased pressure on our nation's soils. Soils are the living resource that support all these functions; they will be increasingly relied on into the future, both here and globally. Yet, amidst these growing pressures and increasing dependence, we are at risk of overlooking elaborately simple solutions to help sustain agriculture and natural resource functions. This is potentially the case with our soils.

As farmers, we understand that soils are our most basic and thus our most valuable resource. Yet, for much of the past, we have not treated our soils as the important commodity that they are. We have been tilling them for many years using nonsustainable practices that have resulted in the loss of over 50% of native soil organic carbon. This, in turn, has led to increased soil erosion and surface water contamination by sediments. We have also allowed our soils to lose much of their most precious function — their fertility. The green revolution of the 1950s and 1960s brought a salvo for the loss of soil fertility: Chemical fertilizers restored yield losses and reversed declining profits, but masked the degradation of soils and the loss of their important ecological functions. We were lulled by clean seedbeds, believing we were doing what was best for our soils and our lands. The green revolution also brought improved varieties of seeds, as well as new equipment and technologies that made farming larger parcels of land easier, thus leading to larger farms. Moreover, we saw yields increase even further.

Nevertheless, we are now learning that there were some unfortunate consequences to chemical fertilizers and intensive tillage technologies. Together, they have allowed us to keep yields up while masking the fact that our soils are losing or have lost their lifeblood — the organic carbon that supports the very structure of soils. This structural decay of soils has been accompanied by leaching and nutrient transport with eroded sediments. Leaching causes groundwater contamination with nitrates, and sediment transport contaminates surface waters with nitrogen and phosphorus. Without knowing it, we changed the very nature of the soils so that they no longer functioned as they should. The natural cycle of plant and litter decay that restored the soils with organic matter, nitrogen, and other nutrients necessary for soil productivity and fertility was replaced with nutrients from a bag. The levels of nitrogen, phosphorus, and potassium from fertilizers helped to maintain plant growth and yields, but at a cost to surrounding water bodies, as well as the loss of both soil and soil functions within agricultural and other ecosystems.

Accepted farming practices left a "clean field" without any residue on the surface; this was based on research from the period 1940–1960. Prepared seedbeds were required to obtain the proper soil–seed contact. The Midwest Corn Belt, an area of high soil fertility due to its historical native grasslands, had high levels of soil carbon, all of which was mined by agriculture for the carbon and nitrogen content, without any consideration for replenishment of these resources for the future.

We, as innovators, switched to no-till farming for many reasons, including reduced fuel costs, reduced wear and tear on our equipment, and an ability to farm more acres with less labor and in less time. We were often called "trash farmers," a reference to the residue left on our fields postharvest. Gone were the clean fields and tilled seedbeds. Our fields were considered eyesores to many, and were thought to give agriculture a bad name. Yet, we persisted, not understanding that what we were doing was managing soil carbon and replenishing our soils, in addition to all the other benefits we sought. What we observed over time was that "the soils were changing." They began to look more like the native soils in undisturbed areas. They were improving as the soil carbon increased, they had better tilth (i.e., structure), and the infiltration rate and water-holding capacities increased. The bottom line overall was that soil quality and sustainability were improved. We encountered less wind and water erosion, improved fertility, and increased yields. The soil, it appeared, was again functioning as a natural system. This is the "new green revolution" that we need: better management of soil carbon to better manage soils — our most precious resource.

This book examines soil carbon management in a different way than we have studied soils in the past, where yield improvements were the primary goal. This book discusses the on-site benefits to the soils, as well as the off-site benefits of good soil carbon management. Typically, yield improvements accompany improvements in soil organic matter, depending on soil type.

The bottom line is that we need to conserve and preserve our soils for future generations, and restore the ecological functions of soils. If we do this, we will have a multitude of benefits both on-site and off-site. We believe that the authors of this book compiled information that will help in the formulation of future agricultural policies that have long-lasting and positive environmental impacts, including improved farm income and conservation of our most vital soil resource. They demonstrate that soil carbon management has economic, societal, and environmental benefits. This book should help in the formulation of future agricultural policies that consider the soil as a precious resource needed to sustain food, fiber, and fuel needs. An underlying reason for the success of the United States as a sovereign state has been the continued availability of an abundant, safe food and fiber supply to underpin economic growth and development. Our soils literally fueled this growth and development. As we move into the future, we will need to be ever more vigilant in managing and renewing this precious resource, as well as replacing the nutrients and life-sustaining matter that we remove for our own needs.

James Moseley
Farmer and Former Deputy Secretary, U.S. Department of Agriculture

William Richards
Farmer and Past Chief of Soil Conservation Service, U.S. Department of Agriculture

About the Editors

John M. Kimble, Ph.D., is a retired research soil scientist for the U.S. Department of Agriculture (USDA) Natural Resources Conservation Service, National Soil Survey Center, in Lincoln, Nebraska, where he was employed for 24 years. He is now a consulting soil scientist working on issues related to global climate change. Previously, he was a field soil scientist in Wyoming for 3 years and an area soil scientist in California for 3 years. He has received the International Soil Science Award from the Soil Science Society of America (SSSA). While in Lincoln, he worked on a U.S. Agency for International Development Project for 11 years, helping developing countries with their soil resources, and he remains active in international activities as a member of the International Union of Soil Sciences. During the last 11 years in Lincoln, he focused more on global climatic change and the role soils can play in these changes. He has worked on issues related to soils and sampling soils, from the high Arctic to Antarctica and most areas in between. Dr. Kimble was an active member of the U.N. Intergovernmental Panel on Climate Change (IPCC) and other groups interested in soils and climate change. His scientific publications discuss topics related to soil classification, soil management, global climate change, and sustainable development. He has collaborated with Rattan Lal, Ph.D., Ronald F. Follett, Ph.D., and other authors to produce 12 books related to the role of soils and global climate change.

Charles (Chuck) W. Rice, Ph.D., is a professor of soil microbiology in the Department of Agronomy at Kansas State University. Dr. Rice teaches courses in soil microbiology and conducts research on microbial ecology and soil carbon and nitrogen transformations, including C and N emissions in agricultural and grassland ecosystems. He is currently advising or has advised 14 M.S. students, 16 Ph.D. students, and 9 postdoctoral scholars. He has authored or coauthored more than 100 publications. He has been elected fellow of the SSSA, America Society of Agronomy, and the American Association for the Advancement of the Sciences. Dr. Rice currently serves as director of the Consortium for Agricultural Soils Mitigation of Greenhouse Gases (CASMGG). This consortium is a 10-institution organization that conducts research on the potential of agricultural soil to sequester carbon dioxide while providing benefits to producers. In addition, Dr. Rice is currently serving on the IPCC to author a report on climate change.

Debbie Reed, M.Sc., is currently a consultant for national environmental and energy policy groups on agricultural mitigation strategies for state and federal global climate change policies. She has particular expertise in the area of agricultural mitigation opportunities as a cost-effective, near-term strategy to help reduce U.S. emissions of greenhouse gases and to provide additional income opportunities for farmers and ranchers.

In her previous position as the legislative director at the National Environmental Trust (NET) in Washington, D.C., Ms. Reed worked on a variety of environmental issues, including global climate change, domestic clean air policies, and energy policy. She participated in U.N. Climate Change negotiations in Buenos Aires, Argentina (2004), Milan, Italy (2003), New Delhi, India (2002), and Bonn, Germany (2001) while at NET.

Previously, Ms. Reed worked at the White House Council on Environmental Quality as the director of legislative affairs and of agricultural policy for the Climate Change Task Force. She participated in the U.N. global warming treaty negotiations as a U.S. Delegate at The Hague, the Netherlands (2000).

Before that, Ms. Reed was a senior legislative assistant for U.S. Senator J. Robert Kerrey (D-NE), where she handled environmental, natural resource/agriculture, and energy issues. While working for Senator Kerrey, she participated as a congressional delegate to U.N. climate change negotiations in Kyoto, Japan (1997), and Buenos Aires, Argentina (1998). In previous positions at the U.S. Department of Agriculture (USDA) and at several public health-oriented institutions, Ms. Reed's work focused on federal agricultural, food safety, and nutrition policy.

Ms. Reed has graduate degrees in human nutrition and communications, and undergraduate degrees in dietetics and chemistry.

Siân Mooney, Ph.D., is an associate professor of economics within the College of Business and Economics at Boise State University. She earned a B.Sc. degree (Hons) in agricultural economics from the University College of Wales, Aberystwyth, an M.Sc. in agricultural economics and farm management from the University of Manitoba, and a Ph.D. in agricultural and resource economics from Oregon State University. Before joining Boise State University, she was an assistant professor at the University of Wyoming and a postdoctoral fellow at Montana State University. Her work has focused on problems and opportunities that arise at the agricultural and environmental interface. Over the past 8 years, she has examined the design of economic incentives, policies, and contracts to support the emerging trade in greenhouse gas credits, as well as simulating the response of agricultural producers to alternative incentive designs and estimating the magnitude of transactions costs associated with trades in agricultural soil carbon credits. Dr. Mooney has also worked on a broad range of other topics, including drought management, environmental impacts of releasing new technologies, incentives for endangered species protection, mitigation of climate change, and, most recently, erosion mitigation and water quality protection in sub-Saharan Africa. Her work is international, national, multiregional, and multidisciplinary. Dr. Mooney is a board member of the American Agricultural Economics Association, Committee for Women in Agricultural Economics, and holds a governor's appointment to the Wyoming Governor's Carbon Sequestration Advisory Committee. She has published many journal articles and reports related to the economics of carbon sequestration, climate change, and environmental economics.

Ronald F. Follett, Ph.D., is supervisory soil scientist and research leader with the Agricultural Research Service (ARS) of the USDA, with 40 years of research experience. For the past 20 years, he has been with the ARS Soil-Plant-Nutrient Research Unit in Fort Collins, Colorado. Dr. Follett previously served 10 years as national program leader with ARS headquarters in Beltsville, Maryland. He has also been a research soil scientist with ARS in Mandan, North Dakota, and Ithaca, New York. Dr. Follett is a Fellow of the SSSA, the American Society of Agronomy (ASA), and the Soil and Water Conservation Society (SWCS). Among his numerous awards, he received the USDA's highest award, the USDA Distinguished Service Award, in 1984 and 1992, and in June of 2000, he received the USDA Superior Service Award — "For promoting sensible management of natural, soil, and water resources for an environmentally friendly and sustainable agriculture." In 2004, Dr. Follett received the U.S. Presidential Rank Award for Meritorious Service, and in 2005 he was the recipient of the ARS Senior Scientist of the Year Award in the Northern Plains Area. Dr. Follett is adjunct professor with Colorado State University and Virginia Tech. He has also organized and written the ARS Strategic Plans for both "Groundwater Quality Protection — Nitrates" and for "Global Climate Change — Biogeochemical Dynamics." Dr. Follett has edited 12 books, including a recent book titled *The Potential of U.S. Grazing Lands to Sequester Carbon and Mitigate the Greenhouse Effect*, coauthored the book titled *The Potential of U.S. Cropland to Sequester Carbon and Mitigate the Greenhouse Effect*, and served as guest editor for the journals *Contaminant Hydrology* and *Soil and Tillage Research*. His numerous scientific publications include topics about nutrient management for forage production, soil-N and C-cycling, groundwater quality protection, global climate change, agroeco-sytems, soil and crop management systems, soil erosion and crop productivity, plant mineral nutrition, animal nutrition, irrigation, and drainage.

Rattan Lal, Ph.D., is a professor of soil physics in the School of Environment and Natural Resources, and director of the Carbon Management and Sequestration Center at The Ohio State University (OSU). He earned a B.Sc. degree from Punjab Agricultural University, an M.Sc. from Indian Agricultural Research Institute, and a Ph.D. from OSU. Before joining the International Institute of Tropical Agriculture as a soil physicist in 1969, Professor Lal worked with the University of Sydney, Australia, as a senior research fellow. From 1969 to 1987 at the IITA, Ibadan, Nigeria, Professor Lal conducted long-term experiments on watershed management, water budget in relation to land use and land use change, erosion control, water conservation in the root zone, no-till farming, and agroforestry. Since joining OSU in 1987, he has worked on soils and climate change, drainage of agricultural lands, soil degradation, and global food security. Professor Lal is a fellow of the ASA, the SSSA, the Third World Academy of Sciences, the American Association for the Advancement of Sciences, the SWCS, and the Indian Academy of Agricultural Sciences. He is the recipient of the International Soil Science Award, the Soil Science Applied Research Award, and the Soil Science Research Award of the SSSA; the International Agronomy Award, the Environment Quality Research Award, and the Carl Sprengel Agronomic Research Award of the ASA; the Hugh Hammond Bennett Award of the SWCS; the 2005 Borlaug Award; and the 2006 Liebig Award of the International Union of Soil Science. Professor Lal is the recipient of an honorary Doctor of Science degree from Punjab Agricultural University, Ludhiana, India, and from the Norwegian University of Life Sciences, Aas, Norway. He is past president of the World Association of the Soil and Water Conservation and the International Soil Tillage Research Organization, and president of SSSA (2006–2007). He was a member of the U.S. National Committee on Soil Science of the National Academy of Sciences (1998–2002) and lead author of the IPCC (1998–2000). He has served on the Panel on Sustainable Agriculture and the Environment in the Humid Tropics of the National Academy of Sciences.

Contributors

Steven I. Apfelbaum
Applied Ecological Services, Inc.
Brodhead, Wisconsin

Dan Devlin
Department of Agronomy
Kansas State University
Manhattan, Kansas

David Douds, Jr.
U.S. Department of Agriculture–ARS
ERRC
Wyndmoor, Pennsylvania

N. H. Euliss, Jr.
U.S. Geological Survey
Northern Prairie Wildlife Research
Center
Jamestown, North Dakota

Karina Fabrizzi
Department of Agronomy
Kansas State University
Manhattan, Kansas

Ronald F. Follett
U.S. Department of Agriculture–ARS
Soil-Plant-Nutrient Research Unit
Fort Collins, Colorado

Robert A. Gleason
U.S. Geological Survey
Northern Prairie Wildlife Research
Center
Jamestown, North Dakota

James Hanson
University of Maryland
College Park, Maryland

Paul Hepperly
The Rodale Institute
Kutztown, Pennsylvania

John M. Kimble
Retired Research Soil Scientist
Addison, New York

Rattan Lal
Carbon Management and Sequestration
Center, SENR/OARDC
The Ohio State University
Columbus, Ohio

M. K. Laubhan
U.S. Geological Survey
Northern Prairie Wildlife Research
Center
Jamestown, North Dakota

Andrew P. Manale
U.S. Environmental Protection Agency
Washington, D.C.

Siân Mooney
Department of Economics, College of
Business and Economics
Boise State University
Boise, Idaho

Jeff Neel
Department of Agronomy
Kansas State University
Manhattan, Kansas

Gary Pierzynski
Department of Agronomy
Kansas State University
Manhattan, Kansas

David Pimentel
Cornell University
Ithaca, New York

Debbie Reed
DRD Associates
Arlington, Virginia

Charles W. Rice
Department of Agronomy
Kansas State University
Manhattan, Kansas

Rita Seidel
The Rodale Institute
Kutztown, Pennsylvania

Paul White
Department of Agronomy
Kansas State University
Manhattan, Kansas

Jeffery Williams
Department of Agricultural Economics
Kansas State University
Manhattan, Kansas

Contents

SECTION III Off-Site Benefits

Section I

Overview, Policy, and Economics

1 Soil Carbon Management: Economic, Environmental, and Societal Benefits

John Kimble, Charles W. Rice, Debbie Reed, Siân Mooney, Ronald F. Follett, and Rattan Lal

CONTENTS

INTRODUCTION

The years 2007–2009 have been designated as the International Year of Planet Earth, with the United Nations declaring that 2008 will be the Year of Planet Earth. For this important endeavor, a brochure was developed by the International Union of Soil Scientists called "Soil — Earth's Living Skin" (Dent, Hartemink, and Kimble, 2005). This brochure and other activities can be found at www.yearofplanetearth.org. The following excerpt is from the brochure:

Earth's Living Skin

Soils are truly wonderful. They are major support systems of human life and welfare. They provide anchorage for roots, hold the water long enough for plants to make use of it, and hold nutrients that sustain life — otherwise the Earth's landscape would be as barren as Mars. Soils are home to myriad microorganisms that accomplish a suite of biochemical transformations — from fixing atmospheric nitrogen to decomposition of organic matter — and to armies of microscopic animals as well as the familiar earthworms, ants and termites. In fact most of the land's biodiversity lives in the soil, not above ground.

We build on soil, as well as in it and with it. And it's not the same out there! The abundance of life, habitats, and opportunities for human occupation mirror the tremendous variety of soils that are the Earth's living skin.

Source: Dent, Hartemink, and Kimble, 2005

Soil is not dirt, and we need to treat it like a valuable resource, which it is. Soil organic carbon (C) is the glue that holds soil in place and helps to prevent soil wind and water erosion, as well as preserve or enhance overall soil quality. Soil C is a very important part of this living skin, and we need to manage it so that it functions to benefit all of society. A good, simple overview of where soils fit into the environment can be found in the *American Geologicl Institute Environmental Awareness Series 9 Soils, Society and the Environment"* (Loynaham et. al., 2005).

This book highlights the importance of soil C management and the significant economic, environmental, and societal benefits that result from land management practices that maintain or increase soil C. Previous books and articles have focused on soil C, but have not comprehensively addressed the societal, economic, and environmental benefits that arise from soil C management. The book is written for a broad audience, including scientists, nonscientists, and policy makers, and brings together the best available current information concerning the multiple benefits of considering soil C management and its importance for developing future land management and global warming mitigation policies. Recently, soil C management has generated considerable interest because increasing soil C via sequestration (i.e., locking or storing C in the soil) has been suggested as a way to reduce greenhouse gases and help offset global warming. Much of the interest in soil C sequestration has been focused on the potential for agricultural soils to generate soil C credits that could be traded in the broader greenhouse gas credit market. This opportunity could make practices that increase soil C more economically attractive to producers and landowners and yield other benefits that are tied to good C management; these will be identified and discussed in this book. These benefits are both economic and environmental, and have highly positive as well as potentially broad societal impacts.

Federal policies are often developed to achieve a single management parameter or a specific environmental outcome. However, the policy prescriptions can often lead to other secondary co-benefits that were not the original focus. For example, inputs to the past farm bills have suggested activities and incentives that would create the opportunity for land managers (e.g., farmers, foresters, ranchers, etc.) to participate in the developing C market and benefit from another income source. Nevertheless, the multiple agricultural (i.e., on-farm) and societal (i.e., off-farm) benefits of managing soil C are not reflected in policy development, which tends to discount the value of these activities and their broad-ranging positive impacts.

We are interested in having policy makers, land managers, and the public understand the multiple benefits attributed to soil C management. These include reduced soil erosion, improved air and water quality, sequestration of C to reduce atmospheric CO_2, increased agriculture sustainability, improved wildlife habitat, increased farm productivity, as well as potential increases in the profitability of production agriculture. However, the economic or societal values of many co-benefits are, for the most part, undocumented.

All too often, policy development lags behind scientific findings. This is partly attributable to the fact that scientists and policy makers often require an intermediary to translate science into the building blocks of policy-based applications. Scientists publish their work in limited distribution scientific journals and texts intended for a distinct audience that tends not to include policy makers or practitioners.

This book is intended to translate available science on the topic for federal-level policy makers and others that influence them, including the targets of federal policies such as farmers and land managers. This book is not written in the same way as a normal scientific paper. Instead, it poses a problem or set of problems of interest to the audience, and then describes how soil C management can help to solve these problems, listing also the multiple benefits that will arise from good soil C management. A major focus will be an attempt to describe or evaluate the co-benefits of soil C sequestration, specifically for policy development purposes. A long-standing doctrine of policy development is the use of cost-benefit analyses. In the case of environmental outcomes, it is frequently difficult to place a value on community or societal benefits because many of these benefits are nonmarket (i.e., they are not sold in the marketplace and do not have an explicit price). This is particularly true for many of the benefits that are discussed in this book. The myriad co-benefits are documented, and the values of societal and economic impacts are described in a manner that is useful to policy development. As with any scientific article, the writing is supported by references.

We present a new conceptual framework upon which to develop policies for managing and enhancing soil C and new approaches to achieving environmental outcomes. Many papers and books on soil C management have addressed specific ecosystems, such as agricultural lands, rangelands, and forestlands. Instead, each chapter in this book begins by addressing a particular concern and potential options to manage it, along with the real and perceived benefits of these options. Wherever possible, the information presented uses an economic metric to determine the broader value of activities that increase soil C sequestration to the land manager, society, and the environment. Scientists may say a practice will reduce soil erosion, for instance, without then stating or documenting the true value of this outcome or its connection with soil C management. This book will address that critical link.

The bottom line is a work that is scientifically sound but one that will be read by both scientists and nonscientists. It will help to show the public the benefits they reap from farm programs that address specific problems such as soil erosion and land degradation. It offers an array of policies that can be implemented at various political scales and tailored to meet local and regional needs and constraints, and shows how they can encourage the management of soil C. Finally, it will show that if we manage soil C, we can achieve improved environment quality concurrently with more sustainable and productive farms and land use practices.

The book is divided into three sections. The first section provides a general overview of farmer perspectives, some policy options and recommendations, and economics. The second section examines on-site benefits from soil C management. Section 3 considers off-site benefits from additional soil C sequestration. Although on-site and off-site benefits are presented in separate sections, they are closely related to each other; it is very difficult to completely separate on-site and off-site benefits.

SUMMARY OF CHAPTERS

Section 1: Overview, Policy, and Economics

This section provides an overview of the purpose of the book and an overview of policy options, the views of farmers, and the economics of soil C and the environment.

Chapter 1 provides an overview of the purpose of the book, including policy options, the views of farmers, and the on-site and off-site economic benefits of soil C sequestration.

Chapter 2 develops and discusses possible future policy options. The elucidation of policy options is given up front because the overarching goal of this book is to present information that will help in the development of polices to treat soil and soil C as the valuable resources that they are. Many of the subsequent chapters also contain policy recommendations, but Chapter 2 connects the ideas and elements from the chapters and presents them in a clear and concise format that gives the "takeaway message" to the readers of this book: Societal goals for environmental protection cannot be achieved without the maintenance or enhancement of soil C. We want to give scientifically supported information as a foundation for future policies that will be developed and will be tied to the management of soil organic C. We need policies to encourage the maintenance or increases in soil C stocks for a variety of societal benefits, including improved soil quality, reductions in both wind and water erosion, and the mitigation of economic impacts of natural events, such as hurricanes and droughts. To these ends, the existing policy structure does not suffice. Interventions have been poorly targeted, funded, and implemented at the necessary levels. However, although some states are developing rules and regulations for the protection of soil resources, little discussion has taken place at the national level — beyond the programs established for agriculture — to maintain agriculture productivity. To date, however, such discussions have not resulted in broad-based effective policies. All too often, farm policies do not consider the soil but instead are tied to the production of agriculture commodities without considering how their production may affect the soil. Subsequent chapters outline the benefits of C management. The policy options presented in this book put the management of soil C in the forefront so that the many benefits detailed in the rest of the book become reality. As a nation, we need policies that put the soil first because the pedosphere is the lifeblood of farming, and it is the link among all the spheres (i.e., pedosphere, hydrosphere, biosphere, atmosphere, lithosphere, and geosphere).

Chapter 3 presents the views and perspectives of farmers from different eco-regions. These farmers are innovators and leaders in no-till farming, which is one of the best ways to manage soil C to obtain the multiple benefits detailed in the other chapters of this book. This chapter is based on the experiences of pioneering farmers, who have taken risks and changed their farming practices — not necessarily to undertake soil C management but for other reasons. What they observed when they switched to no-till farming were reductions in the fuel they use, reduced water and wind erosion, and, in many cases, increased yields. When Drs. Follett and Kimble were working on the benefits of the Conservation Reserve Program in the mid-1990s, a farmer in Texas said he was very happy to see the increase in wildlife as a result of his enrollment as well as the increased C in his soil. Some unanticipated impacts also occur. No-till farming has presented some new problems, such as slugs, in many no-till soils in Ohio. This is bound to happen because as systems change, new problems will develop. These problems will need to be addressed as they occur. Overall, however, no-till farming has been very effective in reducing soil erosion, improving wildlife habitat and agricultural sustainability.

Chapter 4 deals with the economics of C management. It points out that there may be a substantial monetary benefit from increased C sequestration through proper soil management. These benefits result from reductions in sedimentation and nutrient runoff, but activities that increase C sequestration can also add value to recreation and wildlife. Significant problems occur when trying to estimate the value of increased C sequestration. It is also not clear how much additional C can be sequestered in soils, and this will require additional study. The link between physical changes in the environment and economic values is still very weak and is an area in which considerable additional work is needed. One caution in interpreting these values is that they could include value generated by other factors in addition to soil C sequestration. Thus, the benefit of soil C sequestration may account for some portion of the total values reported. However, it is clear that despite these caveats, soil C management yields positive economic benefits to society.

Section 2: On-Site Benefits

On-site benefits of soil C management are discussed in this section. Benefits include increased yields and better utilization of plant biomass. On-site benefits are also directly related to off-site benefits.

Chapter 5 discusses with the economic and societal benefits of soil C management on cropland and grazing land systems. This chapter gives much of the scientific background for soil C sequestration and how it improves soil productivity and the environment. It emphasizes how practices and policies that encourage maintaining and improving soil C sequestration can consistently be associated with improved soil and water quality, reductions in silt loads and sedimentation into streams, lakes, and rivers, as well as improvements in air quality. The Conservation Reserve Program (CRP) and other incentive programs generally improve the aesthetic and economic value of land, while also enhancing the biodiversity of either the immediate land parcel or surrounding areas.

Many proposals have been made to use crop residues for biofuels and to find alternative uses beyond the needs of the soils for the actual biomass production. Policies will be needed to balance these different — and sometimes competing — uses so that agriculture can help to reduce the need for nonrenewable resources and, at the same time, maintain soil quality and health as well as productivity. The need exists for policy intervention, including economic incentives, to sequester more soil C, encourage more efficient use of the nutrients in residues and manures, and to capture those improvements that are consistent with economies of scale and increased efficiencies of production on large areas of U.S. agricultural land. Tillage practices that require less energy inputs and seeding directly into largely undisturbed crop residues is consistent with enhancing soil C sequestration and with reducing fuel costs, as well as recycling nutrients contained within the residues back into the soil for the next crop. These practices also have the potential to conserve soil moisture.

Chapter 6 discusses how organic farming can enhance soil C and the benefits derived from soil C management. Organic farming has been used for over 6000 years to make agriculture sustainable. The practices used focus on enhancing soil C. The benefits of organic farming include increased soil C and nitrogen (N), more drought tolerance in wet years, and reduced leached N from soils because the increased soil C helps prevent the breakdown and loss of soil N. The use of crop rotations helps reduce soil erosion and pest problems. Studies have demonstrated that organic agricultural systems require less "energy" when compared with conventional farming systems. It does need to be emphasized that organic farming practices are typically more labor intensive than conventional farming systems, but because of higher market prices for organic agriculture products, the net return may not differ much from conventional farming systems. Finally, the use of crop rotations, cover cropping, and manure/compost utilization in organic agriculture systems are practices that also have applications to conventional farming systems. They were used in the past and are used in reduced till systems today. Presently, most large-scale operations have moved away from these practices. Perhaps it is time to return to these practices. Several recommendations made for organic farming systems need to be considered in future policy development.

Chapter 7 outlines the benefits of soil organic C to the physical, chemical, and biological properties of soil. Most soil functions are driven by soil organic C, which in turn affect plant production and other ecosystem services. The chapter shows that nutrient supply from soils can be converted to a monetary value as can the integrated effect on crop yield. In many cases, the better the C management increases, the better the net returns. At this time, however, the value of many other ecosystem services have not been adequately quantified, and this need to be fully developed.

Section 3: Off-Site Benefits

Chapter 8 discusses the relationship of soil physical properties and soil erosion. A well-developed soil structure is a major determinant of reduced or avoided soil erosion. Important parameters here are structural stability and infiltration rate. Some scientists consider that soil erosion provides sinks for buried C, while others feel that erosion may increase decomposition of soil organic matter. This chapter presents the strong and weak points of each hypothesis. An objective resolution of these opposing theories requires research data from long-term field experiments conducted — and performed at different scales — to assess the fate of erosion-displaced soil organic matter at all four stages of erosion. In the meantime, a strong rationale exists for adopting effective erosion control measures because of numerous, adverse on-site and off-site effects of accelerated erosion.

Chapter 9 on wetlands and global C deserves special consideration because the United States has lost 95% of its original wetlands and because of the past ecological benefits derived from these wetlands. Converted wetlands that are cropped annually are depleted increasingly more soil C through land tillage, which is a primary source of greenhouse gas emissions. Existing wetlands that occur within the agricultural landscapes are receiving nutrient enrichment from fertilizers from other drained and farmed wetlands. They are also contributors of CO_2 and potential trace gases such as methane. To reduce these emissions, the nation needs to recognize the C-storage and greenhouse gas (GHG)-storage potential of historic wetlands and to develop the policies needed to encourage their restoration. The restoration will also help to buffer against nutrient and fertilizer influxes in existing wetlands where trace gas emissions are occurring. An option is to consider converting current farm program subsidies into "carbon farming" and "water quality and quantity farming" enterprises. This strategy seems more appropriate than continuing to focus on crop failures and subsidies that encourage the farming of soils that are delivering crops and commodities at great "downside" costs to the global atmosphere and environment. The takeaway message is that wetlands are very important to C storage, but we are not adequately managing wetlands today.

Chapter 10 examines the wildlife benefits of soil C. Because farmers, ranchers, and private landowners manage two-thirds of the nation's land, their actions have a major impact on wildlife. Conversely, actions to maximize C storage, which will have a major impact on wildlife, must involve private landowners. Proper C management equates with improved wildlife habitat. Existing Farm Bill programs have been instrumental in providing these simultaneous benefits that help to conserve our nation's natural resources. Given the current political climate, future Farm Bill programs will likely place even more emphasis on C sequestration benefits. However, while it is possible to develop programs that maximize incentives to store C on agricultural lands, the feasibility of such programs is limited by ecological constraints and would compromise the value of other ecosystem services. Given the diverse environmental concerns of modern society, future Farm Bill programs that provide for a full range of ecosystem services likely will be the best alternative.

Chapter 11 addresses soils and water storage by suggesting ways to manage soil C to mitigate runoff and flooding. Floods cause billions of dollars in damages in the United States every year. Between 1965 and 2003, over two-fifths of all Presidential disaster declarations involved floods, and much of this flooding was a result of poor C management. In 2005, much of New Orleans was flooded, and most people blame the flooding and its impacts on the failure of the levees in that city. Yet, if the soils, especially the wetlands, in this region had been managed better, soil water storage would have been increased, and some of the effects would have been better mitigated. After the 1993 floods of the Missouri and Mississippi rivers, towns in the flood plains were moved to higher ground, drained wetlands were restored, and the result was that nutrients and pollution runoff and downstream flooding were greatly reduced.

All too often, science takes a backseat to politics, and the root cause of problems, such as the flooding in New Orleans, is never addressed. We can more effectively manage soils for water storage if we have the political will to do so. Scientists and authorities concerned with flooding and its associated social, environmental, and economic costs have long known that managing landscapes, particularly agricultural landscapes, and implementing certain types of agricultural practices that retard water flow or retain water on the land contribute to effective flood management. Yet, we are still not doing it. Because soil C has a major influence on water storage, any policy that benefits the increase of soil C will have a beneficial effect on soil water storage.

Chapter 12 addresses why we need to consider the environmental and ecological benefits of soil C management as they are related to surface water quality. Efforts to increase soil C concentrations represent a "no-regrets" strategy for mitigating the potential effects of increased atmospheric CO_2 concentrations. In addition to helping reduce the rate of increase in atmospheric CO_2 concentration, increasing soil C concentrations can act to significantly improve surface water quality, primarily by influencing surface runoff characteristics. The most common contaminants found in surface waters in the United States are sediments, nutrients, bacteria, and herbicides, and most of these originate from agricultural lands. Improved soil C management can help reduce the impact of land management practices on these contaminants. Improved soil C management can reduce runoff volume, and it can reduce the concentrations of sediments, nutrients, and herbicides such that the total amounts lost from the landscape are decreased. Most of the research that has been conducted on runoff utilizes an "edge-of-field" approach where the relative differences in runoff volume and composition induced by various practices are measured. We need to go beyond this edge-of-field effect and look at watersheds in their entirety. For a given land use practice, the effects of best management practices are not additive. Watershed modeling can overcome some of these difficulties by attempting to simultaneously consider all land use and best management practices on surface water quality within a watershed. The models can be calibrated against stream monitoring data and then used to predict the impact of best management practices on water quality over a large area. Overall, results from watershed models are still viewed with some skepticism, but they have clearly progressed to the where they can appropriately be used for planning purposes and to reduce the need for direct monitoring of water quality.

Chapter 13 discusses "urban lands" and how the management of soil C on urban lands has not been a major focus for national policy to date. We see increasingly more flooding in urban areas as houses, roads, malls, factories, and parking lots cover more land area. To reference Joni Mitchell's "Big Yellow Taxi" song, released in 1970, we "paved paradise and put up a parking lot." Development has greatly increased urban runoff and contributed to the loss of soil C storage — and water storage in the soil — and the ability of the soil to reduce runoff. Today, major C emissions are attributed to urban lands. Some headway is currently being made in individually addressing some of the key C emission sources. For example, improvements of storm water best management practices are becoming more common, and development strategies are tending toward the consolidation of developed areas, which is freeing up and improving open space lands where soil C can be better managed. In some areas, a marked shift has occurred toward green buildings, clean streams and lakes programs, and ecological restoration projects in urban areas that engage the public in treating urban open lands as treasures that need management and stewardship, not unlike gardens. Despite these wonderful strides, the need is significant for a coherent, consistent, and integrated series of solutions across the landscape to help stem the C emission challenges from urban areas.

Chapter 14 on prairies, savannas, and forests and global C management discusses the challenges that we face in these ecosystems because they have been highly modified and their functions diminished. The lands formerly occupied by enormous areas of these ecosystems for over 100 years have been converted to the economic engine of U.S. agro-industrial operations. They have the potential to be major storage areas for soil C. The development of decisive policy changes and economic incentives, along with integrating ecosystem restoration strategies into agricultural operations and in marginal lands, can be used to again rebuild these soils, reestablish carbon stocks, and create environmental service benefits such as reduced flooding, improved ground water supplies through restoring recharge, and a range of biodiversity and water quality benefits. Because of the extensiveness of these ecosystems, significant soil C management benefits can be derived from policy and restoration programs that can operate at large scales.

REFERENCES

Dent, D., Hartemink, A., and Kimble, J., *Soil — Earth's Living Skin*, Earth Sciences for Society Foundation, Leiden, The Netherlands, 2005.

Loynaham, T.E., Brown, K.W., Cooper, T.H., Kimble, J.M., Milford, M., and Smith, D.B., *AGI Environmental Awareness Series, 9, Soils, Society and the Environment*, American Geological Institute, Alexandria, VA, 2005.

2 Economic and Societal Benefits of Soil Carbon Management: Policy Implications and Recommendations

Debbie Reed

CONTENTS

INTRODUCTION

Soil carbon (C) sequestration represents an existing, low-cost, high-benefit approach to begin dealing with the looming problem of global climate change; it also has multiple ancillary environmental and societal benefits. As one part of a global warming policy response, in particular, it represents a bridge to the future, and can be implemented now while we wait for the cost of some solutions to be reduced and others that may take more time to develop or implement.

WHY SINGLE OUT GLOBAL CLIMATE CHANGE?

The phenomenon of global climate change has become front-page news in this country. Some scientists have been ringing the global warming alarm for decades, but in the past 2–3 years, increasing scientific evidence of the phenomenon of global climate change, as well as evidence of global impacts occurring at rates faster than scientists predicted, have brought the issue to the fore with a far broader audience than climate scientists. Since the beginning of 2006, it appears that new, more alarming studies on the impacts of global climate change are published on a weekly basis. Although some media outlets have been rather alarmist (e.g., *Time Magazine* cover story, March 28, 2006: "Be Worried, Be Very Worried"), many scientists believe the attention is warranted and long overdue. Public concern and engagement may provide the stimulus for federal policy makers, particularly the U.S. Congress, to take some action to begin to reduce U.S. emissions of greenhouse gases (GHGs). The recent media attention might thus prove beneficial to the future of U.S. policies to deal with global climate change.

To date, in this country, federal policies targeted at the abatement of GHG emissions have been primarily voluntary in nature, and of limited impact. As a result, U.S. emissions of GHGs continue to rise steadily, and are currently 16% higher than in 1990 (USEPA, 2006).

A major reason cited by the current Administration and some members of the U.S. Congress for not supporting mandatory GHG reduction policies has been dire predictions of negative economic impacts. However, as economists point out, the costs of inaction are great, and estimates of the costs of climate change to societies are rising. In a survey of climate change in a recent issue of *The Economist*, the notion that global economic development will be negatively impacted by continued temperature increases is discussed (*The Economist*, 2006). Economists cited in that issue predict damage to the global economy ranging from 0.1% a year to 3% (the latter is tied to a 2.5°C increase in temperature).

As other governments have shown, policy changes and market signals are imperative to reducing a country's emissions. According to a research report of the German Federal Environmental Agency, climate-specific policies in both Germany and the United Kingdom have helped keep these countries on track to meet their Kyoto obligations. Both experienced net GHG emissions reductions due at least in part to "special circumstances," such as unification of states, in the case of Germany, and liberalization of the energy and gas markets in the United Kingdom in the early 1990s. Analysis indicates, however, that significant reductions were also due to climate-specific and other environmental policies at the federal, regional, and local levels (German Federal Environmental Agency, June 2001). For example, federal policies in Germany targeted at CO_2 abatement include fuel taxes, support for wind energy, and new building codes (plus refurbishment codes for existing dwellings) for the residential sector. Similar policies in the United Kingdom include increased support for renewable energy, for Combined Heat and Power (CHP), and for improved building standards.

In this country, a Department of Energy (DOE) study on the U.S. potential to reduce GHG emissions using energy-efficient and low-C technologies concluded that the overall economic benefits of efficient, mandatory C-reduction policies appear to be equivalent to their costs, and that "smart public policies can significantly reduce not only C dioxide emissions but also air pollution, petroleum dependence, and inefficiencies in energy production and use" (Interlaboratory Working Group, 2000). The report also found that numerous existing cost-effective, energy-efficient technologies are not being adequately used in the four analyzed sectors: buildings, industry, transportation, and utility. Utilization of these technologies alone could constitute a minimal, least-cost response to a known problem.

Also significantly, a report prepared for the Pentagon, U.S. Department of Defense, for instance, illustrates that the economic impacts of global warming in this nation (and others) is likely far greater in the face of inaction than are policies that begin to address or resolve the issue now, particularly if the gradual warming observed to date leads to abrupt climate change (Schwartz and Randall, 2003). The Pentagon report cites national security issues as a primary concern, with the potential for global conflict if severe weather and climate impacts occur. Specifically, Schwartz and Randall cite potential disruptions to food, water, and energy as reasons for the United States to be concerned, and to further explore the possible consequences of climate change. They conclude: "This report suggests that, because of the potentially dire consequences, the risk of abrupt climate change, although uncertain and quite possibly small, should be elevated beyond a scientific debate to a U.S. national security concern." Among its recommendations to prepare the United States for potential abrupt climate change, the report authors call for the identification and implementation of "no-regrets strategies," among other recommendations.

A U.S. Response Needed

The longer we wait to begin to address the problem of global climate change, the more costly the solutions and the potential impacts, and the more difficult it will be for other governments to impact global emissions. Policies need a lead time for implementation, and gradual implementation is less costly than having to change energy production and GHG emissions abruptly. Delayed action increases the costs of actions and policies largely because, through delay, we prevent market and investment signals from going out to industry and other emitters of GHGs that indicate GHG emissions must go down. Business plans and investment decisions made in the absence of these signals have long-term impacts, due particularly to capital turnover rates. Consider power plants, for instance. If power plants have a 50-year lifespan, and 10 or 100 coal-fired power plants are built in this country in the intervening years before a policy response or a "carbon signal," these high GHG-emitting plants will further exacerbate the problem of global warming. Once a carbon signal goes out (i.e., enactment of a policy that GHG emissions must be reduced), however, these plants must be retrofitted or otherwise overhauled or must find a way to otherwise comply with the new policy(ies). In any event, the enterprise becomes needlessly expensive, economically and environmentally, for investors and for society. Investment decisions would quite likely have been very different if a policy to reduce GHGs had been in effect.

To begin to reduce U.S. emissions, which represent fully 25% of global GHG emissions, a well-planned, balanced mitigation approach that ramps up the response over time is needed. A cohesive, thoughtful long-term approach is consistent with good business management and investment decisions, and is preferable to a wait-and-see approach that ignores long-term investment and strategic decision making. Many existing technologies and methodologies can be employed now to begin reducing U.S. emissions. These include, for example, increased use of renewable energy, increased automotive and appliance efficiency standards, increased efficiency and energy use standards for buildings and manufacturing processes, enhanced use and availability of mass transit systems, and, as documented by the authors, a focus on biological C sequestration strategies. Additionally, innovative actions, products, and businesses typically arise from necessity and entrepreneurial activity when market-based approaches to address environmental problems are enacted.

U.S. Businesses Step Up

Many U.S. businesses acknowledge that mandatory U.S. caps on GHGs are both necessary and inevitable, and are taking action now even in the absence of mandatory government policies. These activities will prepare the companies to participate early in future C markets, and allow multinationals to participate in international GHG reduction schemes and markets, such as the Kyoto Protocol (see the subsection titled "International Response Under Way"). Dupont, Johnson and Johnson, 3M, General Electric, and even Chevron-Texaco represent just some of the U.S. companies taking action to combat global climate change. The synopsis of their actions, below, is not nearly a comprehensive list of what U.S. businesses are doing in the face of climate change, but instead an illustration of some company's actions.

- **DuPont** is one of the country's corporate leaders in GHG mitigation policies, having reduced its GHG emissions 72% since 1990. To help further reduce U.S. emissions, the company has developed climate-friendly products, such as energy-efficient building materials, solar, wind and fuel cell system components, and climate-friendly refrigerants.
- **Johnson and Johnson** adopted company-wide "Next Generation" goals in 2000, committing each of the company's business units to reduce GHGs emissions 4% from 1990 levels by 2005, and 7% by 2010. The company promotes C trading and C sequestration policies as a means of reducing GHG emissions.
- **3M** cites global warming emissions reductions as a company priority, and set a goal of reducing its worldwide GHG emissions by 50% in 2005, from 1990 levels. 3M has also pledged to reduce its U.S. GHG emissions 30% from 2002 levels by 2007.
- **ChevronTexaco**, a major U.S. oil producer, has developed a comprehensive program to manage GHG emissions, and is incorporating GHG emissions reduction activities into all its business decisions.
- **General Electric**, in 2005, launched a project called "ecoimagination," which is the company's plan to double investments in climate-friendly technologies and to reach $20 billion in annual sales of these products by 2010.

International Response Under Way

In February 2005, the Kyoto Protocol took effect as international law for more than 140 of the world's countries that have ratified the treaty; the United States and Australia have not. Under the Kyoto Protocol, countries agreed to reduce global emissions of GHGs an average of 5.2% below 1990 levels by 2012. Internationally, C markets established under the auspices of the Protocol are rapidly growing. The

European Union's Emissions Trading Scheme (EU ETS) began operating in January 2005, with an estimated $40 million worth of trading in its first month (Ecosystem Marketplace, 2005). The *Environmental Business Journal* in 2005 quoted sources as saying the Kyoto emissions trading market is "probably the most explosive market in terms of size and volumes" (*Environmental Business Journal*, 2005). Business certainty and entry into force of the Kyoto Protocol are cited as drivers for these markets. Some predict that trading of GHGs under the Protocol will reach $40 billion annually by 2010 (Point Carbon, 2006).

Although the United States is not a party to the Kyoto Protocol, and thus cannot participate in the markets or trading that are established by the treaty, many multinational corporations that operate in the United States and in "Kyoto countries" will not only have to comply with rules and restrictions of those countries, but will also be able to participate in the C markets in those countries. Most of these businesses, and others who are not participating, believe that it is inevitable that the United States will one day establish its own mandatory GHG emissions reduction policies, and that the United States will participate in some future international treaty, as well. Whatever the case, it is clear that the global market signal from the Kyoto regime will eventually affect nonparticipating U.S. companies. Some, such as the Chicago Climate Exchange (see the subsection titled "Private Sector Efforts") are already intricately involved in these markets.

No-Regrets Strategies

Regardless of the timeframe for inevitable federal-level global warming action in the United States, some policies make sense to promote low-cost, early-benefit options, which can begin reducing GHG emissions as other sectors retool for later, deeper reductions. Some policies make sense even in the absence of global warming benefits — policies that can reduce U.S. emissions of GHGs as an ancillary benefit, but that have inherent societal value above and beyond global warming mitigation. These constitute "no-regrets" strategies.

This book is about one category of no-regrets policies, and is intended to compile and simplify available scientific documentation in support of soil C sequestration. This strategy could have multiple, sometimes incalculable societal benefits, only one of which includes the reduction of GHG emissions and global warming abatement. By packaging GHG mitigation activities with other environmental benefits, as is the case for soil C sequestration, efficiencies are multiplied.

Given the timeliness of the issue of how to address global warming at federal, state, and local levels – as well as in the international arena – the authors hope to influence U.S. policy development at all levels by showing the interrelatedness of C management in the nation's soils with multiple other direct and indirect beneficial outcomes. These outcomes include: public health; air and water quality; flood control; biodiversity; and wildlife. For a more comprehensive compilation of the known direct and indirect benefits of soil C, see Appendix A: Known Benefits of Soil Carbon and Soil Carbon Sequestration.

Soil C, because of its link to air and water quality and its ability to help mitigate other potential ecological disasters — large-scale flooding or drought, for instance

— can provide an immediate response to existing, visible woes visited upon our society and others. Managing lands for soil C and soil "health" at larger geographic scales than the field or farm is one of the lessons to be learned from such recent disasters as Hurricane Katrina, in which the large-scale loss of wetlands and marshlands and inadequate land management contributed greatly to the devastating impacts of the hurricane in the Gulf Coast.

A landscape-scale approach to managing soil C need not imply or necessitate a return to a preindustrial or predevelopment state. Instead, the approach can be used to manage lands so that the ecological *functions* of soils and landscapes that have been lost or impaired can be restored for the benefit of society. As the authors of this book have indicated, such an approach will entail greater planning and coordination, at many levels, and over longer timeframes than is currently the norm. Soil C is the common currency underpinning this approach, whether the desired outcome is climate change mitigation, water quality, water quantity, wetland integrity and function, or agricultural productivity and sustainability.

Specific policy recommendations are presented, with attendant economic benefits where those benefits have been calculated or studied. As the authors documented, however, some benefits have not been valued in monetary terms. In these instances, the authors state the case for benefits using available supporting data.

The Value of Soil Carbon

The value of soil C, when direct and indirect effects and impacts are considered, is substantial. As the authors show, the wide range of beneficial impacts of soil C – or soil organic matter content – goes far beyond on-farm or on-site impacts. The true impacts of improved soil organic matter content are far ranging and valuable – perhaps even invaluable – to society. As cited in a recent report, the value of natural capital is "far in excess" of what can be reported or calculated, due largely to the many "intangible benefits of nature – attributes we know exist, but cannot place a value on" (Olewiler, 2004). If a ton of emitted C has a price of $10 (a hypothetical assumed by the researchers), these researchers calculated the value of incremental soil C sequestration at $17.90/hectare/year. Researchers in New Zealand, attempting also to put a dollar figure on the value of soil organic matter and soil C, concluded that, although soil C is valuable to agricultural production, it is 40–70 times more valuable to environmental protection than to crop production (Sparling et al., 2006).

This book shows the interdisciplinary nature of soil C solutions to a wide range of problems and the need for greater integration of policies and approaches to enhance soil C stocks. Current approaches to achieving an integrated program present many barriers. For example, societal understanding of ecosystem services and their value is relatively low, with the result that a general lack of support exists for such programs within current policy constructs; soil C sequestration and the problem of global warming generally fall into this category. Additionally, the U.S. federal program funding mechanism (i.e., the annual appropriations process in the U.S. Congress) makes coordination of funding across federal agencies difficult, and makes continuity of funding for programs uncertain from one year to the next. A proper, coordinated federal response will require Congress and multiple federal and state agencies to work together for extended periods toward a common goal or set of goals.

Others have suggested national-level agendas to tackle the problem of global warming (Pew Center on Global Climate Change, 2006a); the authors concur that a national agenda and a crosscutting set of policies is necessary to approach this issue and the issue of soil remediation in a cohesive yet efficient manner. The recommendations in this book stand alone, but are designed as the framework for a national agenda.

No Silver Bullets

Soil C sequestration is not a silver-bullet approach that will work anywhere, anytime. Regional and climatic differences, differences in soils, crops, and farm activities will require different practices and techniques in different situations. In fact, the term "soil C sequestration" does not refer to any one particular practice or technology but instead to a suite of management practices and methodologies that may result in improved soil organic matter content. See Appendix B, Methods and Technologies to Enhance and Improve Soil Carbon Sequestration, for a listing of these practices.

The technologies and practices recommended herein can be implemented now with a net economic benefit to practitioners and the nation as a whole, and provide a "quick-start" option to begin dealing with a large existing problem. The overarching societal and GHG benefits demand that soil C sequestration be considered as a beneficial policy response aimed at reducing U.S. GHG emissions as well as to improving agricultural sustainability, farm profitability, and environmental quality, all with positive societal impacts. Through this book, it is envisioned that policy incentives to promote soil C sequestration will be enacted in recognition of these multiple and ancillary benefits.

Although the issue of transferability of the practices and policies to other nations and global regions is not discussed in this text, improved soil C sequestration is a proven means to reverse land degradation and desertification that plagues many countries, particularly many arid and developing countries (SSSA, 2001; Lal, 2001).

Finally, additional research is always needed to fill the gaps between the knowledge base and the best means to address a problem. In some instances, the need exists for better assessment of the unintended consequences of policies aimed directly at improving soil organic matter content. For instance, organic matter, such as corn stover, is beneficial to boosting or restoring soil C content, but the removal of corn stover for biofuel production can compete with soil C sequestration as well as soil stability and productivity if too much stover is removed. The proper balance must be struck. Scientists and policy makers are already grappling with the issue of total GHG accounting, and the need to better understand impacts of actions to increase soil C on the emissions of other gases, such as methane and nitrous oxides, particularly in reference to cropping systems. Some high-priority research recommendations by the authors are included for the area, overall. Optimally, policy constructs should include a feedback mechanism to allow for new research findings as they emerge.

OVERVIEW OF EXISTING POLICY CONSTRUCTS AND OPPORTUNITIES FOR BIOLOGICAL SOIL CARBON SEQUESTRATION

PUBLIC EFFORTS

U.S. Department of Agriculture (USDA) and the "Farm Bill"

In the public policy arena, there has been some acknowledgment of soil C sequestration as a potential means to help mitigate global climate change. This has resulted in some explicit programs and incentives for sequestration, as noted below. Additionally, many on-farm conservation programs administered by the USDA can directly and indirectly benefit soil C sequestration by promoting management practices that enhance sequestration. Most of these opportunities pertain to agricultural programs and policies laid out in the Food Security Act of 1985, and its successor, the Farm Security and Rural Investment Act of 2002. These acts are popularly known as the Farm Bill(s).

Beginning in the 1985 Farm Bill, conservation programs received increased support with the creation of the Conservation Reserve Program (CRP). CRP is a land set-aside program intended to protect and restore degraded and environmentally sensitive lands. Lands in the CRP program cannot be farmed, and must be maintained for conservation purposes. Carbon sequestration is a subcategory criterion factor in the Environmental Benefits Index (EBI) of the CRP.* The EBI is used to rank land offers for acceptance into the CRP program.

A new program enacted in the 2002 Farm Bill, the Conservation Security Program (CSP), specifically intended to create incentives for soil C sequestration, among other conservation measures. The CSP was crafted to provide financial and technical assistance to agricultural operations to promote conservation and improvement of the quality of soil, water, air, energy, plant and animal life, and other conservation purposes. The program is meant to apply to all working lands, including cropland, grassland, prairie land, pasture and range land, and forested lands that are part of an agricultural operation.

Under CSP, a three-tiered payment system was established to reward farmers for good conservation practices. Unlike other farm programs, however, farmers are to be rewarded for maintaining conservation practices on their farms. The authors believe this type of reward system is justified given the multiple societal and off-farm benefits generated by conservation and C sequestration activities.

It is noted, however, that the potential impacts of the CSP are currently limited due to underfunding of the program, and, presumably due to this lack of funding, implementation of the program within a limited geographical range. Some conser-

* Though carbon sequestration is a subcategory under the "Air Quality Benefits from Reduced Wind Erosion" category of the EBI, carbon sequestration could be included in virtually all the EBI categories as a criterion. These categories include wildlife habitat benefits, water quality benefits, on-farm benefits from reduced erosion, enduring benefits factors, and costs (i.e., the cost of environmental benefits per dollar expended).

vationists feel that the program has not been implemented as intended, further undermining its potential benefits. Full funding of the program, as intended, is necessary to achieve the program's conservation goals.

The Value of Conservation Programs Recognized

Federal policy makers have recognized the value of multiple-benefit farm conservation programs. The 2002 Farm Bill (aka the farm Security and Rural Investment Act of 2002) significantly increased funding for conservation programs, nearly 80% over the previous Farm Bill. In recognition that these programs have ancillary environmental benefits additional to the known soil conservation, wildlife habitat, and reduced nutrient runoff impacts, for instance, USDA has undertaken a broad-scale evaluation of conservation program effects on watershed-scale environmental impacts, including water quality. Annual data from USDA's Natural Resource Inventory (NRI) will be used for this national assessment, known as the Conservation Effects Assessment Program (CEAP), targeted at quantifying the natural resource benefits from conservation practices on private lands. Recognition of the water quality and other ecological benefits of agricultural practices and programs is a signal that the Congress and USDA recognize the need to better evaluate agricultural impacts beyond the farm gate. Ongoing collection of these data at regular intervals can help to improve policy responses over time.

Beyond the CRP and CSP, several other Farm Bill programs also promote practices that enhance soil C sequestration, including:

- The Agricultural Management Assistance (AMA) Program
- The Conservation Reserve Enhancement Program (CREP)
- The Environmental Quality Incentives Program (EQIP)
- The Grassland Reserve Program (GRP)
- The Wetland Reserve Program (WRP)
- Wildlife Habitat Incentive Program (WHIP)

AMA provides cost-sharing and incentive payments for voluntary adoption of conservation practices on farms, mainly to address water quality and erosion control issues. It is limited to 15 states with historically low participation in the federal crop insurance program. The states include Connecticut, Delaware, Maine, Maryland, Massachusetts, Nevada, New Hampshire, New Jersey, New York, Pennsylvania, Utah, Rhode Island, Vermont, West Virginia, and Wyoming.

CREP is a component of the CRP Program, but is a cost-share program (up to 50% federal CRP funds, balanced with nonfederal funds) targeted at high-priority geographic regions with specific local environmental concerns.

EQIP provides incentive payments and cost-share funds to livestock or crop producers to implement conservation practices on eligible working lands.

GRP is a voluntary program to assist landowners in protecting, restoring, or enhancing grasslands, including rangelands. Lands can be permanently enrolled in easements, or under 30-year or rental easements. The program allows each state to prioritize the enrollment of working grasslands in the program.

WRP is a voluntary program to assist landowners in protecting, restoring, or enhancing wetlands on their property. The program provides both technical and financial support.

WHIP provides technical and financial assistance to landowners to develop, restore, and enhance fish and wildlife habitats on private lands.

The USDA: Beyond the Farm Bill

Some particularly useful (for the purposes of C sequestration) USDA policies go beyond the Farm Bill. One major USDA achievement is the development of a user-friendly database that farmers and ranchers can use to evaluate on-farm C management activities. The Carbon Management Evaluation Tool for Voluntary Reporting (COMET-VR) is an online Web tool that allows any practitioner to consider how changes in management and conservation practices will affect soil C and fuel and fertilizer use on their lands. The COMET-VR uses an agricultural model of present soil C levels. After data is imported into the program by the user, including locale, soil type, history of management practices on the land, and proposed changes in management practices, the program calculates changes in soil C levels. The reports provide estimated 10-year averages, complete with uncertainty values. A fuel and fertilizer report can also be prepared by COMET-VR to report changes in fuel use and fertilizer, based on changes in practices.

The USDA created the database not only as a decision-making tool, but also to allow farmers and ranchers to record emissions reductions (or increases) in the voluntary GHG emission reduction registration program established in the 1605(b) program of the Energy Policy Act of 1992.

U.S. Department of Energy

The DOE operates a C sequestration program focused on technologies that capture or permanently sequester C. Terrestrial sequestration activities, including in forests, plants and soils, are a focus of this program, which the DOE operates in concert with the USDA Forest Service. Under this program, the DOE has established several regional partnerships that are currently identifying potential sequestration opportunities specific to their geographic areas, for implementation in the event that the United States determines that deployment of a wide-scale sequestration strategy is necessary in the future. The program concentrates heavily on geologic sequestration, a less-proven technology than terrestrial sequestration, perhaps to the detriment of future policy and funding support for terrestrial sequestration activities.

The U.S. Congress

In the legislative arena, increased attention to the issue of global climate change has become evident as each successive Congress since the 105th has introduced or considered a growing number of bills to tackle the problem. According to one source, the number of climate change-related legislative proposals increased from seven in the 105th Congress (1997–1998) to 25 in the 106th Congress (1999–2000), and over 80 in the 107th Congress (2001–2002), and nearly 100 in the 108th Congress (2003–2004). The recently concluded 109th Congress saw 106 climate change-related bills introduced (Pew Center on Global Climate Change, 2006c). Soil C sequestration and other agricultural mitigation activities, which received only moderate support in the past, are also gaining attention as credible, realistic options to help this country begin to address the problem. Some of the bills that were introduced or considered in the 109th Congress proposed specifically to reward the agricultural sector for mitigation activities, and many of the global warming-related bills introduced early in the 110th Congress contain agricultural mitigation provisions. To date, none of these bills have passed or appears close to adoption, but the federal legislature should be watched for progress on this issue, as the evidence continues to build, and as public pressure mounts.

Other Public Activities: State Programs

Some states have undertaken far-reaching programs to assess soil C sequestration potential within their borders, and/or to promote soil C sequestration practices and activities. Four states in particular have undertaken policies to assess and increase their soil C stocks: Indiana, Nebraska, Iowa, and Oklahoma. These states have positioned themselves to take advantage of future GHG markets and to reap the rewards of the ancillary environmental benefits of soil C sequestration.

The Natural Resource Ecology Laboratory (NREL) at Colorado State University (CSU) assisted Indiana, Nebraska, and Iowa in their county- and state-wide assessment activities. NREL's Agroecosystems Research Group developed a two-phase, state, and county level approach that is documented in reports for each state (Brenner 2002a, 2002b; Brenner, 2001). The fourth state, Oklahoma, undertook a similar activity on its own, as called for by that states' legislature.

Other states have undertaken more limited programs that have had the impact of directly or indirectly increasing soil C stocks. Some of these programs are summarized next, by state.

Idaho

Idaho formed a Carbon Sequestration Advisory Committee in 2002 in response to concerns about rising GHG emissions, and also to possibly create or participate in GHG emissions trading markets. In a 2003 report of the Advisory Committee, a comprehensive set of options was presented to the state legislature for consideration of how Idaho's agricultural and forested lands and landowners can participate in C trading markets.

Illinois

The state of Illinois was an early innovator among the states, and authorized and formed a Task Force in 1991 to recommend and later implement state strategies to combat climate change. The Task Force included representatives from the agricultural sector, and when the Task Force's first report, A Climate Change Action Plan for Illinois, was issued in July 1994, it included recommendations for tree planting and afforestation activities as one potential mitigation strategy. The state later inventoried its own GHG emissions and developed pilot mitigation projects, including forestry programs and energy efficiency policies, to begin to reduce its emissions. In 2006, Illinois announced the formation of the Illinois Conservation and Compliance Initiative (ICCI), together with the Illinois-based Delta Institute (a nonprofit group dedicated to environmental quality improvements and community and economic development). ICCI awards payments for approved C sequestration practices, such as conservation practices, no-till farming, grass and tree plantings, and for the use of methane digesters. Like some other states in the region, the ICCI trades C tons sequestered by the program in the private Chicago Climate Exchange (see the subsection titled "Private Sector Efforts").

Indiana

Indiana, together with the states of Iowa and Nebraska, has been involved in a unique Carbon Storage Project. Together with USDA/NRCS, the National Association of Conservation Districts, and state and local conservation districts (and with state support and some funding from the DOE), these states undertook initiatives to assess the soil C storage potential on a county-by-county basis. The goal of the Project was to assess existing rates of sequestration on agricultural lands, and to help landowners predict changes in sequestration rates that would occur if new conservation or other management practices were adopted. For each of these states, the county data that was collected was linked to the CENTURY Model (Century Eco-System Soil Organic Matter Computer Model), an agricultural model developed at Colorado State University that predicts changes in C and organic matter in response to management and climate changes on grasslands, agricultural lands, prairies and savannas (Parton et al., 1987). The data were then combined in state-specific software programs known as the COMET programs (for CarbOn Management Evaluation Tool), which allow individual landowners to predict changes in soil C content on their lands before making management changes, and to estimate C sequestration rates after the management changes are undertaken. A report on the Indiana Carbon Storage Project is available (Brenner, 2002a). The work of the Carbon Storage Projects set the stage for the later development by USDA of the nationwide COMET-VR software described elsewhere in this chapter.

Iowa

The Iowa Farm Bureau Federation is an innovator in the realm of agricultural sequestration options and GHG trading. The Iowa Farm Bureau, operating as an aggregator, has organized farmers in Iowa, Kansas, and Nebraska to sell their soil C credits within the Chicago Climate Exchange (see the subsection titled "Private Sector Efforts"). Additionally, Iowa was an original participant in the Carbon Storage Project (see the previous subsection on Indiana), and served as the pilot project. A report on the Iowa Carbon Storage Project is available (Brenner, 2001).

Nebraska

In 1998, the Nebraska state legislature passed a bill mandating a study of the potential for a C credit trading system, as well as a statewide assessment of soil C sequestration potential, for possible C credit trading. That bill, LB 957, marked the start of the Nebraska Soil Carbon Sequestration Project. LB 957 required the state to form an advisory committee of scientists and land users to study the potential for a C credit trading program and to conduct a soil C assessment to establish a baseline for possible C credit trading activities. The Committee also encouraged awareness and educational activities regarding soil C sequestration and the role of C in global warming, and sought to identify and recommend areas of research needed to better understand and quantify the process of soil C sequestration. Nebraska was thus an early adopter of agricultural mitigation opportunities for future C markets, and is a participant in the Chicago Climate Exchange (see the subsection titled "Private Sector Efforts"). Nebraska, along with the states of Iowa and Indiana, was an original participant in the Carbon Storage Project (see the previous subsection on Indiana). A report on the Nebraska Carbon Storage Project is available (Brenner, 2002).

A state-based public–private partnership in Nebraska, called the *Forever Habitat Equipment Program*, encourages conservation tillage by leasing no-till equipment to the state's farmers. Pheasants Forever operates the program, and has received $670,500 in grants from the Nebraska Environmental Trust since 1996 to purchase 60 no-till grass drills and other habitat equipment for lease at low- or no-cost to Nebraska landowners.

North Dakota

The North Dakota Farmers Union (NDFU) announced in Spring, 2006, a new Carbon Credit Program, operated by NDFU in conjunction with the Chicago Climate Exchange (CCX) (see the subsection titled "Private Sector Efforts"). Under the program, NDFU acts as an approved aggregator of agricultural C credits for sale or trade in the CCX. Farmers and ranchers can bank credits by undertaking no-till practices or by grass seeding of lands. NDFU has extended participation beyond North Dakota farmers and ranchers to include those in four other states involved in a regional Farmers Union Enterprises group, which includes Montana, South Dakota, Minnesota, and Wisconsin.

Oklahoma

Oklahoma has been active in seeking to prepare the state's agricultural interests for participation in future GHG trading schemes. The Oklahoma state legislature passed the Oklahoma Carbon Sequestration Enhancement Act of 2001, which called for the creation of an Advisory Committee by the Governor, and required a state-wide assessment report of the state's soil C sequestration potential, to be completed by the Advisory Committee and the state's Conservation Commission by December 1, 2002 (Oklahoma Law 27A O.S., 2001). The resulting report, submitted to the state legislature in November, 2002, considered several state level options for pursuing soil C sequestration provisions, and made a set of additional recommendations to the Oklahoma Legislature (Oklahoma Conservation Commission and Carbon Sequestration Advisory Committee, 2002). The participants in the Oklahoma project followed procedures similar to those undertaken in the states of Indiana, Iowa, and Nebraska, including working with USDA/NRCS staff to undertake a baseline survey of soil C in the state's soils. A second report by the Oklahoma Conservation Commission and the Carbon Sequestration Advisory Committee was submitted to the state legislature on January 6, 2003 (Oklahoma Conservation Commission and Carbon Sequestration Advisory Committee, 2003). This report presented the findings of the baseline soil C survey, and estimated potential soil C gains and losses through management practices on agricultural lands.

Oregon

Oregon has also been an innovator in the realm of promoting terrestrial sequestration activities as means to mitigate rising GHG emissions. Spurred on by progressive state policies to reduce net GHG emissions, two of Oregon's power providers — Portland General Electric and PacifiCorp — developed a public–private partnership in 1999, the Klamath Cogeneration Project, to invest in forest sequestration offsets for GHG emissions. Forest and soil conservation practices are a focus of the sequestration activities of the Klamath Project.

Wyoming

Wyoming created a Carbon Sequestration Advisory Committee, which produced a lengthy report to the Wyoming Legislature in 2001. The report documented agricultural and forestry activities in the state that could reduce GHG emissions, and identified potential existing markets for the sale of credits. This led in 2002 to an Advisory Committee report that recommended state actions to benefit from C sequestration activities, including a state GHG registry of emissions, sequestration pilot projects, and potential C trading. The state has since undertaken various sequestration projects on rangelands, croplands, and forested lands, and is seeking to facilitate the marketing of the resulting C credits.

Private Sector Efforts

As is typically the case when entrepreneurial opportunities arise, the private sector has created additional ventures in support of soil C sequestration activities. Most have been created to test the viability of the concept of soil C tons or credits being traded in a C market, and to enable the agricultural sector to participate in such systems. Though operating outside the public sector, these ventures are not mutually exclusive from public policy, and may prove invaluable in helping to inform future policies. It is also hoped that the experience will give producers and farm organizations an advantage when C markets inevitably begin in the United States, or when the United States becomes a participant in international markets.

Some of these projects and initiatives to reduce GHG emissions include direct incentives for activities that enhance soil C sequestration, whereas others promote the concept indirectly.

The **Chicago Climate Exchange (CCX)**, which describes itself as "North America's only, and the world's first, greenhouse gas (GHG) emission registry, reduction and trading system for all six greenhouse gases (GHGs)," is a voluntary program whose members make legally binding commitments to reduce GHG emissions from their operations, or through the purchase of offsets (see www.chicagoclimatex.com). Currently, members of the Exchange span diverse business interests, including, for example, the Ford Motor Company, Motorola, Inc., Bayer Corporation, IBM, DuPont, American Electric Power, the Roanoke Electric Steel Corporation, public and private universities, and several municipalities. Members agreed to reduce direct emissions by December 2006 to 4% below a baseline period of 1998–2001. In a second phase of commitments, CCX members will reduce GHG emissions 6% below baseline by 2010.

CCX issues C credits for certain forestry and agricultural practices. For the latter, committing land to continuous no-till, strip-till, or ridge-till cropping in the central United States and other regions (or countries) can earn credits, as can initiating grass cover planting in specified states, counties, and parishes.

In addition to its U.S. operations, CCX operates the European Climate Exchange (ECX), where it says it handles 85% of the total exchange volume in the European Union Emissions Trading Scheme (www.europeanclimateexchange.com/index_flash.php). ECX members include Goldman Sachs, Merrill Lynch, Morgan Stanley International, Shell Energy Trading Company, and Shell International.

PowerTree Carbon Company LLC is an initiative in the lower Mississippi River Valley to manage terrestrial C dioxide, mainly by planting trees (PowerTree Carbon Company LLC, 2006). The initiative restores bottomland hardwoods on marginal agricultural lands while also improving wildlife habitat, air and water quality, and C sequestration. Twenty-five participating power companies have contributed $3 million for projects in three states (Louisiana, Arkansas, and Mississippi). The projects are scale-variable, with the largest at 1100 acres, and the smallest, 200 acres, and involve private landowners, business entities, state and federal partners, and nonprofit groups.

With five planting projects to date, PowerTree estimates that 1.624 million tons of CO_2 will be sequestered in a 100-year period because of these projects, with additional environmental and wildlife benefits for the duration.

Direct Sale of Carbon

In what is believed to be the first such trade of its kind, the Pacific Northwest Direct Seed Association (PNDSA) contracted with Entergy Corporation (www.entergy.com) in 2002 to lease sequestered soil C credits. Entergy leased offset credits for 30,000 tons of sequestered C dioxide from PNDSA over a 10-year period. Entergy described the venture as a means of offsetting its emissions from power plants. Each participating farmer in PNDSA, who agreed to use no-till or direct-seed agricultural practices for at least 10 years, received a check for their C sequestration efforts. The agreement is providing valuable marketplace experience in C trading for both PNDSA and Entergy, giving them an advantage when C markets take hold in this country – a future that many businesses and policy makers agree is a certainty. The deal also helped to show the potential value of businesses such as Entergy looking to the agricultural sector to provide valuable, low-cost GHG emissions reductions — a bridging strategy to future mandatory emissions reduction policies.

POLICY RECOMMENDATIONS

Broad-Based Policy Recommendations

1. USDA should change its policy of managing the nation's soils for structural erosion, or soil loss tolerance rates ("T") and should instead manage soils for soil organic C content, with a focus on maintaining existing C stocks and on increasing soil C levels. The concept is familiar to USDA; on its Web site, and in literature developed by NRCS, the notion of managing soils for C is widely promoted (NRCS/USDA, 2003). USDA currently estimates that 1.8 billion tons of soil is still lost from U.S. croplands annually, and that if all cropland were managed for soil organic C instead of structural erosion (T) losses, as per current policy, then annual soil loss would decline by 1.29 billion tons, at a savings of $8.2 billion annually (USDA/NRCS, 2006).
2. Congress should fully fund all existing Farm Bill programs that contribute to enhanced soil C sequestration. Programs authorized by the Farm Bill(s) are frequently underfunded in the annual appropriations (federal funding) process, thereby limiting their value and benefits to the agricultural sector and society, despite having been identified as top priorities by policy makers. Compliance checks for all programs are essential to ensure attainment of intended benefits. Conservation programs that should receive full funding as authorized include (but are not limited to) the following:
   - The Agricultural Management Assistance (AMA) Program
   - The Conservation Reserve Program (CRP) and Conservation Reserve Enhancement Program (CREP)
   - The Conservation Security Program (CSP)
   - The Environmental Quality Incentives Program (EQIP)
   - The Grassland Reserve Program (GRP)
   - The Wetland Reserve Program (WRP)
   - The Wildlife Habitat Incentive Program (WHIP)

   The Conservation Technical Assistance Program provides on-the-ground technical assistance to land managers, including for reduced soil erosion and sedimentation, and should be fully funded and expanded to protect this valuable, nationwide infrastructure and delivery support program. (See also Recommendation 6).
3. The federal government should create an integrated, National Soil Carbon Sequestration Agenda, working in concert with the existing North American Carbon Program (NACP).* The Agenda should require all Federal land-management agencies to manage the nation's soils and lands at the landscape scale (as opposed to the farm or field scale), seeking to restore or enhance the functioning of soils and lands — and their ecosystem

* The North American Carbon Program (NACP) is a scientific project created to measure and understand carbon stocks and the sources and sinks of carbon dioxide (CO_2), methane (CH_4), and carbon monoxide (CO) in North America and in adjacent ocean regions. Participating federal agencies include NASA, NIST, NOAA, NSF, DOE, EPA, and USGS (see NACP Web site: www.nacarbon.org/nacp/index.html).

services — while retaining existing critical policy objectives of these agencies. The goals of the agenda should be to:

- Coordinate federal agency activities, policies, programs, regulations, and research on soil C and land management into a cohesive approach to maximize soil C sequestration on agricultural, forested lands and urban landscapes, thereby maximizing ecosystem services and environmental outcomes on these lands.
- Establish quantifiable goals and milestones for the Agenda, with outcomes to be assessed at regular intervals (e.g., 3–5 years), as appropriate. This should include mechanisms for mid-course corrections in the event of unanticipated, inadequate, or inappropriate impacts, to enhance beneficial impacts, or to take into consideration new research findings.
- Prioritize policies and activities that accomplish the greatest net benefits to both agriculture and society, taking into account the following factors: water and air quality issues on-farm and off-site, particularly at the urban/agriculture interface; restoration of marginal and highly erodible lands; promotion of organic farming systems technology and methods; restoration and replacement of wetlands; flood control and mitigation; wildlife habitat maintenance and restoration; recreational opportunities; and other multibenefit outcomes, as appropriate.
- Increase the focus on and funding for biological C sequestration activities as part of the Department of Energy's Carbon Sequestration Project and Carbon Sequestration Regional Partnership. Since 2001, the Carbon Sequestration Project has devoted relatively large sums to geological and ocean sequestration projects, despite both being unproven technologies with certain associated risk (such as changing the pH of ocean waters or disrupting geological formations or leaking from the formations). Soil C sequestration promises proven GHG reduction and ancillary societal benefits, but receives little attention in these projects, with some exceptions, such as a handful of narrowly focused projects on reclaimed or abandoned mine lands. Some beneficial projects on soil C measurement and monitoring technologies do exist. One project In particular is examining soil sequestration on degraded lands, such as lands disturbed by mining, highway construction, or poor management practices. The USDA, EPA, and other land-management agencies should work cooperatively with the DOE to identify and fund high-priority soil C research projects to maximize soil C sequestration opportunities across the broader U.S. landscape.
- Plan and coordinate with other federal agencies and state and local authorities, as appropriate, as well as private sector stakeholders, to maximize federal and private sector dollars (and leverage additional private-sector funding), avoid duplication of effort, and build on existing research, data, C sequestration, and C market efforts. In addition

to the agencies included in the NACP, planning and coordination efforts should include the Department of Interior, Fish and Wildlife Service; and the Department of Defense, Army Corps of Engineers. In addition to coordinating and prioritizing this work through annual appropriations bills for these agencies, funding priorities and recommendations in the biannual Water Resources and Development Act and the 5-year Farm Bill should be included. Private sector entities involved in planning and coordination efforts should include national, state, and local conservation groups, and those involved in GHG trading and markets.

- Take into account total GHG accounting, seeking to better understand impacts of soil sequestration on other GHG emissions, and to achieve net reductions of GHG emissions.
- Be crafted in anticipation of future mandatory, market-based nationwide policies to reduce U.S. GHG emissions. These policies should utilized market-based trading and other market efficiencies, but should be crafted now, as a bridging strategy to the future, to:
  a. Begin to reduce net U.S. GHG emissions now, using currently available, multibenefit activities and technologies.
  b. Prepare farmers and agricultural producers for participation and financial remuneration in existing and future C markets.
  c. Provide low-cost C (and other nutrient) "credits" in existing markets as well as future mandatory system(s). The USDA's COMET-VR program (see list item 4) offers a basis for future documentation and awarding of such credits or offsets, and should be enhanced to include all applicable GHG and total system GHG accounting and to make the data in the program more robust.

4. Increased funding should be provided to USDA to continue to strengthen and expand the Carbon Management Evaluation Tool for the Voluntary Reporting of Greenhouse Gases (COMET-VR), an online C management program developed for farmers, ranchers, and foresters. For example:
   - Additional GHG should be added to the software program; USDA is working to add nitrous oxide and (potentially) methane to the system now.
   - Current proposals to use the tool as part of the Conservation Security Program (CSP) should be supported, and with time, expanded to include additional USDA conservation programs.
   - Besides procuring feedback from producers and land managers, as it is currently doing, USDA should consult with the private sector entities that are already operating in emerging C markets, both to prepare the system for future markets and to ensure maximum compatibility of the system and transferability of certified credits within future C markets.
   - USDA should make the COMET-VR system and its components more robust by improving and enhancing the data in the program and in the

programs' underlying datasets, such as the CENTURY* model (Parton et al., 1987). The goal would be to make COMET-VR reports acceptable in market-based GHG trading systems, if producers undertake the management changes as documented in the reports. COMET-VR reports currently include an uncertainty estimate, which might be used to estimate or apply a discount rate to C credits used in a market-based trading program, for instance. This would overcome the criticism that the COMET-VR program is not as accurate as direct measurement of soil C — which is true — but COMET-VR reports are much less expensive, time-consuming, and less disruptive than direct soil measurement of C content. Improvements to the COMET-VR databases to make the data more robust and to decrease the uncertainty rates can improve the market value of C credits generated by the program. (See also list item 7, which includes a recommendation to enhance the CENTURY model.)

- The USDA should work with the DOE, the National Aeronautics and Space Administration (NASA), and relevant private entities to further develop, test, and calibrate remote-sensing and direct soil C measurement technologies to develop standardized, cost-effective monitoring technologies for market-based applications. At present, remote sensing technologies report data at rather coarse scales and do not penetrate soils to measure C content. Direct soil sampling technologies are costly, labor-intensive, and slow. The DOE's Terrestrial Carbon Processes Program has developed new, more rapid instrumentation for measuring soil C. A concerted effort is necessary to further develop these technologies. Pilot projects to test, scale, and calibrate measurement technologies as a means of allowing the agricultural sector to participate in GHG markets are recommended. A next step would be to ground truth, cross correlate, and calibrate the technologies, linking them as appropriate to the COMET-VR program. The end-goal would be the development of standardized measurement and monitoring technologies that are appropriate for multiple scales and acceptable for market-based trading and the sale of soil C credits.
- The USDA should work with state and private-sector entities and within existing federal programs such as the Conservation Technical Assistance Program to educate potential users of the existence of COMET-VR, and of emerging market opportunities that could utilize COMET-VR reports, particularly reports based on an enhanced COMET-VR program (see the previous bullet in this list).

---

* Century Ecosystem Soil Organic Matter Computer Model (CENTURY Model), developed by CSU and the USDA Agricultural Research Service, is a land-based ecosystem model that simulates nutrient dynamics (e.g., carbon and nitrogen) on cropland, grasslands, forest, and savanna ecosystems, as well as land use changes between these different systems. CENTURY is used by the United States and several other countries to estimate national soil carbon inventories, and is linked to the COMET-VR database. See also: www.nrel.colostate.edu/projects/century/.

5. USDA should tailor incentives for C-specific conservation policies to different farm circumstances to maximize adoption of voluntary programs. For instance, variations of conservation programs should be crafted with specific incentives tailored to farm size (small farms versus industrial sized or factory farms, for instance) and circumstances and motivations (e.g., farmers dependant solely on income from the farm versus farmers with off-farm incomes and work). According to research conducted by USDA's Economic Research Service (ERS), the determination of who participates in voluntary conservation programs is dependant on factors such as farm size, commodities produced, and "operator motivation" (e.g., profit, on-farm or off-site environmental impacts) (Lambert et al., 2006). ERS data also show, for example, that on more than half the farm acres where conservation programs are utilized, only three conservation practices are used, singly or in combination: conservation cropping, conservation tillage, and/or crop residue use (Claassen, 2004).

   The USDA should explore targeting additional, less-utilized conservation practices as appropriate to encourage maximum adoption rates of conservation programs. For each conservation program operated by USDA, two versions should be considered, as appropriate: one targeted to small, family-owned and operated farms with little additional income to contribute or dedicate to conservation practices and programs; and one targeted to larger, industrial-sized farms with typically larger profits, but also larger impacts on the environment. The policies should consider ERS's findings and other appropriate findings as necessary, but should ensure that by tailoring the programs, C sequestration and conservation impacts are maximized. The USDA should also collaborate with other federal land-management agencies to identify and adapt conservation programs and practices for nonfarm private land use and participation.
6. The USDA and the U.S. Congress should make it a priority to increase and maintain funding and infrastructure of the NRCS Conservation Technical Assistance (CTA) Program. The localized delivery infrastructure of the program is essential to the goals of a nationwide soil C sequestration program, as well as to most land conservation programs. In fact, the purposes of the CTA are entirely consistent with the known benefits of soil C sequestration (NRCS/USDA, 2006).
7. States should adopt state "carbon sequestration enhancement plans" to assess soil C stocks within their boundaries, and then develop landscape-scale policies to build C stocks and the associated ecosystem services. Proven methodologies (as well as worksheets, databases, and other necessary documents, such as the Carbon Sequestration Rural Appraisal, a survey instrument) have already been developed, and are available (Brenner, 2002a; Brenner, 2002b; Brenner et al., 2001; Oklahoma Law 27A O.S., 2001). The Colorado State University (CSU) Natural Resources Ecology Laboratory (NREL) has conducted state and county-level assessments of soil C and changes in soil C in three states (Iowa, Indiana, and Nebraska), and a fourth state (Oklahoma) has done the same, following

a similar process. State and federal funding should be made available to assist states in these efforts. Data collected through this assessment process can be fed back into the COMET-VR and CENTURY Model databases, to enhance the robustness of these datasets. As more states follow this process, the datasets will be improved.

8. USDA should make soil C sequestration a specific criterion or goal of all conservation programs where such criteria are weighted as part of the selection process for acceptance into the programs. For instance, C sequestration is a subcriterion for the Environmental Benefits Index (EBI) of the CRP program — but only as it relates to the air quality criterion (Farm Service Agency/USDA, 2006). Carbon sequestration has known benefits for these additional EBI factors as well: wildlife habitat benefits; water quality benefits; on-farm benefits from reduced erosion; benefits that will likely endure beyond the contract period; and cost. As such, soil C sequestration should be a single, stand-alone criterion or category of the EBI, in addition to the other criterion.
9. USDA should analyze and carefully track and assess the impacts to soil quality of future increased renewable fuel production from agricultural biomass. Title XV of the Energy Policy Act of 2005 (P.L. 109-58) has provided a much-needed boost for renewable fuels development and utilization in this country. Some analysts are predicting land-use competition for renewable-fuels biomass and rising food prices due to fewer cropped acres devoted to food crops in the future. Greater attention must be paid to the potential impact of these policies to the nations' soils and water quality, however. The removal of agricultural biomass, such as corn stover, for fuel production may be a boon to renewable fuels, but the loss of the organic matter and soil quality can further degrade soils in this country. At present, chemical fertilizers are utilized to boost soil productivity, often masking deteriorating soil quality and productivity due to low organic matter content. Increased chemical fertilizer use to boost biomass production for fuel production could exacerbate exiting nutrient runoff and leaching from the fertilizers, and mask continued land degradation and soil quality losses. Continued health of the nations' soils will be necessary to produce the biomass needed to fuel the increased renewable energy needs.

   A potential biomass utilization process that captures C and can renew and boost soil productivity may provide an appropriate policy response to this situation. The utilization of agricultural biomass to produce a charcoal (aka "biochar" or "agrichar") soil amendment and a bioenergy coproduct thus deserves further research and development consideration (see the subsection titled "Research Policy Recommendations" for more information).
10. The Federal government, through all land-management agencies, should expand public education efforts pertaining to the multiple benefits of soil C sequestration and of soil and C management at landscape scales, instead of field or farm scales. The education should target all stakeholders, including: farmers, the public (including the public education system),

policy makers, and state and local governments. The goal should be to educate all stakeholders of the value of well-managed, high organic matter content soils, and the benefits of public support for proper soil management. These efforts should include print and electronic materials, including online materials.

Ecosystem-Specific Policy Recommendations

Wetlands

- Develop a **National Strategic Vision and Implementation Plan for Watershed and Wetland Restoration** and focus financial and technical expertise from complimentary programs to undertake and implement the plan. Plan priorities include:
  - To map, watershed by watershed, all wetland restoration opportunities;
  - Develop a general performance-based watershed condition model to apply to each watershed to determine the acreage of wetland restoration that would be targeted to achieve the desired performance outcomes; and
  - Create a Carbon Utility, similar to a potable water supply utility, to provide incentives to private landowners for wetland restoration on their lands in return for annual payments for supplying C management services. The service could provide a proven public benefit at much lower cost than, for example, an industrial strategy to remove C from the atmosphere.

Expand and encourage continued use of wetland mitigation banking in urban areas and other areas where wetlands are subject to development pressure. Wetland mitigation banking is a primary compensation tool used by the Army Corps of Engineers and the U.S. Environmental Protection Agency (US EPA) in the Clean Water Act Section 404 regulatory permitting process. Wetland banks are restorations that are constructed to compensate for the losses of small, often isolated wetlands directly through filling or dredging and indirectly through hydrological and water quality changes and sedimentation from development activities. Wetland banks are constructed for public good, and provide benefits in strategic locations where they will be most successful and effective. Privately constructed wetlands often become public parks and urban open spaces, ensuring their protection and use for the public good.

Urban Ecosystems

- Establish integrated development and conservation policies wherein new and existing developments are co-located to retain or create open space systems. Utilize ecological design principles by managing storm water in these open space systems, with the goal of eliminating storm sewers. The resulting savings are substantial for both developers and in terms of eliminating long-term maintenance by the responsible public agency (e.g., municipality, county, or state). In addition, by keeping water within the soil system instead of in subterranean pipes, local hydrology and groundwater supplies are regenerated, and soil systems are protected from deterioration typically associated with dewatering. Downstream flooding is also reduced because the water does not run off as fast.

- Perform a Metropolitan Area Natural Resource Inventory (MANRI) of each metropolitan area greater than 25,000 persons, and incorporate the findings into each area's Master Plan. The goal of the MANRI would be to provide each community with the data and mapping to prepare conservation plans showing protected areas, and to prioritize areas for restoration, land management, and the protection of natural resource capital which provides C management, storm water management, improved human and wildlife habitat, cleansing functions, and other benefits. Important areas should be protected through permit and entitlement easements negotiated with developers by the metropolitan area, and the proceeds of permit sales would be utilized to fund the stewardship of protected areas.
- Develop metropolitan regional economic incentives to restore degraded lands and lands existing in developments that were created without ecological design principles. A primary goal would be to create networks of protected parcels in each development, to maximize storm water and soil C management within open spaces. The quantity of retained open lands would contribute to other regional values, including recreation, habitat, and park uses.
- Encourage the use of native and low maintenance landscaping in urban areas to reduce or eliminate the need for watering, herbicides, pesticides, or mowing. Use of these landscaping planting strategies in parks, right of ways, vacant lands, and on school and municipal building properties (e.g., public works, fire departments, and sanitary facilities) can serve as regional demonstrations for private landowner models and conversions.

Prairies, Savannas, and Forests

- Encourage intensive short rotational grazing on national forest and grassland grazing allotments, which enhances soil C sequestration and reduces soil degradation, sedimentation, and nutrient loading in the nation's waterways.
- Establish a new federal perpetual conservation easement tax benefit program for landowners willing to permanently retire CRP lands, providing the lands continue to be managed for soil C content. The program should be created with the goal of working in concert with existing state programs that remove lands from development, to optimize conservation opportunities and efforts.
- Expand on the Wetland Reserve Program (WRP) to promote restoration of native grasslands, savannas, and forests by removing or disabling drainage ditches and tiles that dewater millions of acres of historic prairie,

savanna, and forested lands. Specifically, WRP should target restoration of these transitional ecological systems that supported the functions and health of the historical wetlands. To accomplish this, USDA should establish a database of these lands within the National Resources Inventory (NRI), prioritizing areas for restoration under the WRP.

- In former forested floodplain environments now converted to agricultural cropping uses, establish a program to provide incentives to adopt soil and crop management strategies that maximize soil C sequestration (e.g., use of no-till soil preparation and management, reforestation, game, and non-game wildlife activities in lieu of cropping).
- Encourage and provide incentives for sound forestry practices that do not disturb or degrade forest soils and their productivity.

Flood Control and Abatement

- Create regional initiatives to link landscape-scale management for soil C sequestration and flood control mitigation, prioritizing within each region those watersheds that are prone to repeated flooding, including agricultural watersheds. The cost of practices to restore soil C can be more than compensated through the savings from reduced flood damage and other outlays associated with excessive runoff.

Research Policy Recommendations

The Federal government should develop and continuously update a National Soil Carbon Research Agenda to optimize soil C sequestration activities and efforts at a landscape scale. High-priority research areas include the following:

- Increase research and development support for innovative biological technologies to accelerate an increase in total C sequestration in cropping systems, agricultural lands, and nonagricultural lands. The use of charcoal produced from agricultural biomass — termed "agrichar" or "biochar" — as a soil amendment is a high-priority area. The conversion of biomass to biochar via pyrolysis can capture up to 50% of the biomass C, with bioenergy coproducts. The system is C-negative, by virtue of capturing up to half the C biomass, which can be stored in soils. The addition of the biochar to soils can potentially create a large, long-term C sink, far in excess of natural state soils, or even primary forests (Lehmann et al., 2006). Additionally, displacement of chemical fertilizers by the biochar can reduce associated nutrient runoff and leaching that hampers water quality. Soils produced in this manner, such as the Terra Preta (i.e., "dark earth") soils of the Amazon Basin may help to prevent soil degradation, which will be particularly important if biomass is increasingly used to fuel the nation's energy needs. Stable, high-quality, fertile soils can provide the underpinnings of a biomass-based energy system. Bioenergy coproducts from the system may take the form of bio-oils or syngases, contributing to renewable energy production. Research and development as well as commercial projects on various aspects of this system are under way in several countries, including Australia, Brazil, Canada, China, Italy, the United States, and New Zealand, among others (*Nature*, 2006; Marris, 2006).

- Mycorrhizal fungi may be associated with higher soil C accumulation rates. Understanding their nature and role in soil C sequestration will be important to understand the optimization of soil C in organic and traditional agricultural systems.
- Cost-effective, accurate monitoring and verification methodologies are needed to measure changes in on-farm GHG emissions or emissions reductions, including on-site, rapid assays of soil C and changes in soil C. This will be particularly important to insure maximal sale value for soil C credits in a market-based GHG system.
- Assess potential for leakage due to activities and/or emissions increasing elsewhere when mitigation activities are adopted.
- Improve on-farm whole-system GHG accounting systems.
- Improve knowledge and monitoring of nitrogen oxide emissions, particularly from croplands and agricultural operations, and mitigation options.
- Improve knowledge and monitoring of methane emissions from agricultural operations and mitigation options.
- Develop and promote regional and state-specific approaches to enhance soil C stocks, tailored to the climate, soils, crops, and farm circumstances particular to each.
- Assess how climate change impacts may alter soil C sequestration opportunities; for example, whether soil sequestration potential or rates of accumulation or soil C stocks will be impacted by changed climatic conditions in affected regions.

SUMMARY

Soil C sequestration poses unique, multi-benefit opportunities for the United States to help reduce net GHG emissions while developing other, more costly or longer-term solutions to the problem of global climate change. The varied benefits to society of the conservation and land management practices that lead to soil C sequestration represent win-win solutions to multiple problems that plague this nation at local, state, and regional levels.

The authors document additional impacts beyond GHG reductions from soil C sequestration, in an attempt to show that traditional cost-benefit analyses shortchange the true societal and environmental benefits of soil C sequestration, and recommend multiple policy goals to maximize the benefits on a large scale.

Although multiple federal, state, and private entities are engaged in a network of activities aimed at maximizing soil C sequestration for various outcomes, including financial and environmental outcomes, a cohesive federal framework and agenda for these activities is lacking, but would greatly benefit both spending and impacts. The basis of a comprehensive policy is recommended, with the USDA as the lead agency, together with specific additional policy recommendations that would benefit society.

REFERENCES

Brenner, J., Paustian, K., Bluhm, G. et al., Quantifying the change in greenhouse gas emissions due to natural resource conservation practice application in Iowa: The Iowa Carbon Storage Project, March 2001. Available at: www.nrel.colostate.edu/projects/agroeco/projects/statelevel/ia.html. Accessed on 03/15/06.

Brenner, J., Quantifying the change in greenhouse gas emissions due to natural resource conservation practice application in Indiana: The Indiana Carbon Storage Project, February 2002a. Available at: danpatch.ecn.purdue.edu/ncstore/comet/doc/indiana_final_report.pdf. Accessed on 1/10/07.

Brenner, J., Paustian, K., Bluhm, G. et al., Quantifying the change in greenhouse gas emissions due to natural resource conservation practice applications in Nebraska: The Nebraska Carbon Storage Project, March 2002. Available at: www.casmgs.colostate.edu/vignette/399_7_nebraska_ final_report.pdf. Accessed on 03/15/06.

Claassen, R., ERS analysis of USDA conservation compliance data, reported in "*Have Conservation Compliance Incentives Reduced Soil Erosion?*" Amber Waves, the newsletter of the U.S. Department of Agriculture Economic Research Service (USDA ERS), June 2004. Available at: www.ers.usda.gov/AmberWaves/june04/features/have conservation.htm. Accessed on 04/14/06.

The Economist, Climate change: the heat is on, September 7, 2006.

Energy Information Agency (EIA), Office of Integrated Analysis and Forecasting, U.S. Department of Energy (DOE), Washington, D.C., October 1998. Available at: www.eia.doe.gov/oiaf/kyoto/pdf/sroiaf9803.pdf. Accessed on 02/09/07.

Farm Service Agency, USDA, *Conservation Reserve Program Sign-Up 33: Environmental Benefits Index,* March 2006. Available at: www.fsa.usda.gov/pas/publications/facts/ebiold.pdf. Accessed on 04/11/06.

German Federal Environmental Agency (Umweltburdesant), Greenhouse gas reductions in Germany and the U.K. — Coincidence, or policy induced? *Research Report 201 41 133 UBA-FB000193*, Berlin, June 2001. Available at: www.umweltdaten.de/publikationen/fpdf-l/1987.pdf. Accessed on 04/11/06.

Hasselknippe, H., Carbon market update 2006: towards a truly global market. Point Carbon, February 28, 2006. Available at http://www.pointcarbon.com/getfile.php/fileelement_74094/carbon_2006_final_print.pdf. Accessed on 01/15/06.

Interlaboratory Working Group, Scenarios for a clean energy future (Oak Ridge, TN: Oak Ridge National Laboratory, and Berkeley, CA: Lawrence Berkeley National Laboratory), ORNL/CON-476 and LBNL-44029, November 2000. Available at: www.ornl.gov/sci/eere/cef/. Accessed on 03/15/06.

Lal, R., Potential of desertification control to sequester carbon and mitigate the greenhouse effect, *Climate Change*, Springer, Dordrecht, The Netherlands, 51(1), 35–72, October 2001.

Lambert, D., Sullivan, P., Claassen, R., and Foreman, L., *Conservation-Compatible Practices and Programs: Who Participates?* USDA Economic Research Service, February 2006. Available at: www.ers.usda.gov/publications/ERR14. Accessed on 04/11/06.

Lehmann, J., Gaunt, J., and Rondon, M., Bio-char sequestration in terrestrial ecosystems — a review, *Mitigation Adaptation Strategies for Global Change*, 11, 403–427, 2006.

Markets for emissions credits showing strength on all fronts. *Environmental Business Journal,* XVIII, 5/6, 2005. Available at www.ebiusa.com/News/ArtV18No5.htm.

Marris, E., Black is the new green, Sequestration News Feature, *Nature,* 442, 624–626, 2006.

Natural Resources Conservation Service (NRCS)/USDA, *Managing Soil Organic Matter: The Key to Air and Water Quality (Erosion Control Is Not Enough)*, 2003. Available at: www.nm.nrcs.usda.gov/technical/technotes/soils/soil2.pdf. Accessed on 04/14/06.

Nature, Correspondence, Letters to the editor submitted by M.H.B. Hayes and a second submitted by W.I. Woods, N.P.S. Falcão, and W.G. Teixeira, 443, 144, 2006.

Nicholls, M., EU carbon prices crash as trading begins. *Ecosystem Marketplace,* February 2005. Available at http://ecosystemmarketplace.com/pages/article.news.php?component_id=647&component_version_id=886&language_id=12.

NRCS/USDA, *Conservation Technical Assistance, Natural Resources Conservation Service Fact Sheet*, 2006. Available at: www.nrcs.gov/programs/cta/. Accessed on

Oklahoma Law 27A O.S. Section 3-4, Oklahoma Carbon Sequestration Enhancement Act, 2001. Available at: www.oscn.net/applications/oscn/DeliverDocument.asp?CiteID=212414. Accessed on 05/14/06.

Oklahoma Conservation Commission and Carbon Sequestration Advisory Committee, Oklahoma Carbon Sequestration Enhancement Act: A Report to the Oklahoma State Legislature Relating to the Requirements of Oklahoma Law 27A O.S. 2001, Section 3-4-102, November 29, 2002. Available at: www.okcc.state.ok.us/Publications/Carbon_Sequestration_Act_Report.pdf. Accessed on 05/14/06.

Oklahoma Conservation Commission and Carbon Sequestration Advisory Committee, Oklahoma Carbon Sequestration Enhancement Act: An Assessment Report to the Oklahoma Legislature Relating to the Requirements of Oklahoma Law 27A O.S. 2001, Section 3-4-102, January 6, 2003. Available at: www.okcc.state.ok.us/Publications/Carbon_Assessment_Report.pdf. Accessed on 05/14/06.

Olewiler, N., *The Value of Natural Capital in Settled Areas of Canada*, Ducks Unlimited Canada and the Nature Conservancy of Canada, 36 pp., 2004.

Parton, W.J., Schimel, D.S., Cole, C.V., and Ojima, D.S., Analysis of factors controlling soil organic matter levels in Great Plains grasslands, *Soil Science Society of America J.*, 51, 1173–1179, 1987.

Pew Center on Global Climate Change, *Agenda for Climate Action*, February 2006a. Available at: www.pewclimate.org/docUploads/PCC%5FAgenda%5F2%2E08%2Epdf. Accessed on 02/09/07.

Pew Center on Global Climate Change, *Learning from State Action on Climate Change*, updated March 2006b. Available at: www.pewclimate.org/docUploads/UpdatePew-States BriefMarch2006%2Epdf. Accessed on 03/14/06.

PowerTree Carbon Company LLC, 2006. Available at: www.powertreecarboncompany.com. Accessed on 06/14/06.

Schwartz, P. and Randall, D., *An Abrupt Climate Change Scenario and Its Implications for United States National Security*, a report prepared for the Pentagon, U.S. Department of Defense, Washington, D.C., October 2003. Available at: www.grist.org/pdf/Abrupt ClimateChange2003.pdf. Accessed on 06/14/06.

Soil Science Society of America (SSSA), *Carbon Sequestration: Position of the Soil Science Society of America*, approved by SSSA Board of Directors, October 25, 2001. Available at: www.soils.org/pdf/pos_paper_carb_seq.pdf. Accessed on 03/15/06.

Sparling, G.P., Wheeler, D., Vesely, E.-T., and Schipper, L.A., What is soil organic matter worth? *J. Environmental Quality*, 35, 548–557, 2006.

U.S. Environmental Protection Agency (US EPA), *The Inventory of U.S. Greenhouse Gas Emissions and Sinks: 1990–2004*, April 15, 2006. Available at: www.epa.gov/global-warming/publications/emissions. Accessed on 05/15/06.

APPENDIX A: KNOWN BENEFITS OF SOIL C AND SOIL C SEQUESTRATION

Agricultural, On-Farm Benefits and Societal, Off-Site Benefits

A host of direct and indirect benefits from soil C sequestration has been documented. However, as shown in the text, the findings for some of the benefits are mixed. Additionally, as also noted, some of the benefits may also be attributable to management practices associated with soil C sequestration, instead of the sequestration itself. The spectrum of on-farm, direct benefits is listed next, followed by a list of off-site, indirect benefits that are enjoyed by society.

Agricultural/On-Farm Benefits of Soil Carbon Sequestration

- Improved soil structure and stability leads to reduced soil erosion, improved soil fertility, improved water infiltration and retention, improved aeration and root growth, and reduced environmental degradation
- Reduced production costs
- Labor cost savings (reduced labor needs, reduced labor hours, reduced labor costs)
- Reduced energy costs
- Reduced machinery repair and maintenance costs
- Reduced equipment ownership costs
- Overall cost efficiencies
- Organic matter is a source of plant nutrients (nitrogen, phosphorus, sulfur)
- Improved soil cation exchange (i.e., the ability to store nutrients such as potassium and calcium), which is important for plant growth
- Improved plant production, crop yields
- Improved soil biodiversity
- Reduced bulk density, which improves the plant-rooting environment
- Improved net returns/profitability in some tillage systems/some areas
- Potential federal program payments
- Potential marketability of sequestered tons of C

Off-Site, Societal Benefits of Soil Carbon Sequestration

- Reduced wind erosion of soil reduces sedimentation, improves air quality
- Soil organic C binds contaminants such as petroleum products, pesticides, and heavy metals, thereby reducing toxicity and minimizing leaching
- Reduced water erosion and runoff reduces nutrient loading and thus improves water quality, improving wildlife habitat and human health impacts
- Flood mitigation and control (reduce the magnitude and impact of floods from extreme weather events or excessive snowfall via "temporary water storage")

- Wildlife and recreational benefits (wildlife viewing, pheasant hunting, and freshwater-based recreation and fishing)
- Restored wetlands provide C storage, mitigation of various GHGs, improved water quality, attenuation of floodwater, reduced topsoil loss into river systems, and provide critical habitat to support biodiversity

APPENDIX B: METHODS AND TECHNOLOGIES TO ENHANCE AND IMPROVE SOIL CARBON SEQUESTRATION

On-Farm Management Practices

- Organic farming
- Crop rotations
- Cover crops
- Reduced-till, no-till, and strip-till
- Organic solids management
- Use of biochar/agrichar from pyrolysis of biomass

Prairies, Savannas, and Forests

- Short rotational grazing
- Silviculture
- Restoration of floodplain forests, or changes in cropping strategies and/or soil management strategies in former forested floodplain environments now converted to croplands
- Changing forest rotations, such as by lengthening the time between harvests
- Reduced impact logging strategies (reduced impacts to stored C in biomass and soils)
- Sound forestry practices that do not degrade soils

Wetlands

- Wetland preservation
    - Cease artificial drainage of wetlands and hydric soils
    - Reduce mechanical disturbance of soils (reduced or no-till operations)
- Wetland restoration (particularly lands that have been drained with ditching and drain tile systems, but which are not predictably dry enough for stable cropping, or for which restoration of drainage infrastructure is costly)
- Buffer wetlands from fertilizer and nutrient runoff/migration

Urban Carbon Management (Urban Planning)

- Integration of storm water management into urban planning
- Use of native and low-maintenance landscaping that does not require irrigation or fertilizer
- Minimum seasonal mowing
- Reduced disruption of ecosystems
- Use "smart growth" principles in urban planning and development

Flood Control Strategies

- Landscape-scale management for flood mitigation and flood control (via soil C sequestration)

3 On-Farm Benefits of Carbon Management: the Farmers' Perspectives

John M. Kimble

CONTENTS

INTRODUCTION

This book is about the management of soil carbon (C) management and the benefits to the environment and society in general. In a recent paper about the worth of soil organic matter, Sparling et al. (2006) found that C management had an environmental protection benefit that was 40–70 times the productivity (yield) benefits. The benefits of soil cover and soil organic matter have been covered in many books and papers and will not be reviewed here other than to point out the review by Bot and Benites (2005) was recently released by Food Agriculture Organization. Scientists talk about benefits of soil C management many times in abstract terms based on work done on small research plots or by modeling studies, and many of these are discussed in other chapters in this book. The purpose of this chapter is to look at both the benefits as well as the negative impacts of soil C management as identified by farmers using no-till and other conservation practices. Farmers from different regions of the country that have been involved in work with Natural Resources Conservation Service and

Agriculture Research Service over the years, as well who are innovators, were asked to describe their reasons for making management changes. These changes are often ones that increase soil C. Their comments are based on whole farm management and are not replicated field studies that would allow statistical analysis but they report what they observe and measure as the benefits or drawbacks of changes in farm management practices that increase soil C. First and foremost, the farmers look at yield but they also see changes in erosion, equipment needs, time required for planting, and fertilizers and chemicals usage. Farmers from Georgia, Virginia, Ohio, Pennsylvania, Kansas, Saskatchewan, and Idaho submitted write-ups. These write-ups form the bulk of this chapter.

Farmers are interested primarily in their bottom lines, but many also are very concerned about the environment and leaving the land in a better condition for the coming generations.

The following is from Gene Johnston, *Successful Farming* Managing Editor and contains tips for growing corn as submitted to him by farmers.

Agro-Connect Survey, sponsored by Dow AgroSciences and Mycogen Seeds, brings together farmers and agronomists to share sound science and practical field experience, 3/17/2005.

Want to lower your costs of growing corn, and perhaps improve yields at the same time? Here are some of the tips submitted by 250 top corn growers who responded to Successful Farming *magazine's Agro-Connect survey, sponsored by Mycogen Seeds.*

1. *Reduce tillage.* Fully half of the farmers say that the biggest money saving change they've made in recent years is reduced tillage. Some say they've gone to complete no-till, while others have become more selective in performing tillage operations. Typical comments are like this one from an Illinois farmer: "I have gone nearly 100% minimum tillage or no-till from conventional tillage. I have cut many passes, saving lots of hours on equipment and fuel." And from Nebraska: "We've gone to a total post herbicide program, and do no cultivating anymore. It saves soil moisture."
2. *Use higher planting populations and narrower rows.* More than 10% of the farmers in this survey said they have increased planting populations recently to increase yields. Some say that with the improved genetics on the market today, they now plant for stands of 29,000–33,000 plants per acre. "Going to 20-inch rows gives us a fast canopy, so the soil does not dry out as fast," says a South Dakota farmer who has also increased his population stands. "Fast canopy also helps with weed control, so the corn itself gets all the moisture."
3. *Get bigger equipment.* Several farmers say they have moved up to 16-row planting equipment, and 8-row combines, to get over ground more quickly and reduce fuel costs per acre. Says a Wisconsin farmer, "We also bought a seed tender to fill our big planter faster."
4. *Make in-season fertilizer applications.* Several farmers say they apply at least part of their fertilizer in-season, when the crop needs it the most. "Our plan is to spoonfeed nitrogen," says an Illinois farmer. "We apply some at planting, then side dress more after the crop is growing." A Wisconsin farmer says, "We split the nitrogen in half; 50% before planting, 50% when the crop is up."

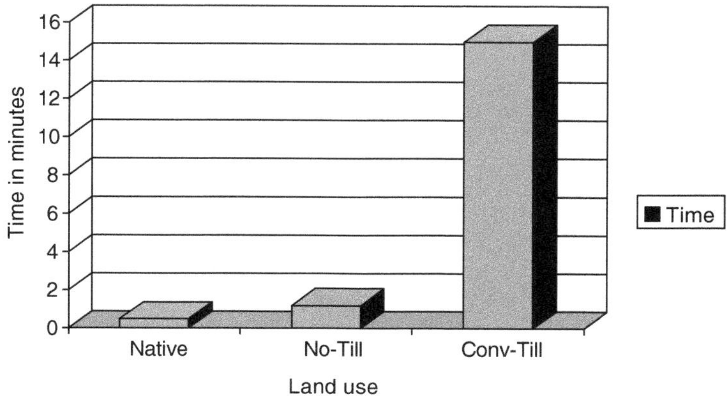

FIGURE 3.1 Time needed for 1 inch (2.54 cm) of water to infiltrate into the soil, South Central Ohio.

The biggest money saving change came from reduced tillage, and this is really soil C management.

Water is the controlling factor in crop growth. Without water, you have no growth. In the irrigated region of western Nebraska, farmers involved in a U.S. Department of Energy (DOE) project have seen a reduction in the number of irrigation passes needed when they go to strip tillage and leave most of the biomass on the surface. They are reducing watering needs — a reduction of two irrigation cycles. As the organic matter improves, more reductions may be expected. In addition, by leaving the residue on the surface they are seeing less wind erosion.

Figure 3.1 from work on the Richards farm in Ohio illustrates the effects of no-till on water infiltration. If you are able to get more water into the ground with less runoff, then moisture from rainfall or irrigation is much more effective. The greater infiltration observed on the no-tilled field is a result of improved soil tilth with more micro- and macropores as well as more material on the soil surface. Figure 3.2, also from Ohio, depicts the aftermath of a 2-inch rain on a long-term no-tilled field (a) as well as a conventionally tilled field (b). A massive amount of soil erosion occurred on the conventionally tilled field and next to none on the no-tilled field. Both fields were on the same soil series and on the same geomorphic landform.

When many farmers began to see the changes that conservation tillage, especially no-till, had on the soils, they did not look at the C in the soil. Instead, they saw more earth worms, more infiltration, less need to clean sediment ponds, improved crops in dry years, and a great deal of "trash" on the soil surface. Many of the farmers we visited had not even looked at the changes in their soil C. When they did, they were pleasantly surprised to see that it had increased. Even on the Richards farm in Ohio, they had not really looked at or considered the change in the soils — just the ancillary benefits they were getting from the change in the practices. When work was being done on the farm, we looked at material from different fields; Figure 3.3 shows material from the soil surface of a no-tilled field on the right and from a conventionally tilled field on the left. The material on the right had a more granular

FIGURE 3.2 Effects of a 2-inch (5 cm) rain event on no-till (top) and conventionally tilled fields in South Central Ohio (bottom).

structure and visible earthworm holes versus the much more massive compacted structure of the material from the tilled field on the left both soils were on the same soil series (Miamian). All too often, the soil all too often is taken for granted and, in the past, was considered inexhaustible.

On one of the projects in Central Nebraska, we spent a day with several local farmers looking at the soil pits from the sampling work. The conventionally tilled field had lost about 90% of the A horizon (i.e., the thick dark mollic layer high in soil C) when compared with a native site. The site under about 10 years of no-till was recovering from past erosion with better structure and a deeper, darker surface horizon. The farmer who owns the conventionally tilled field made the comments that the soil was still black when it was tilled, and he said he had never looked at the soil in a pit before. The interesting thing was that this impressed him so much

FIGURE 3.3 Comparisons of the surface soil of a conventionally tilled field on the left and a no-tilled field on the right.

that he switched to no-till the next year. He started looking at the soil with a different perspective and saw that he needed to do more for his soil.

CASE STUDIES: COMMENTS BY FARMERS IN THEIR OWN WORDS

Elmon Richards, Richards Farm, Circleville, Ohio

The emergence of no-till farming: In the early 1970s, my father, William Richards, (now considered "the father of No-Till"), decided to develop a planter that would plant seeds without the need for any kind of soil preparation or tillage. We found this to be possible after constructing a small test planter capable of planting through various ground conditions of soil and cover. Having achieved this, we also found that we could cover a substantial amount of ground with greatly reduced fuel, labor, and equipment costs.

In the mid-1970s, we decided to try no-till on larger amounts of acreage. To do so, we built a 24-row, 30-inch spacing (61 and 76 cm), 60-foot (18 m) wide planter with complete no-till capabilities. After building and using this planter in no-till and minimum-till conditions over a greater amount of acreage, we found that after the first year of no-till, the yield remained the same. However, after the second and third years, weed control problems, seedbed compaction, and subsoil compaction problems resulted in reduced yields. To eliminate these problems, we changed to a rotation of minimum till for corm and no-till for beans while studying different types of farming throughout the country.

While attending Michigan State University, I went on a work-study program on a vegetable farm located outside of Stockton, California. I studied the concept of permanent beds for irrigation and water control. Even though they were called permanent beds, they would still cut the beds down and rebuild them each season.

After returning home to the farm, we thought over the permanent bed concept and came up with permanent tramlines in an attempt to stop soil compaction. In order to leave two skip-row spaces for the tractor tire traffic, we needed to make up for the lost seed population. To raise the seed population, the planter was completely rebuilt to a 31-row unit with 20-inch (50-cm) row spacing, with two skip-rows for permanent wheel tracks. We made planters this way for beans and com. To keep with the tram-line idea, we made an anhydrous ammonia applicator to fit the same concept. Now, whether we planted com or beans, everything was to go down the same wheel tracks year after year, thus stopping all soil compaction on the seedbed areas.

Along with stopping soil compaction, the 20-inch (50-cm) rows add to earlier development of a close-crop canopy, thus cutting down on the use of herbicides for weed control. With this new complete planting package, we converted to a 100% no-till farming program.

After approximately five years of complete no-till, yields increased back to normal or higher levels than had been obtained with tillage. This is mainly due to increases in soil quality, organic matter, and improved water retention.

No-Till Benefits: The most noticeable results of continuous no-till are the need for fewer, smaller tractors, a need for a fraction of the fuels, more time to conduct other activities, and the ability to farm more land with less workers. In no-till, the smaller tractors use an average of 0.3–0.4 gallons of fuel per acre (2.8–3.7 L ha^{-1}) for planting and spraying. In conventional tillage, the average tractor is 300–400 horsepower with a fuel consumption of 3–4 gallons per acre (28–37 L ha^{-1}) for chiseling, disking, field cultivating, planting, and spraying.

On a national scale, the fuel-savings that could be achieved from converting all conventional-tillage to no-till farming would be dramatic. With an average of 240,000,000 cropped acres (97.2 million ha) in the United States, it takes approximately 720 million gallons (2.7216 million L) of fuel to till and plant, using conventional tillage methods and equipment. It would take approximately only 96 million gallons (362.880 million L) of fuel to no-till the United States, on the other hand. The difference is a saving of 624 million gallons (2.35872 million L) of fuel using no-till farming, and since each gallon of fuel burned releases 6.1 pounds (2.77 kg) of C to the atmosphere, reduced fuel use of 96 million gallons (363 million L) would drastically reduce greenhouse gas emissions.

Although C credits are not likely to provide a substantial income to producers and landowners, the combined benefits and possible savings of fuel, equipment, labor, C credits, and increased land values can amount to a substantial amount of income. Not to mention that very few industries can say they turned their livelihoods over to the next generation in better condition than when they started.

In closing, when I personally ask other farmers why they're still conventionally tilling their farm, most claim the loss of yield between the second and fifth year is too great, and they cannot afford the loss. To promote a greater adoption of a no-till farming program, producers need to be compensated for the low yield years while their soils are rebuilding. They will most likely continue with the no-till program due to the cost savings and an improved lifestyle. No-till is a "no-lose" program for all. It cuts costs, improves soil quality, improves water quality, and helps the environment in general. It is an absolute win-win deal for every aspect of an environmentally sound life.

Bill Jones, Saskatchewan, Canada

I first contemplated changing my farming methods when I purchased a half section of sandy loam farmland that had considerable topsoil loss from wind erosion in the previous years. This land had been farmed conventionally before I purchased it, which meant that half was in summer fallow and half was seeded to a crop every year. The previous owners had worked the land with the plow in early years but progressed to a cultivator and disk in later years. This left the surface free to blow every spring and in open winters with little snow. In 1988, the wind blew all year long with such fury that the air was full of dirt morning, noon, and night. It was obvious to me that we were losing our precious top soil due to our farming methods. We tried every method we could devise to stop this from happening but to no avail. I finally decided that I had to look for a better way to farm. I would either have to seed this ground back to grass or try continuous cropping. Because grassland wasn't too lucrative, I had no choice but to resort to continuous crop. I started to seed the same ground every year to a different crop but my biggest problem was that the hoe drill I was using because it left clumps of residue from the previous crop on the soil's surface. I did not have a very effective method of spreading or breaking up the residue of straw left on the field at harvest time. I looked at a Haybuster® no-till drill but the price was way out of my reach at the time, so I struggled on.

In the early 1990s, I started attending every no-till meeting that was within driving distance. I also watched my neighbor who was using minimum-tillage and it seemed to be closer to what I should be doing to save the land. In those years, I worked at a local seed plant in the winter to supplement my farm income. While there, our coffee time conversation often turned to seeding methods and how to better handle the straw left after harvest so spring seeding didn't turn into a nightmare of straw dragging up into piles. We talked about various kinds of disks and after considering many designs and tests, I felt the angle disk would be my best option. The end result was the single shoot disk opener called the Barton Opener®. As I had a 24-foot (7.3-m) John Deere® air seeder that disks would fit right on; this subsequently became the implement design featured at zero-tillage demonstrations. Once the disk openers were on the machine and I witnessed the results of my initial tests, a whole new way of farming began for me. This was a brand new experience as my father and his father before him had never done anything but farm conventionally.

For me, zero-till farming has many benefits. The first being the wear and tear on my machinery, which is considerably less because I put fewer hours on my tractors and implements compared when I farmed conventionally. My farm fuel consumption was also cut in half and there was less soil compaction as the crop residue, which has made the soil surface mellow. After a few years of no-tillage, the soil felt soft and spongy when you walked in the fields compared to when I conventionally farmed and the ground was rock hard and cracked.

When we used conventional farming methods, we had to "crust-bust" in early spring to seal the ground and retain as much moisture as possible. If we did not work the fields in the spring, the ground would crack down at least four or five inches. Also, if not worked early enough, the soil would often work up in lumps so hard that small seedlings had trouble emerging from the seed bed. In these situations, by the time small plants struggled to the surface, they were yellowed and in a weakened state.

Since using zero-till, our seedlings emerge from a nice warm moist seed bed enhanced by the stubble and are protected from winds that often cut off young tender plants. We also found that the stubble collects more snow and retains this moisture better because the cover from crop residue and standing stubble reduces evaporative water loss from our winds. The soil also soaks in moisture more readily as it is considerably less compact from the rotted straw layered on the surface of the soil. If you drove your truck right after a rain, you could drive on prairie with no problem, but on summer fallow, your vehicle would be instantly become stuck. This happens because the rain infiltrates into the undisturbed prairie more readily than in soil that is summer fallowed. The summer fallow seals over and lets the rain runoff in lower areas where it is lost to evaporation instead of providing soil moisture to benefit our crops. Another plus of no-till is the return of wildlife to our farm over the last few years. I believe this is due to the fact that there is more and constant vegetation on zero-till that provides food and cover. We also find we have fewer problems with some of the weeds we were plagued with before. We can now control wild oats and millet much better than when we first switched to zero-tillage farming. The reason is that the nice finely tilled soil and plentiful sunlight that these weeds prefer is reduced by the stubble and residue cover under no-tillage management. The zero-tillage process can also benefit from using a stripper header. This leaves the straw two to three feet tall at harvest so it collects more snow that winter and therefore collects more moisture for crops planed in spring. Our biggest drawback to zero-tillage, however, is the cost of chemicals but, at least, the chemicals we use are largely environmentally friendly ones such as glyphosate. All in all, I believe our no-tillage farming practices are better and will leave our land in much better condition for our next generation.

Now that we have made such progress in saving the land by conserving our soil, we have to try to save the farming way of life by enhancing profits for the farming community. The farm economy has severely plummeted over the past ten years in Canada. This has caused our next generation of farmers to consider many options, including taking a part-time job to support their families while working the farm or to forget the farm entirely and pursue a different career altogether. Therefore, we are in jeopardy of losing our next generation of farmers.

In the 1930s, during the dust-bowl days of the depression, some of the people from this area around Conquest were determined to stop the severe erosion of the soil by planting tree rows. Tree plantings consisted mostly of carriganas planted in strips every 600–1000 feet (540–914 m) apart in the fields. My grandfather was in charge of a lot of these plantings along with my father and many others from the community working hand in hand with Indian Head Nursery and the government. These tree strips have diverted the wind from our soil for many years; they have also housed much wildlife and birds. Hence, some infrastructure is already in place to conserve our valuable top soil in Saskatchewan, and when coupled with the benefits of no-till farming, I believe that we are doing the very best we can to ensure the sustainable productivity of our lands for the foreseeable future.

Lamar Black, Tilmanstone Farms, Millen, Georgia

Strip till and heavy cover crops in East Central Georgia: My main reason to begin strip till was to control water and soil erosion. Since the beginning, I found many benefits that I did not expect. On Tilmanstone Farms, we have to subsoil to break up a hard pan that is 10 to 14 inches (25–34 cm) deep; therefore, we use strip till to allow for the subsoiling.

I plant winter cover crops every year, mostly rye. With a long-term use of cover crops, we are increasing our organic matter in the top half inch of soil from less than 1% to 3% in 4 years. This has translated into much better water retention from rainfall or irrigation. This was brought to my attention dramatically in 1993 in an irrigated field. In 1990, we planted corn and had to make 41 trips over the field to irrigate. We could not apply over half inch (1.27 cm) per application because of severe runoff in a field with only 3–4% slope. In 1993 we planted cotton back in the same field but used the corn residue from 1992 as our cover. I found that I could apply 1.5 inches (3.61 cm) of water per application without any runoff. With the addition of a rye cover crop to the field, we can apply 2.5 inches (6.35 cm) plus per application without any runoff. This makes our use of water much more efficient by using less irrigation water to make the same yields.

In our nonirrigated fields with use of a heavy cover crop, we can get through a 3 to 4 week drought which we commonly have. By heavy, I mean 6-foot (1.8-m) tall rye at burn down. Also, the use of a heavy cover crop is providing some weed control, particularly in the row middles. With some soil disturbance in the row at planting we still have some weed problems, but not enough that I need to use a preemergence herbicide. All of my herbicides are applied postemergence.

I have reduced my preplant trips over the field from 4 tillage trips to only 2 nontillage. These are a spray trip with a 60-foot (18-m) boom and the other with a grain drill. Our fuel use has been reduced from 4.5 gallons per acre (42 L ha^{-1}) to 1.25 gallons per acre (11.60 ha^{-1}) up to planting.

Other benefits are less equipment required, less hours on equipment, and less labor cost. Strip till and cover crops in Georgia give us cleaner water, no sand blasting of young crops, and better use of fertilizer because it stays in the field after a heavy rain. The organic matter helps reduce leaching. I have been using strip till and cover crops for 12 years and will not go back to conventional tillage. My initial reason to strip till was to control water erosion, which it has done completely. That was the only reason in the beginning. I was the first in my area to strip till. It was not new because I tried strip till in the 1970s with corn.

Yields have stayed the same in years of rainfall. In a drought year, yields are better, as much as 6 bushel per acre (376 kg ha^{-1}) with soybeans. Cotton yields are 75–100 pounds (34–45 kg) better. All corn is irrigated. 2004 was the best corn yield ever with an average of 220 bu per acre (22,062 kg ha^{-1}) Before Bt (Bacillus thuringiensis) cotton, we sprayed less insecticide, usually two less applications per year with a savings of \$12 per acre (\$29 ha^{-1}) for the insecticide. Now we spray for different insects, and we use the same for conservation till or conventional till. The savings in trips across the field will be \$20 per acre (\$49 ha^{-1}), taking into account the cover crop and burn down herbicide; \$20 is a net reduction. According to soil tests, the fertilizer usage is the same. Nitrogen applied to the cover crop is recycled the next year.

Kenneth J. Cain, Cain's Homelike Farms, Darlington, Indiana

The following is a short description of Cain's Homelike Farms (CHF) and our relationship to the land, and some information about things we are trying as they relate to soil C and no-till. CHF is a family farm comprised of two brothers, their wives, and families. We are located in west central Indiana, in the Sugar Creek watershed, not too far from Purdue University, West Lafayette, Indiana. Our family operation consists of approximately 3000 owned and rented acres (1250 ha) in a 50/50 corn/soybean rotation, on land that was once home to the Shawnee and Miami Indians. The farm covers a wide range of soil types, from the Stark-Mahalasville association, a nearly level poorly to very poorly drained soil, to the Miami-Fincastle association, a nearly level to strongly sloping somewhat poorly drained and well-drained soil. Farming this land is a privilege, not a birthright, and we are challenged to leave it better than we found it.

The success of generational farming relies solely on the stewardship of the present generation. Soil health is our primary concern, and we are constantly looking for better methods of "growing" our soils to higher levels of biodiversity. In other words, to work in harmony with natural processes, versus trying to subdue and control them with commercial heavy iron, will keep more of our natural resources in place for the next generation. We must keep in mind that our soils are not just inanimate clays, sands, silts, and minerals, but a living, breathing community of interrelated organisms from microbes to earthworms.

In our corn and soybean cash grain farming operation with its constant pressure on the soil, versus our father's livestock farm with diverse cropping rotations, we have found that the less we disturb the soil the better our bottom line. A no-till cropping system has kept the farm competitive in the consolidation phase of today's agriculture. Economic health follows soil health. Since our conversion from a conservation tillage mode in the 1970s to a total no-till mode in the 1980s to present, we have experienced both. No-till has its challenges, but with today's herbicides and seed delivery/placement technology, it has become easier.

The future production efficiencies will not come only from genetics, equipment, and synthetic chemicals, but from a renewed awareness of carbon's role in soil health. Biological management of the soil profile will play an important role in soil health. We are currently trying annual rye grass as a cover crop to sequester carbon and scavenge soil nitrogen to hold it over for the next crop. Manure from neighboring hog farms is also an integral part of our fertility and carbon needs. We are also working with Purdue University, Cinergy, a local utility, and Pheasants Forever to establish some baseline data on carbon sequestration in the different warm season grass stands we have had under cultivation since 1997.

Dave Hula, Renwood Farms, Charles City, Virginia

Benefits to no-till:

1. Reduced labor — Renwood Farms is able to crop 4500 acres (1822 ha) with 2 people operating the equipment and 2 support personnel.
2. Reduced fuel cost — Prior to no-till our tillage system required about 250 gallons (945 L) of fuel to plant 60 acres (24.3 ha) of grain and with no-till one can plant about 300 acres (121 ha) on 120 gallons (453 L) of fuel. (*Editor's Note: Before no-till, 4.16 gallons per acre were used. After no-till, 2.5 gallons per acre were used; this is a 40% reduction in fuel usage.*)
3. Timing (able to cover more acres per man-hour) — With a 30-foot (27-m) no-till drill, one can plant 120–175 acres (49–70 ha) a day depending on field size, and with a 30-foot (27-m) corn planter, one can easily no-till 100+ acres (40.5 ha) a day. Under conventional tillage our operation was limited due to the time required for tillage, which the production was about 60–80 acres (24–32 ha) per LONG day.
4. Timeliness (spraying, planting, harvesting) — No-tilling allows one to enter the field much sooner after a wet period.
5. Yield increase for corn/soybeans — Our soils have low water-holding capacity, and no-till allows for less water loss and over time increases storage as the organic matter increases.
6. Less equipment — Prior to no-till we required 2 large tractors and the tillage equipment, then a tractor to plant with. Today, we crop with only one 225-hp tractor.
7. Tax credit availability on equipment — We take advantage of Virginia tax credits on conservation tillage equipment (i.e., no-till planters and drills).
8. Long-term fertility programs become more stable — Lime requirements are reduced, nitrogen needs are maintained stable based on crop removal, P applications are more specific, and K requirements become more cyclical.
9. Reduced soil erosion — There are visible improvements of reduced soil erosion not only noted in the fields but also in the drainage ditches, as they need to sediment cleaned out less often.
10. Improved water quality — The water runoff is not only minimized but much cleaner.
11. Moisture conservation — Based on our irrigation-scheduling program under an extended dry period the irrigation begins 1–2 days later and the infiltration rate is superior in a no-till situation versus the conventional system.

Problems to no-till:

1. Tire wear — There is an increase in tire wear particularly behind corn and soybean stubble.
2. Yields on small grain (scab potential) — At the beginning of the continuous no-tilling process, the wheat yields seemed to lag behind but our long-term average. The no-till wheat has only averaged 1.7 bushels (106 kg) less as compared with the conventional tillage system. The corn yields have increased over time to where

our long-term (11-year) average is 131 bu/acre (8215 kg ha^{-1}). No-till does contribute greatly to this increase. As for the soybean crop, our yields have not increased like the corn production has, but the no-tilling practices have not hindered the production. The few years we incorporated cotton in our rotation, we no-tilled the crop and our production average was 890 pounds/acre (997 kg ha^{-1}). We have no experience with conventionally tilled cotton, but our observations were that the lack of tillage created no additional problems.

3. Weed shifts — An increased amount of perennial weeds (Johnsongrass, pokeberry, milkweeds, field bindweed, horsenettle, and wiregrass) and hard-to-control annuals (Italian ryegrass, morning glory, and mustards).
4. Compaction on certain soils — On the sandy soils there is evidence of a hard pan where some subsurface tillage is helpful. The challenge is that the economical return from the subsurface tillage is inconsistent.
5. Fertility — The need to manage nitrogen does not change, but with the lack of incorporating nutrients we see a greater need to manage P through the use of starter, since this is the only time of application until the soil levels fall into the M-H (medium to high) range. And at that time a maintenance application is used in front of a small grain crop.
   - Need to increase nitrogen at beginning of continuous no-till
   - Need to place P more efficiently in starter, due to the lack of incorporation and starter P is more expensive
6. Cooler soils — In the corn production, soil temperature becomes critical because as we try to plant earlier, one must select a variety that has good seedling vigor. In the small grain production, if possible, earlier plant is an option but one must watch for other challenges (heisson fly, mildew, aphids).
7. Pests — With the lack of tillage, certain animals become problems (ground hogs, skunks, and fox dens).
8. Disease — With the abundance of residue, there is an environment for high levels of disease spores, which can affect several crops in the rotation (scab, the fungi that aids in breaking down the corn residue, is harmful to the wheat crop in the pollination process). Also, there is an increased presence in both corn and soybeans of diseases that primarily affect seedlings.
9. Insects — Hession fly becomes a concern in the wheat production.

Observed changes and improvements since the early days of no-till:

1. Pesticide availability
   - Herbicides to control problem weeds
   - Insecticides to control insects due to the continued live soil residue
   - Fungicides to control most disease less scab
2. Equipment enhancements and availability
   - Prior to the Midwest implementing no-till, all planting equipment had to be modified, but today one can purchase proven no-till equipment.
   - Minimum soil disturbance deep tillage
   - Improved floatation equipment despite the size to reduce compaction
   - GPS to offer controlled traffic patterns or tram lines
3. Technology
   - Genetically modified crops (Roundup Ready® for weed control and BT for insect control)

- Plant breeding (development of varieties specifically for no-till [i.e., varieties that emerge in a cooler seed bed and have greater disease resistance or tolerance])

Why did we start no-tilling? We started no-tilling corn and soybeans in the 1970s just to speed up the planting process and conserve moisture. While back in 1987 we conducted our first experiment of continuous no-tilling or "NEVER-TILL" ON 67 acres (27 ha), this was our first exposure to no-tilling wheat behind corn. The experiment soon became the idle conservation and little did we know that it is becoming more accepted as a production practice. Today, the majority of our land has been in continuous no-till since 1991, and the only application for tillage in our operation is for sludge, subsurface tillage of filling in ruts. The application of no-tilling has allowed us to focus more on the planting and spraying applications of the cropping system. When correctly managing this system, one will be pleased with the results.

Why don't others practice no-till?

1. Cost — Having to buy a no-till planter.
2. Change — FARMERS DO NOT LIKE CHANGE.
3. Appearance — No-till farming is ugly, and some landowners as well as producers do not like the appearance.
4. Recreation — Some farmers just enjoy tillage, and they have the equipment, so why not use it?
5. Yield — Potential for wheat yield reduction, therefore some hesitation.

Table 3.1 summarizes some of the effects observed with the use of no-till. A large reduction in runoff, as well as sediment and reduced nitrogen and phosphorus losses occurs. When hurricane Hugo struck with about 20 inches (50 cm) of rain, almost no erosion occurred on the no-till fields, while extensive erosion occurred on conventionally tilled fields.

TABLE 3.1
Renwood Farm Data by the Innovative Cropping Systems Group in Virginia[a]

Treatment	Runoff Cu.Ft/Ac	Sediment lb/ac	Nitrogen lb/ac	Phosphorus lb/ac
Plowed + fert	6506	3176.3	9.17	3.65
No-till + litter	1547	30.5	0.54	0.38
No-till	2014	18.5	0.49	0.27
No-till + fert subsoil	1537	5.4	0.47	0.26
No-till + fert	1373	16.0	0.46	0.25
			%	
Average reduction by no-till	74.9	98.2	94.7	92.0

[a] 7.5% slope and 5-year storm event.

Gordon Gallup, Ririe, Idaho

Dry land wheat and barley: In 1984, we were trying to produce with a conventional tillage approach on 900 acres (365 ha); we had just changed from a summer-fallow rotation to a continuous cropping program. The planting process at that time was a minimum of 5 passes with tillage equipment, putting 1300 hours on two tractors both over 300 hp. In 1985, we purchased our first no-till drill, and that spring we put under 90 hrs on one tractor, seeding the same 900 (365 ha) acres. It's hard to put a value on the reduced time on the tractor because you also have less fuel consumption, longer life of the tractor, less annual maintenance, and longer intervals in the major eminence. We now seed around 3500 acres (1418 ha) each year with one tractor and just less than 180 hrs, so the progress in the no-till equipment has helped once again to improve our productivity with even less cost per acre as far as fuel and fertilizer inputs are concerned. With the rise in fuel costs this winter, I am glad that I'm not still trying to make the 5 pass operations we were making in the early 1980s.

We were fortunate in that our first years we actually outyielded the conventional tillage neighbors from the word go; we had some dry summers that cut into conventional tilled crops because of the moisture loss from the preplant tillage. We were able to conserve all the winter moisture, and it held our crops until some late summer rains saved us. We always said that the no-till was the last to die in those years. This year, 2004, will go down as an all-time record yield for the whole farm, with barley coming at 104 bushels (6522 kg) and spring wheat at 89 (5581 kg); the 10-year average is 75 bu (4703) on barley and 67 (4201 kg) on wheat.

The biggest plus that I like to talk about is the erosion factor; we have seen our erosion virtually stop. The small and large gullies have now healed, and we can farm whole fields without cutting them into small pieces going up and down deep ruts cut by spring runoff and heavy summer storms. The stubble is also a great catch for snow in the windy winter season. The overall health of the soil just seems to be better.

Soil quality goes hand in hand with the erosion: Leaving the residue on the surface has played an important role in both erosion and soil quality. The residue seems to break down better in the sun light and doesn't rob the soil of important nutrients that the crop needs that would otherwise be used up breaking down the residue that is traditionally turned under.

The no-till experience for me has been a positive one from the first year till now. I at this time have a 10-year average of 75 bu (4703 kg) barley and 65 bu (4201 kg) of wheat with 4 of the driest years since we started no-till. As for my neighbors' averages, he is reluctant to share. The best I can do is give my yields before no-till. In my pre no-till years, it was good to get 75 bu (4703 kg) of barley and unheard of to get 65 bu (4201 kg) of wheat. I always figured my average barley was about 55 bu (3449 kg) and 40 bu (2500 kg) on wheat, and not much difference when we were in summer-fallow rotation. *(Editor Comment: That is a 36% increase in barley and 37% increase in wheat for the 10-year average. Figure 3.4 presents the benefits of no-till on tractor hours and yields.)*

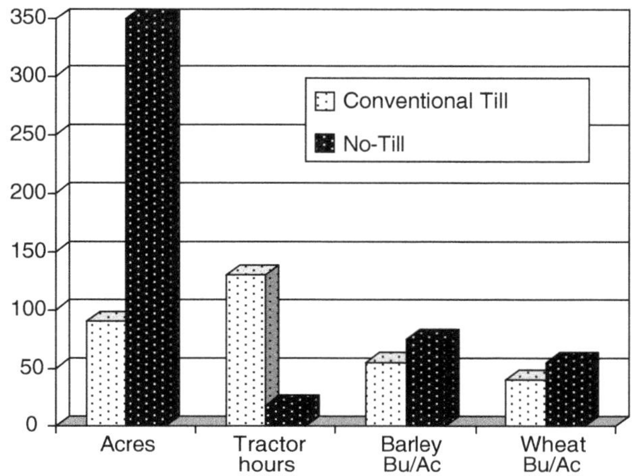

FIGURE 3.4 Data showing changes in acres farmed, tractor hours, and barley and wheat yields for conventional tillage versus no-till. The acres farmed and tractor hours need to be multiplied by 10.

We use about 5000 gallons (18,900 L) of fuel per year on 3500 acres (1417 ha), including harvest and spraying most of the bigger engines will burn about 12 gallons 45 L) per hour under light to moderate load; this is 90% of what I do. The conventional tillage use of fuel is extremely high with sometimes as many as six trips across the ground by the time seed is put in the ground. My friend to the south of me still uses conventional tillage and tells me that his fuel consumption is in the 6000 to 7000 gallon (22,680 to 2611 l) range for his 1000 acres (405 ha). *(Editor Comment: This works out to about 1.43 gallons per acre for no-till and 6.5 gallons per acre for conventional tillage, which works out to about a 78% reduction in fuel usage per acre. This alone will lead to a major reduction in greenhouse gas emissions as well as a major monetary savings.)*

There were two reasons that we looked into no-till. The first was the time it was taking in the spring to get the grain planted. We needed to get the crops in within 10 days, and it meant extra long hours to get that accomplished. The second was the amount of erosion that we were getting by working the ground in the fall. It seemed to loosen the ground too much, and with the right snowmelt we could get some severe erosion in the spring.

Merle Holle, M&K Farms, Marysville, Kansas

Why I no-till a partnership with Mother Nature: No-till farming appears today as a coined concept of what Mother Nature has practiced for centuries. It was not until recent history of mankind's existence that the perceived need to upset the soil structure, stir it up, pack it down, and then seed into a "good" seedbed was deemed necessary to get a good crop yield. We did not have the technology, equipment, or know-how to achieve a healthy growing crop free from weeds and pests without tilling until recent years. It was a proven, economical way to produce food and fiber from the rich soils we had the opportunity to be stewards of.

However, in the last 40 years, many innovators have tried new concepts of production agriculture by mimicking Mother Nature. These innovators reasoned that by cooperating with "her" we could improve the soil structure and provide a seedbed that is better than anything man can establish, provide cover to retain moisture in limited rainfall areas, and this cover also retains soil when we do get a heavy rainfall event. Mother Nature is such a willing partner in this concept that she even provides us with the underground microflora, bacteria, and invertebrates to make this soil a living, breathing, and water-absorbing medium. We only have to observe and understand what she needs and work with her, and not try to make her do things our way.

I was trained in production agriculture at being an astute manager of the land I farmed. By that, I mean we would work the ground often to make sure no weeds were allowed to make seed, provide a seedbed with "good," moist, seed–soil contact, and would walk the growing fields roguing any weeds that escaped the cultivation. We were taught that the only good weed is a dead weed. I still believe in the "dead" weed concept, but thanks to the innovators I have learned from, and from my own experience, I've discovered that some of the concepts I previously thought mandatory are not necessarily so.

The economics of production agriculture have changed tremendously in the past few years. The price of fuel, fertilizer, machinery, and labor has increased dramatically, and the grain prices vary but overall have stayed fairly constant. The cost–price or margin we receive is narrowing at a steady rate. In order to survive this highly competitive business, we must look for ways to either "cut cost," or "improve income." Either one is very difficult unless we are willing to "change." I truly believe that by changing our mindset of production agriculture and adopting the no-till methods, we not only can cut costs, but also improve sales by increased yields due to better soil structure, soils that are more active, retaining more moisture, lessoning soil erosion, less root pruning, etc. I have several charts showing the savings of fuel, machinery, and labor on our farm, comparing "no tilling," to "conventional tilling."

Our yields have increased steadily since we have not tilled, not so much because of the "bumper" yields, but because our yields have evened out and are more consistent than before. Many have said that by less tillage they have more time to do other things they enjoy, such as family outings, recreation, etc. This is true, however, you can use this time for machinery maintenance, livestock duties, management planning, increasing acreage, etc., the point being you are spending less time doing the "unnecessary" tillage, and you can be more productive with your efforts. Crusting of the soil is no longer an issue because of the partially decomposed residue on the surface. Most times, residue is visible from the two or three previous crops.

There are, however, some tradeoffs or disadvantages to not tilling the soil. Adequate fertilization is a key issue with no-till; by disturbing the soil, there is increased decomposition of organic matter and residue which gives off carbon dioxide and leaves available nitrogen for the crop. And by so-called "mining" the soil of organic matter, it gives a short-term boost of plant nutrients, especially nitrogen. This is quite noticeable at the growth or vegetative period. In no-till, starter fertilizer and a balanced fertility program are a must because the bacterium of the soil, as it decomposes the residue, is a strong competitor with the crop for available nitrogen. This nitrogen in addition with C and other nutrients combine to add organic matter to the soil, which will be released at a later time. Additional nitrogen needs to be placed, so the plant gets what it needs at the proper time. By adding organic matter, instead of mining it, the soil texture and structure improve, the beneficial organisms increase, and after a few years, it repays the manager with better yields.

In extremely heavy residue situations, *without proper equipment*, it may be hard to establish a good stand if planted in that residue. It doesn't take a lot of sophisticated machinery to provide a good seedbed. Most times, a good set of residue managers is quite adequate; the secret is not to try going too deep with the managers, just move the residue.

Perhaps the most challenging of all is to convince yourself, and then your landowner, that the no-till concept does work. It isn't the only way to raise a good crop, but is a proven, reliable, environmentally friendly practice that pays good dividends to you, if you are willing to change the way you are now doing things.

The first 3–4 years are the hardest; the soil *appears* solid, and small rills appear in the field. The fertilizer costs increase the first 4–5 years as you are building organic matter, you are the subject of "coffee talk gossip," you see the neighbors tilling, you smell that clean fresh soil, and almost have to tie your hands behind your back, so you don't grease up the tillage machine and join them.

At first (depending on weed–seed concentration), it may take some more pesticide applications. But if you stay after the weeds, in a few years, your herbicide application will be less because you're not "planting" the weed seeds when tilling. Your partner, Mother Nature, will decompose most of the seeds as they're not in the germination zone.

In summary, if there is a better way to farm with less cost, more yield, improved environment (air, soil, water), and less labor, is it not a great opportunity to change to that program?

Roy Henry, Clay Center, Kansas

Being mostly involved in the livestock, specifically swine, I have been proactive with waste management for environmental reasons. Societal concerns about concentrated livestock production have increased with time. Some concerns are without a doubt justified, but many other accusations are not accurate. The same statement could be said about all of agriculture as well as most industries in some stage of their maturity. With the change in tillage made possible, no-till erosion is controlled. When animal nutrients are applied to crop land with residue, the nutrients will stay in place. Fields that have been in no-till for some time have levels of residue and percolation rates that are very receptive to the surface application of animal nutrients for the use of plant growth, without the risk of erosion into our streams and rivers. This is usually the method used for dewatering a lagoon where the nutrient load is relative low. The value of this to the livestock sector in many parts of the U.S. could very often be measured by the ability to raise livestock. In the near future, it will be very difficult to produce livestock unless waste application fields are able to take the nutrients without concern of loss to erosion.

To the pure no-till person, the injection of waste under the surface destroys some of the benefits of no-till. This could be true, but we need to evaluate the benefit versus the cost before we draw too many conclusions. When injecting manure, we typically only apply to a specific field once every four or five years. The reasons are that many times the conditions are not suitable for application when it needs to be done, and the fields are unable to take additional nutrients in order to stay in compliance with regulations. The issues we have to consider are crop rotation, compaction, when last applied, and the crop to be planted. Some crops seem to do better when one or two years removed from the actual application of manure. Wheat is likely to lodge the year following manure application, which can cause reduction in yield. When manure is applied, it is not as predictable as commercial fertilizer as to availability and content of nutrients. Because of this, I like to space out applications. The disturbance to the soil caused by the application varies greatly with the soil type and the moisture conditions. Fall application for the spring crop works the best on fallowed ground.

Because of the use of no-till and the injection of animal waste, I have been able to improve the soil fertility greatly, especially with regards to P, with no concern of erosion causing problems for our lakes and streams. The injection of manure 6 to 8 inches below a surface that has a good covering of residue is a sustainable system with the safeguards in place to protect the environment and raise crops to a high level of productivity. There really is very little disturbance to the soil, as can be seen in Figure 3.5. It is a system that gets us closer to being totally independent. The pig produces the nutrients that the crop needs in an environmentally friendly manner. I realize that it is not quite that simple, but it is a good start.

FIGURE 3.5 Swine manure injection with sweep blades 8–10 inches (20–25 cm) below the soil surface.

Dan Bower, Pennsylvania

Farm history: We have a 500-acre (202 ha) hilly farm in north central Pennsylvania, of which 300 acres (122 ha) are tillable, 50 acres (20 ha) untillable pasture, approximately 10 acres untillable unused open land (log landings, electric and gas line right of ways, and small and extremely steep areas), with the remainder woodland. Up until 2003, we had a 60-cow dairy and 60-head heifer replacements, with equipment for all phases of tillage and harvesting by husband and wife with no outside labor. Our growing season is relatively short, so timeliness is important for a quality crop.

Priority issues were:

1. April and May plantings
2. August 1–15th for alfalfa grass seeding after rye and oat harvest
3. Open silage ground needed fall cover crop ASAP
4. Replacement cost of extensive line of equipment was aging
5. Chance of thundershowers (erosion)

Lesser issues were:

1. Steeper fields were left in hay too long (declining yields)
2. Permanent untillable pastures needed reseeding
3. Cost of fuel kept rising
4. Untillable small plots were unused and eyesore
5. Tax break for new equipment purchase

Solution: We felt that a no-till drill would address all the issues, the cost being $18,000 less trade in of $3000 — a final cost of $15,000. The main impact was saved time, saved energy, and saved soil, along with improved wildlife habitat.

Results: We have planted every open area (acre), whether steep or untillable; this is mostly planted for wildlife. It's hard to put dollar values on the benefits of this no-till drill from the saved time, less fuel, smaller tractors, saved soil and improved soil quality, less stones to pick up on surface, and improved wildlife areas. A forester walked our woodland ground and commented that our hardwoods were regenerating, which he said was not happening much elsewhere, which could be a benefit from planted areas for wildlife.

The many benefits of no-till drilling may not readily *add* dollars to your income. I believe it will reduce costs and add quality stewardship on our farm. I feel that many farms are no using no-till because they will not see income rise, and are wondering whether the reduced costs and better stewardship will cover the drill costs in such a tight farm economy.

SUMMARY AND CONCLUSIONS

What can be gleamed from what these farmers have observed? Leaving biomass on the surface reduces both wind and water erosion. It increased infiltration and the water-holding capacity of the soil. It reduces the time needed for planting, allowing the farmer to do other things. It can result in up to a 90% reduction in tractor usage and allows for the use of smaller tractors. Over time, some farmers show a yield increase of up to 37%.

Management of phosphorus may be more of a problem because it is not incorporated into the soil. For example, in Virginia, tillage is done when sewage sledge is applied. Roy Henry in Kansas incorporates his manure underground, which reduces the possibility of runoff and the resulting contamination of water bodies. This creates very little disturbance and allows for no turning of the soil, along with many other environmental benefits.

On the Richards Farm in Ohio, where one field was committed to growing of corn for silage for 10 years (and even after the first year without residue being left on the field), a decline in the soil health/quality was observed. To overcome this soil quality change, a cover crop was planted; this improved the soil quality. The main point here is that modifications will be needed as cropping systems change. However, new problems are developing, and they will need to be addressed. In Ohio, major slug problems are currently occurring, and the cost of even partial control is very high. In the past, tillage and removal of the biomass from the surface controlled the slugs. Research is needed to identify economical method to control the slugs. William Richards, from Knox County Ohio, said that many farmers are switching away from no-till because of the slug problem. Concerns, such as slugs, need to be addressed so that fields in long-term no-till can remain in no-till.

Nitrogen (N) management is still needed; most farmers have continued to apply N at recommended rates. This is an area that needs more work. The need for more studies is clearly pointed out by Mulvaney, Khan, and Elsworth (2006). In the past, which was an era of relatively cheap N fertilizers, N was applied many times without regard to how much nitrogen was in the soil or was really needed. This has, in many areas, led to excess application, which resulted in ground and surface water contamination. Nitrogen was applied as an insurance policy without any consideration for the excess application. Hopefully, this will change in the future. As soil quality improves, more biological release of N occurs from the crop residues, and this should, over time, reduce N fertility needs. One farmer in Illinois, however, has not seen any N response on a 12-year, no-till field with applied N. Whether this situation will continue is unknown because prior to N fertilization soil testing is needed and N needs to be applied based on the soil tests. All too often, N is applied based on an expected yield at a rate of 1.2 pounds (1.34 kg ha^{-1}) of N per bushel of corn. This needs to change to help reduce the overapplication of N on many fields.

Some farmers have found the need for more herbicides, while others use less. Going to 20-inch (50-cm) rows in corn may help to reduce weed growth because the crop provides a quicker canopy cover for the soil. In addition, the seedbed with a lot of mulch on the top is not as conducive to seed germination as a well-prepared (i.e., tilled) seedbed. Nor are buried seeds brought to the soil surface where they will germinate. Overall, there would appear to be a reduction in the need for herbicides other than a "knock-down" just before seeding.

Most of the farmers who have gone to no-till say they would never go back to traditional tillage. Changes do need to be made in farming practices, however. In some high-yield areas, some farmers feel there is too much stubble to plant into without some tillage. As Merle Holle pointed out, all that is needed is a good set of residue managers that create a minimum disturbance to the soil surface. If this is done, it should be kept to as little as possible and as shallow a depth as possible. In high-residue areas, heavier planting equipment may be needed with the ability to cut through the residue, or row cleaners may be needed to move the residue to the side without disturbing the soil.

No single system will work everywhere, but the less tillage, the better the long-term soil quality. Strip tillage is also used to manage the residue and leave a cleaner seedbed as well as a strip that will warm faster in areas where the soil temperature for germination may be lower than desired under the crop residue.

Many of the farmers see the multiple benefits of no-till, which are really soil C management. Most have had increased yields, but in the wetter areas, it took about 5 years to see the crop yields stabilize or increase when the switch to no-till was made. However, in drier areas where moisture is the major limiting factor, the effects on yield were seen even in the first year. These farmers are the innovators who are willing to try new things and were able to make the needed investments in equipment, as well as take a yield decrease for a couple of years. They were in it for the long run, and they are also very good stewards of the land. They want to pass it on to the next generation in better shape to ensure the sustainability of our farming systems.

The bottom line is that soil C management will give us more productive soils that are less prone to erosion and of much better quality. There will also be many off-site benefits that will benefit society as a whole. This will allow the future generations to be more productive and help the overall environment.

REFERENCES

Bot, A. and Benites, J., The importance of soil organic matter key to drought-resistant soil and sustained food and production, *Food and Agriculture Organization of the United Nations*, FAO, Rome, 2005.

Mulvaney, R.L., Khan, S.A., and Elsworth, T.R., Need for a soil-based approach in managing nitrogen fertilizers in corn production, *Soil Science Society of America. J.*, 70, 172–182, 2006.

Sparling, G.P., Wheeler, D., Vesely, E.-T., and Schipper, L.A., What is soil organic matter worth? *J. Environ. Qual.*, 35, 548–557, 2006.

4 Private and Public Values from Soil Carbon Management

Siân Mooney and Jeffery Williams

CONTENTS

INTRODUCTION AND SUMMARY

Soils store a significant quantity of carbon (C), and have been identified as one possible sink (i.e., repository) for atmospheric CO_2. The opportunity to increase C within agricultural soils arises, in part, because conversion of native forests and grasslands to production agriculture during early settlement of the United States reduced the quantity of C contained within these soils.* Many traditional agricultural practices, such as moldboard plowing and removal of plant residues, historically lowered C storage, but considerable expertise and effort is currently being applied to change agricultural management practices to increase soil C levels. Soil C sequestration has received increased scrutiny of late because of the role C sequestration could play in reducing atmospheric concentrations of greenhouse gases (GHGs) such as CO_2. Many articles have addressed the benefits of mitigating global climate change. This chapter focuses solely on the other additional benefits to private individuals and society as a result of adopting management practices that sequester additional soil C. For example, soil C sequestration can help mitigate soil erosion by wind and water that has been exacerbated by some farming practices. Erosion can reduce the profitability of farming operations and can also have negative consequences for areas surrounding the farm; for example, erosion contributes to siltation, which increases flood potential, raises dredging costs on navigable waterways, and can lower water quality, which negatively affects aquatic life, recreational opportunities, and increases water treatment needs before human consumption. Erosion increases the quantity of soil particles suspended in water and can increase water treatment costs.

This chapter quantifies some of the private and public benefits that could result from adopting farming practices that increase the rate of soil C sequestration. Private benefits are those that accrue directly on-farm and are enjoyed only by the agricultural producer for example; changes in production costs, crop yields and overall profitability. Public benefits are additional benefits that accrue off-farm as a result of additional soil C sequestration and can be enjoyed by society as a whole for example; reduced soil erosion and increased water quality.

Private, on-farm benefits to land managers are discussed in the first section of the chapter. Considerable research has demonstrated that adopting management practices that increase the rate of soil C accumulation can reduce labor costs, fuel costs and machinery repair, and ownership costs (Table 4.1). In some cases producers also experience increased crop yields and net returns (Table 4.2). In addition, increased rates of soil C accumulation provide an opportunity for producers to generate revenue from the production of a new commodity, C, that has value in the developing market for GHG credits. As with all technologies, these management practices are economically superior for some land managers and have already been adopted. Other land managers may face barriers to adoption because of the physical, biological and economic characteristics of their specific location, their management capability and risk preferences. Implementing these management practices is not economically viable for these managers without additional incentives. Improving

* For more details, see Chapter 8, "Soil Physical Properties and Erosion," by Rattan Lal.

soil C can provide society with additional economic benefits that cannot be captured by individual land managers at present. If it were possible to capture some of this economic value, it is likely that more land managers would find it economically viable to change their management practices to increase soil C.

The second part of this chapter addresses the economic values enjoyed by society because of improved soil C management. These benefits include reduced water and wind erosion, sedimentation, nutrient runoff, and improved water quality, recreation and wildlife. These are environmental services for which property rights (i.e., ownership) are not well defined. Because of this, the land managers cannot sell these benefits to society and have an incentive to under produce these services. However, well-functioning natural ecosystems are an important asset to society, which benefits from the many services that ecosystems provide (Stevenson, 1994; Heal, 2000). Many of the services do not have explicit prices because we receive them free; however, the absence of a market price should not suggest that these services are without value (Heal, 2000; NRC, 2005). In fact, ecosystem services are often very costly and difficult to replace using engineered solutions.

It is difficult to put a dollar value on these services; however, several studies have shown that their value is considerable — at a minimum, worth many billions of dollars to the United States alone (Table 4.3 and Table 4.4). It is clear that the development of policies that provide incentives for land managers to increase soil C management could have considerable benefits to society. Only a small portion of the values presented in this chapter are due solely to better soil C management. Unfortunately, sufficient information is not available to estimate societal benefits that are solely attributable to improved soil C management.

PRIVATE BENEFITS FROM SOIL CARBON MANAGEMENT

Agricultural management practices that increase soil C sequestration can be economically beneficial to producers because they can increase yields and potentially reduce production costs. The rate of soil C sequestration can be increased in many ways. The following sections consider some of the private benefits that could accrue as a result of sequestering additional soil C by reducing tillage. The impact of these changes on input costs, crop yields, net returns, government payments, and the possible joint production of new commodities, such as C credits, are examined. The potential private economic values of management changes that increase soil C sequestration are summarized in Tables 4.1 and 4.2.

PRODUCTION COSTS

Labor Cost Savings

Several factors can decrease the labor requirements under reduced and no-till systems, in comparison with a conventional till system. Less labor time is needed in the field under a reduced tillage system because fewer tillage trips are required for seedbed preparation. In addition, fewer cultivation operations are required and, in some cases,

TABLE 4.1
Summary of Possible Production Cost Changes from Adopting Practices to Sequester Additional Soil C

Studies	Range of Findings
Labor Costs – Field Operations	
Aller et al. (2001); Harper (1996); Massey (1997); Olson and Senjem (2002); Yin and Al-Kaisi (2004).	Labor savings for field operations due to tillage reduction range between \$2.47/ha (\$1.00/acre) and \$19.13/ha (\$7.74/acre).
Fuel Costs	
Williams et al. (1990); Parsch et al. (2001); Massey (1997); Aller et al. (2001); Williams et al. (2004).	Fuel cost reductions from reducing tillage range between \$3.58/ha (\$1.45/acre) and \$28.29/ha (\$7.63/acre).
Fuel Use	
Williams et al. (1990); Williams et al. (2004).	Fuel use reductions as a result of reducing tillage range between 26.2 l/ha (2.8 gal/acre) and 35.70 l/ha (3.8 gal/acre).
Bashford and Shelton (1981); Schrock, Kramer and Clark (1985); West and Marland (2002).	These studies provide a range of estimates indicating that fuel use is lower for operations that do not involve tillage.
Machinery Repair Costs	
Johnson et al. (1986); Mikesell, Williams, and Long (1988); Williams, Llewelyn, and Barnaby (1990); Weersink et al. (1992); Parsch et al. (2001); Williams et al. (2002).	Machinery repair costs are lower for no-till when compared with conventional till. Estimates of cost savings range between \$2.77/ha (\$1.12/acre) and \$27.92/ha (\$11.30/acre).
Lazarus and Selley (2005).	Repair costs are lower for equipment not used in tillage.
Equipment Ownership Costs	
Llewelyn, Williams and Thompson (1988); Williams et al. (1989); Weersink et al. (1992); Ribera, Hons, and Richardson (2004); Aller et al. (2001).	Equipment ownership costs can be reduced by between \$10.13/ha (\$4.10/acre) and \$117.96/ha (\$47.75/acre) as a result of switching from conventional to no till.
Lazarus and Selley (2005); Schnitkey, Lattz, and Siemens (2003).	Higher ownership costs for no-till planting equipment than conventional till.
Overall Cost Effects	
Dhuyvetter et al. (1996); Williams et al. (2000); Parsch et al. (2001); Epplin et al. (2005).	No-till systems have higher overall costs than conventional till. Labor and machinery cost reductions can be offset by increases in chemical costs.
Williams et al. (1990); Weersink et al. (1992); Mitchell (1997); Yin and Al-Kaisi (2004).	No-till per hectare costs found to be less than those for conventional till.
Langemeier (2005).	Farms that use less tillage have lower per-unit production costs.

the equipment used is wider (resulting in fewer field passes) and can be driven at higher speeds. Chemical applications substitute for some tillage operations previously used to control weeds. Each of these factors can contribute to reduced labor costs and benefit producers by lowering production costs. In no-till systems, planting occurs directly into the crop residue left on the field from the previous crop. Chemical applications are used for all weed control before and after planting.

Several studies have examined the labor time or labor cost changes resulting from a switch from conventional to no-till and their results are summarized in Table 4.1. The number of field operations required to perform tillage and to apply chemicals for wheat declined from 8.4–5.6 per hectare and decreased from 8.3–6.2 operations per hectare for grain sorghum when no-tillage was used instead of conventional tillage (Aller et al., 2001). This resulted in labor savings of $13.32/ha ($5.39/acre) for wheat and $11.14/ha ($4.51/acre) for grain sorghum (Aller et al., 2001).

Labor savings normally result from substituting herbicides for tillage. The typical field speed for a boom-type chemical sprayer is 10.5 km/h (6.5 mi/h), whereas a row crop cultivator is 8.0 km/h (5.0 mi/h), a field cultivator is 11 km/h (6.8 mi/h), a chisel plow is 8 km/h (5.0 mi/h), and a heavy-duty disk is 7.0 km/h (4.3 mi/h) (ASAE, 2003). Chemical application equipment is normally wider than tillage implements, so more land area can be covered per hour if tillage is reduced or totally replaced by chemical applications for weed control.

Several studies have indicated that labor hours, and thus labor costs, are lowered as a result of switching to a no-till system. Harper (1996) estimated that labor requirements for producing corn in Pennsylvania declined by 2.13 h/ha (0.86 h/acre) if producers switched from reduced to no-till. At a wage rate of $9.00/hour, this reduction cuts labor costs by $19.13/ha ($7.74/acre). Massey (1997) estimated that moving from a chisel plow and disk system to a no-tillage system reduced the labor needed to produce corn in Missouri by 0.74 h/ha (0.30 h/acre); a savings of $6.67/ha ($2.70/acre) at a wage rate of $9.00/h. Olson and Senjem (2002) report that reducing tillage for producing corn and soybeans on two different soils in two different regions of the Minnesota River Basin reduced labor costs by $2.47/ha to $7.41/ha ($1/acre to $3/acre). Similarly, Yin and Al-Kaisi (2004) reported labor cost savings ranging from $3.35/ha to $7.71/ha ($1.36/acre to $3.12/acre) for no-tillage compared with conventional tillage for soybean production at various locations in Iowa.

Energy Costs

The potential to reduce the number of field operations as tillage is decreased and can also create fuel cost savings to producers as they sequester additional soil C. A large number of studies have estimated the potential fuel cost savings as a result of reducing tillage or switching to no-tillage. The results are summarized in Table 4.1.

Williams et al. (1990) found that no-tillage reduced fuel use in a corn–soybean rotation by 35.70 l/ha (3.81 gal/acre). At a diesel fuel price of between $0.53/l and 0.79/l ($2.00/gal and $3.00/gal), this savings equals between $18.85/ha and $28.29/ha ($7.63/acre to $11.45/acre). Reduced tillage lowered fuel costs in rotations of soybean, grain sorghum, and corn from $5.61/ha to $4.22/ha ($2.27/acre to $1.71/acre) (Parsch et al., 2001) compared with conventional tillage. Massey (1997) also reports fuel savings of \$3.58/ha (\$1.45/acre) for producing corn using no-tillage instead of chisel plowing and disking. The fuel cost savings would be considerably higher today because diesel fuel prices have more than doubled since 1997. Aller et al. (2001) reported that diesel fuel costs for no-till wheat and no-till sorghum were \$13.27/ha and \$9.74/ha (\$5.37/acre and \$3.94/acre) less than for conventional tillage, respectively. Williams et al. (2004) estimated that no-till reduced diesel fuel usage in wheat production by 35.5 l/ha (3.7 gal/acre) and in sorghum production by 26.2 l/ha (2.8 gal/acre) when compared with conventional tillage. At a diesel fuel price of between \$0.53/l and 0.79/l (\$2.00/gallon to \$3.00/gallon), these savings amount to between \$18.75/ha and \$28.12/ha (\$7.59/acre to \$11.38/acre) for wheat and between \$13.83/ha and \$20.75/ha (\$5.60/acre and \$8.40/acre) for sorghum.

Bashford and Shelton (1981) estimated fuel use for individual field operations from a 2-year study that collected fuel usage from farmers throughout Nebraska. They found that a field operation with a chemical sprayer had the lowest fuel usage of any operation. Tillage operations that the sprayer operation replaced used 3.2–10.7 times more diesel fuel. They also found that fuel use for tillage operations to produce row crops was higher than that for wheat and fallow, so fuel savings for adopting no-tillage for row crops should be higher than that for small grains. Schrock, Kramer, and Clark (1985) collected on-farm fuel consumption data and found that diesel fuel used for a sprayer to apply herbicides averaged 2.9 l/ha (0.31 gal/acre), whereas tillage implements had higher fuel requirements ranging between 34 to 252% greater. West and Marland (2002) reported that energy used in application of chemicals was less than that for the tillage operations they replaced. They estimated the diesel fuel used for tillage operations, such as disking, was 6.70 l/ha (0.71 gal/acre) and 3.26 l/ha (0.35 gal/acre) for cultivation, whereas pesticide application for no-till was only 1.22 l/ha (0.13 gal/acre).

Repair Expenses

Generally, no-tillage and reduced tillage systems have lower machinery repair costs due to less use of tillage implements. In most no-tillage systems, tractors are used fewer hours and, in some cases, make fewer trips over the cropland. If one or fewer herbicide spraying trips substitute for each tillage operation, machinery repairs are less. Lazarus and Selley (2005) calculated the repair costs of numerous agricultural implements. They estimated the repairs for a boom sprayer to apply herbicides was \$0.64/ha (\$0.26/acre), whereas tillage implements including a row crop cultivator, field cultivator, chisel plow, and tandem disk were \$0.40/ha, \$0.91/ha, \$1.01/ha, and \$1.61/ha (\$0.16/acre, \$0.37/acre, \$0.41/acre, and \$0.65/acre), respectively.

Experiment field research studies show that repair and maintenance costs are typically less for reduced and no-tillage systems. Johnson et al. (1986) reported that reduced tillage systems for wheat and grain sorghum had machinery repair costs that were \$2.77/ha (\$1.12/acre) less than conventional tillage. This reduction in cost was strictly due to less use of equipment for field operations in the reduced tillage system. It was not due to owning less equipment because the equipment complement assumed for the study could be used for either conventional or reduced tillage. Mikesell, Williams, and Long (1988) reported that repair costs for no-tillage systems for producing grain sorghum and soybeans in rotation were $8.70/ha ($3.52/acre) less than conventional tillage and $3.53/ha ($1.43/acre) less than reduced tillage. Williams, Llewelyn, and Barnaby (1990) reported repair cost savings for no-till versus conventional tillage for a corn–soybean rotation was $4.35 ($1.76/acre). Weersink et al. (1992) estimated costs of conventional and no-tillage systems for a corn soybean farm in southern Ontario. They found that fuel, lubrication, repair, and maintenance costs for no-till were on average $18.52/ha ($7.49/acre) less than a moldboard plow system, $18.87/ha ($7.64/acre) less than a chisel plow system, and $5.28/ha ($2.14/acre) less than a ridge-till system. Parsch et al. (2001) estimated costs of production for conservation tillage versus conventional systems including soybeans, grain sorghum, and corn. Repair costs were always lower for the reduced or conservation tillage system. The savings ranged from $21.99/ha to $27.92/ha ($8.90/acre to $11.30/acre). Williams et al. (2002) found that repairs for machinery were $17.64/ha and $12.28/ha ($7.14/acre and $4.97/acre) less, respectively, for no-till than conventional tillage in wheat and sorghum production. A summary of these results is found in Table 4.1.

Equipment Ownership Costs

The number of field operations needed in no-till is generally less because weeds are controlled by the application of herbicides instead of tilling with a chisel plow, disk, undercutter, cultivator, or other tillage implement. Because of this, no-tillage potentially reduces the need for owning many pieces of equipment. If no-tillage is adopted across the entire farm, the machinery investment and resulting ownership and repair costs are lower because less machinery is needed. Equipment investment and resulting costs such as depreciation, interest, insurance, and property taxes will only decrease if the adoption of no-tillage is accompanied by the sale of unneeded tillage implements and possibly extra tractors. According to Massey (1997), farm managers may find the resale value of used tillage implements to be sufficiently small that they prefer to keep the implement in case it is occasionally needed. If tillage equipment is still needed for some cropped hectares, machinery costs may not be significantly reduced.

Llewelyn, Williams, and Thompson (1988) reported that no-tillage used in a wheat-fallow system in the semiarid region of Kansas reduced machinery ownership costs $10.13/ha ($4.10/acre). They were also $17.99/ha ($7.28/acre) less for sorghum fallow and $18.26/ha ($7.39/acre) less for a wheat-sorghum-fallow rotation. Williams et al. (1989) found that equipment costs used in a no-till corn-soybean rotation in northeast Kansas was $22.07/ha ($8.93/acre) less than that for a conventional system and $12.16/ha ($4.92/acre) less than a reduced tillage system. Weersink et al. (1992) derived the best machinery complement for four tillage systems for production of corn and soybeans. Machinery fixed costs, including annual depreciation, interest, insurance, and housing costs, were lower for no-tillage than any other system. These costs were $113.92/ha ($46.10/acre) less than a moldboard plow system, $117.96/ha ($47.74/acre) less than a chisel plow system, and $3.22/ha ($1.30/acre) less than a ridge-tillage system. Aller et al. (2001) found that depreciation and interest expense for machinery in a no-tillage wheat system and no-tillage sorghum system were less than conventional tillage by \$38.99/ha (\$15.78/acre) and \$37.36/ha (\$15.12/acre), respectively. Ribera, Hons, and Richardson (2004) evaluated no-tillage versus conventional tillage for sorghum, wheat, and soybeans in Texas. They reported that no-tillage machinery depreciation was less than conventional tillage machinery depreciation. The cost reduction ranged from \$18.24/ha to \$23.90/ha (\$7.38 to \$9.67/acre). Schnitkey, Lattz, and Siemens (2003), and Lazarus and Selley (2005) both reported higher ownership costs for no-till planting equipment than conventional planting equipment. However, the research reviewed previously indicates other machinery cost reductions in no-tillage will generally outweigh this increased expense for planting equipment. A summary of these results is presented in Table 4.1.

The results of the cost analysis from research studies are generally based on estimated costs of owning and operating only enough tillage implements and tractors needed to conduct the field operation for each tillage system, and, as a result, may overstate the machinery ownership cost reductions. Farm managers may prefer to keep some additional tillage equipment and tractors for flexibility and risk management reasons.

Field Operation Costs for Herbicide versus Tillage Operations

A study by Beaton et al. (2005) estimates that the field operation cost of herbicide application (excluding the herbicide cost) under no-till is less than that for a typical tillage operation. Application costs in Kansas average \$12.60/ha (\$5.10/acre) under no-till in comparison with \$18.41/ha (\$7.45/acre) for sweep tillage or undercutting, and approximately \$32.12/ha (\$13.00/acre) for deep tillage with a chisel or disk.

Overall Cost Efficiency

Although the research literature clearly identifies cost savings due to reduced labor, energy, repairs, and equipment, the results are mixed concerning overall costs. Some authors report that increased costs of herbicides outweigh the other cost savings. Dhuyvetter et al. (1996) reviewed several studies that compare tillage systems for the Great Plains. They found that several studies indicated that labor, repair, fuel, and oil costs declined as the number of operations decreased. However, no-tillage systems were often higher-cost systems because herbicide expense increased more than associated tillage costs decreased. Williams, Roth, and Claassen (2000) reported that no-tillage systems for wheat and grain sorghum had higher total costs than conventional or reduced tillage. Parsch et al. (2001) reported that although conservation tillage had lower labor, fuel, and repair costs, they were more than offset by higher herbicide costs. Epplin et al. (2005) reported that for the two large size farms in their study of wheat production, no-tillage required \$27.80/ha (\$11.25/acre) more for herbicides and saved \$14.83 to \$17.30/ha (\$6/acre to \$7/acre) in fuel, lubrication, and repairs. For the two small farms, no-till required \$27.80/ha (\$11.25/acre) more for herbicides and saved about \$17.30/ha (\$7/acre) in fuel, lubrication, and repairs.

Other studies have found that no-tillage systems had total costs per hectare that were lower. Williams et al. (1990) reported that no-tillage had lower costs for corn and soybean production in northeast Kansas. Weersink et al. (1992) reported that total costs for no-tillage were less than for a moldboard plow or chisel plow system, but not less than ridge-tillage used for corn and soybean production in Ontario. Mitchell (1997) estimated cost of production system budgets for 19 regions in 12 states, including the Corn Belt, Lake States, and Northern Plains. These budgets included conventional and no-tillage for dryland crop rotations including corn, sorghum, soybeans, and wheat. Most but not all no-tillage systems had total costs that were lower than conventional systems. Yin and Al-Kaisi (2004) found that no-tillage systems for soybean production were generally the lowest cost system at five sites in Iowa. Pendell et al. (2005) also reported that no-tillage had lower costs than conventional tillage for corn production with either commercial nitrogen or manure used as fertilizer.

Langemeier (2005) studied the impact of tillage on farm efficiency, using data from 681 farms that had 5 years of data in Kansas. Efficiency was measured by the ratio of total economic cost to gross farm income. The amount of tillage was measured with an index that was calculated by dividing herbicide and insecticide expenditures by total crop machinery cost. Farms with less tillage had a higher index. The results suggested that farms that have adopted practices that use less tillage have had lower per-unit production costs. The result was more important for western Kansas where reducing tillage may help conserve soil moisture, which can contribute to increased crop yields. A summary of the findings presented within this section is presented in Table 4.1.

Crop Yields

Although no-tillage systems have many on-farm economic benefits in the form of cost savings, yield benefits also occur in some situations. However, the resulting yields from no-tillage, versus systems with more tillage, are highly dependent upon location due to climate and soil conditions. No-till could be advantageous in dryer regions of the country, such as the Great Plains, because a decrease in tillage can conserve soil moisture, which is beneficial to the crop as long as excess soil moisture is not present. Williams, Llewelyn, and Barnaby (1990) reported that no-tillage in wheat-fallow, continuous-wheat, and wheat-sorghum-fallow increased soil moisture at planting time. Yields in the wheat-fallow system were also higher. Williams (1988) found that reduced tillage systems for wheat-fallow and wheat-sorghum-fallow had higher soil moisture at planting time and higher yields than conventional tillage. However, no-till systems that contribute to retaining excess soil moisture may not be an advantage in some locations. Yin and Al-Kaisi (2004) reported that no-tillage might have poorer yield performance relative to other tillage systems in wet seasons.

Soil C sequestration resulting from reduced tillage can reduce soil erosion from cropland and soil-attached contaminate losses compared with conventional and reduced tillage (Devlin, Regehr, and Shroyer, 1999). Although the largest impacts of soil erosion are offsite (see discussion in later sections of this chapter), cropland erosion can also be costly to farm mangers because of the loss of productive soil that can eventually reduce yields and increase production costs. Yield reductions are generally long-term in nature and, in some cases, can be economically reduced with the use of inputs such as additional fertilizer (Williams and Tanaka, 1996; Smith et al., 2000). In other words, the economic benefit of soil conservation increases when managers have a longer planning horizon, soil loss is larger, and technological changes will not make up for the loss in soil productivity.

Dhuyvetter and Kastens (1999) reviewed several field experiments in Kansas that examined the impact of no-tillage on yields. They concluded that yields of dryland corn, sorghum, soybeans, and sunflower in western Kansas were higher in no-till than reduced or conventional tillage systems. In central Kansas, yields of wheat, grain sorghum, and soybeans have generally been unaffected by tillage systems. In east central and southeast Kansas, no-till had a detrimental effect on yield for claypan soils with poor internal surface drainage. Llewelyn, Williams, and Thompson (1988) found that yields for no-tillage wheat-fallow and no-tillage wheat-sorghum-fallow in central Kansas were slightly higher than for conventional tillage. Williams et al. (1990) found no-till yields for corn in a corn-soybean rotation in northeastern Kansas were slightly higher than conventional tillage, whereas soybean yields were equivalent. Yields for no-till continuous corn and continuous soybeans were slightly lower than for conventional tillage. Ribera, Hons, and Richardson (2004) found that average yields from conventional versus no-tillage were not statistically different for sorghum, wheat, and soybean grown either continuously or in rotations. Bushong and Peeper (2004) examined the potential of no-till to increase forage and grain yields of winter wheat in Oklahoma. They reported that forage production was higher for no-till, but grain production was lower.

In other areas outside the Great Plains, the results are mixed as well. Harper (1996) reported that data from 340 farms producing corn during the period 1987–1995 indicated that 177 were using conventional tillage, 108 were using reduced tillage, and 55 were using no-till. Conventional tillage had the lowest yields of 9181 kg/ha (146.3 bu/acre), whereas reduced tillage and no-tillage yields were 9426 kg/ha and 9319 kg/ha (150.2 bu/acre and 148.5 bu/acre), respectively. Parsch et al. (2001) found that conventional tillage resulted in higher yields for soybean and grain sorghum, but cotton and corn had higher yields under conservation tillage in Arkansas. However, the differences were not statistically different, except for grain sorghum. Denton and Tyler (2002) provided a review of work on no-tillage in Tennessee and concluded that on cropland with higher yield potential, yields of major crops for no-tillage are about the same as conventional tillage. Yin and Al-Kaisi (2004) found that average soybean yields for no-tillage were no different from yields for other tillage systems at two of five sites in Iowa. At the other three sites, yields for no-tillage were less than at least one other tillage system.

Hellwinckel, Larson, and De LaTorre Ugarte (2003) examined data from fourteen long-term side-by-side experiments from nine research farms throughout the United States. For all six of the corn–soybean rotations, the yield for no-tillage was less (most of them statistically different). All three of the continuous corn systems had lower yields for no-tillage. However, all the rotations involving wheat and or sorghum had higher yields from no-tillage. They also reported that in the majority of the long-term, side-by-side yield comparisons there was no significant increase in no-till yields over time. Wilhelm and Wortmann (2004) reviewed several studies involving the production of corn and soybeans in the Corn Belt region. Their review

TABLE 4.2
Summary of Yield and Net Revenue Changes from Adopting Practices to Sequester Additional Soil C

Studies	Range of Findings
Yields	
Dhuyvetter and Kastens (1999); Llewelyn, Williams, and Thompson (1988); Williams et al. (1990); Williams (1988); Williams, Llewelyn, and Barnaby (1990); Ribera, Hons, and Richardson (2004); Bushong and Peeper (2004); Harper (1996); Parsch et al. (2001); Denton and Tyler (2002); Yin and Al-Kaisi (2004); Hellwinckel, Larson and De LaTorre Ugarte (2003); Wilhelm and Wortmann (2004); Devlin et al. (1999).	Yields are highly dependent on climate and soil conditions. The impact of no-till and reduced till on yields varies by crop and location. Some studies have found an increase in crop yield resulting from the adoption in no-till, some have found that yields decrease, and others find no statistical difference between yields. In general, additional soil C sequestration is expected to increase yields in areas where soil moisture is limiting or soil has been previously degraded by erosion.
Net Returns	
Dhuyvetter et al. (1996); Williams et al. (1990); Harper (1996); Katsvairo and Cox (2000); Parsch et al. (2001); Williams et al. (2004); Ribera, Hons, and Richardson (2004); Yin and Al-Kaisi (2004); Pendell et al. (2005).	Several studies found no-till to be more profitable than conventional till, whereas others found the opposite result. Results vary by crop and by region.

demonstrates that producing corn and soybeans without tillage increased yields in some environments, was the same in others, and less in some. They concluded that the probability of a corn yield increase with no-till is greatest on well-drained soil when grown in rotation in the southern latitude of the United States. Results from this section are summarized within Table 4.2.

Net Returns

An examination of the research literature also provides a range of results for the profitability of no-tillage systems. These results are summarized in Table 4.2. Although conservation tillage systems have been reported to be generally more profitable than conventional tillage systems, no-tillage has not always been the most profitable of the various conservation tillage systems.

In a review of 9 studies that included a total of 23 comparisons of no-tillage systems to other systems for winter wheat, spring wheat, sorghum, corn, hay, and millet crops in various rotations in the Great Plains, Dhuyvetter et al. (1996) found that no-tillage had higher returns than conventional tillage in eight comparisons. Most of these were for either wheat-fallow or wheat-sorghum-fallow systems. Reduced tillage was also studied in 14 of the comparisons with conventional and no-tillage, and, in seven instances, it had higher returns than both conventional and no-tillage systems. Most of these were in wheat-fallow systems.

Williams et al. (1990) found that net returns from no-tillage for continuous corn and corn–soybean rotations were higher than conventional tillage under several government commodity program designs. No-till continuous sorghum had lower average net returns than conventional tillage. Harper (1996) reported that no-tillage corn producers in Pennsylvania made higher average profits than reduced tillage or conventional tillage over the period 1990–1994. Katsvairo and Cox (2000) reported results for an economic analysis of corn, soybeans, wheat, and red clover rotation for three tillage systems and two different input levels. They found that the net returns for ridge-tillage generally had lower net returns than either moldboard plow or chisel-tillage systems that left less residue on the soil surface and were less suitable for highly erodible land. Parsch et al. (2001) reported that conservation tillage had lower net returns than conventional tillage for five of the six cropping systems they studied in Arkansas. This was due to slightly lower yields in conservation tillage and higher overall costs. Lower labor, fuel, and repair costs were more than offset by higher herbicide costs.

Williams et al. (2004) found that average net returns from no-tillage for wheat and grain sorghum in the central Great Plains were less than for conventional tillage, but the no-tillage system sequestered more C. Carbon credits ranging from \$8.62/tonne C to \$64.65/tonne C (\$7.82/ton C to \$58.65/ton C) would be needed for no-till systems to have the same net returns as conventional systems. Ribera, Hons, and Richardson (2004) examined the net returns for conventional and no-tillage systems for cropping systems producing sorghum, wheat, and soybeans in Texas. No-tillage had higher net incomes than conventional tillage for three of the five cropping rotations; sorghum-wheat-soybean, wheat-soybean, and continuous wheat. The conventional tillage system had higher returns than no-tillage for continuous sorghum and continuous soybeans. Yin and Al-Kaisi (2004) found that net returns for no-tillage systems for soybeans in Iowa were greater than those for any other tillage system at only one of five sites. At two sites, the net returns were slightly lower than a chisel plow system, and at another, they were higher than all systems with the exception of one. A study of tillage systems for continuous corn in northeast Kansas showed that net returns for no-tillage were higher than conventional tillage (Pendell et al., 2005).

Government Programs

A final private producer benefit resulting from soil C sequestration is the ability to receive payments from government programs for engaging in activities, such as no-tillage, that sequester additional soil C. Conservation compliance provisions implemented in the 1985 Food Security Act were retained in the Farm Security and Rural Investment Act of 2002 and have been a part of agricultural policy for 20 years. The objective of the Highly Erodible Land Conservation provisions of these acts is to maintain soil productivity by maintaining soil depth and to reduce off-site damages due to sediment loads (Claassen et al., 2004). Practices, such as no-till, that sequester additional C can reduce erosion and subsequent sedimentation in addition to runoff of nutrients and pesticides. Producers that violate conservation requirements may be denied benefits from a variety of Federal agriculture programs including commodity and disaster programs. Eligibility for Federal agriculture related loans or loan guarantees can also be denied. No-tillage systems also contribute to producers being eligible for incentive payments from the Environmental Quality Incentive Program (EQIP) and the Conservation Security Program (CSP). Claassen et al., (2004) also reported that based on 1997 data, 83% of highly erodible cropland is located on farms that receive farm commodity program payments, disaster payments, or conservation payments from the Conservation Reserve Program (CRP), Wetlands Reserve Program (WRP), or EQIP.

Carbon Credit Generation

Additional C sequestered in agricultural soils, as a result of changing existing management practices, could represent a new saleable commodity for agricultural producers. Although the United States has not ratified the Kyoto protocol, several state-led initiatives have encouraged reductions in the emissions of GHGs and additional C sequestration (Pew Center, 2002, 2004). If the costs of changing management practices to sequester additional soil C are less than the revenues that could be received from selling the C into a C credit market, the ability to sell C credits could provide new economic opportunities for agricultural producers. Agricultural soils could be a part of U.S. efforts to reduce atmospheric concentrations of GHGs (EPA 2005), potentially offsetting up to 9% of U.S. GHG emissions (Lal et al. 1998). The Illinois Conservation and Climate Initiative is one example of an existing program that encourages the production of C credits through changes in agricultural management practices (Illinois Department of Agriculture 2006). The C credit prices on the U.S. Chicago Climate Exchange are in the range of $2 per tonne of CO_2 ($7.32/t C) sequestered, whereas prices on the European exchange are within the range of $30 per tonne of CO_2 ($109.80/t C) as of January 2006, reflecting different supply and demand conditions within the two markets (Williams et al., 2005). Several studies have shown that at C-credit prices less than $30 per tonne CO_2, many agricultural producers could benefit financially from producing C credits (Antle et al., 2003; Pautsch et al., 2001; Mooney et al., 2004). Current prices on the European exchange are more than sufficient to make no-till systems competitive with conventional tillage systems in many cases, and prices on the Chicago Climate exchange are approaching the low end of the range suggested by the work of Williams et al. (2004). The degree of financial benefit accruing to producers will vary depending on the physical characteristics of their land and their management expertise.

SUMMARY OF PRIVATE BENEFITS

Although adopting management practices that sequester additional soil C provides many benefits, as with all technological adoption, some producers will find that additional C sequestration is economically beneficial, whereas others will not. However, as input costs rise, cost savings from no-till increase, enhancing its profitability and potentially resulting in wider adoption in future years. In addition, it is likely that as a market for C credits matures within the United States, these prices will also rise, providing producers with an additional commodity to market that could supplement revenues received from crop production, further increasing the financial attractiveness of managing soil C.*

The Conservation Technology Information Center (CTIC) provides data on the use of no-tillage from crop production. The use of no-tillage as a percent of total planted cropland has increased from 6% in 1990 to 22% in 2004 (CTIC, 2004). However, these data are just a snapshot of the number of acres in no-till for a given year. The number of continuous no-till acres that are important for C sequestration would be less. The increase of 2.9 percentage points from 2002 to 2004 was the largest 2-year increase since the 1992–1994 period. From 1990–2004, conventional tillage declined from 48.7% to 37.7%, and reduced tillage declined from 25.3 to 21.5%. Dhuyvetter and Kastens (2005) provide a summary of CTIC no-tillage adoption data for Midwest crop production. They report that no-tillage use in 2004 was highest for soybeans at 36.9%, up from 26% in 1994. Grain sorghum was next highest at 33.2%, up from 13.6% in 1994. Corn and fall small grains were 17.8% and 18.5%, respectively. They found that overall adoption of no-till was slightly less in Kansas compared with the Midwest region (21.2% versus 24.8%). They also conclude that no-till is increasing in Kansas, primarily due to lower costs in central and eastern Kansas, but higher yields and the associated greater revenue is a major factor in western Kansas. Their conclusions point to the variety of reasons that were discussed previously, which explain why managers throughout the Midwest are adopting no-tillage and sequestering additional soil C.

Prices for inputs used for no-tillage systems have been increasing at a slower rate than those for conventional tillage systems over the last 10 years. Herbicide applications are a substitute for tillage operations and use relatively less fuel, machinery, repairs, and labor than tillage operations. Herbicide costs in no-tillage systems have, in some cases, been demonstrated to offset the savings in labor, energy, repairs, and machinery costs. However, information on agricultural input prices compiled by USDA (2004) show that herbicide prices have increased by an average of less than 1% per year from 1994 to 2003. Alternatively, prices of those inputs used more intensively in conventional tillage systems including diesel fuel have increased at an annual average rate of 7.3%, whereas wage rates, machinery, and repairs have increased at 3.9%, 3.2%, and 2.4%, respectively. If these trends continue, the cost savings from using no-tillage practices should continue to increase.

Although the research literature clearly identifies cost savings due to reduced labor, energy, repairs, and equipment, the results are mixed concerning yields and profitability. Numerous studies show increased, decreased, or similar yields. An examination of net returns yields similar results. Dhuyvetter and Kastens (1999) suggest that traditional research and analysis may not be appropriate for uncovering the actual gains associated with less tillage. This research is often conducted using a traditional enterprise budgeting approach that does not totally capture the ability of a farm manager to adjust the whole farm operation to take advantage of reductions in labor, energy, and machinery use. Increased cropping intensity due to reducing

* Chapter 7 in this book, "Benefits of Soil Organic Carbon to Physical, Chemical, and Biological Properties of Soil," by Rice, Fabrizzi, and White, estimates the value of nutrients provided by soil to farm managers. Similar calculations are presented in Pimentel et al. (1995).

tillage and an increase in crop hectares made feasible because of less tillage may play an important role in adopting no-tillage. However, no-till adoption on a large scale is still relatively slow, so this would indicate that many farm managers regard it as unprofitable or that high transaction costs are associated with changing tillage practices. The types and magnitude of incentives required to encourage farm managers not already using no-tillage practices to adopt them are an important question.

The relative costs of creating C credits using no-till vary across farms because of the difference in profitability that manager's face. Different managers will supply a quantity of C credits at different prices because biological, physical, and economic characteristics vary by location, and thus managers face different opportunity costs for sequestering C (Antle et al. 2003). Therefore, different managers will produce and supply various quantities of C credits at a wide range of market prices. It may be feasible for managers in some regions of the county to adopt no-tillage techniques and increase cropping intensity to increase C sequestration and supply C credits at relatively low credit prices, whereas others require higher prices to do so. One must conclude that some farm managers already find no-tillage profitable and sequester additional C with this practice, whereas others will need incentives to adopt no-tillage.

PUBLIC BENEFITS FROM SOIL CARBON MANAGEMENT

As discussed previously, there can be many private benefits accruing to land managers from soil C management, but as with other technologies, not all land managers benefit because of heterogeneity in their resource endowments and management capabilities. However, the public benefits from soil C management are unequivocally positive; for example erosion reduction is estimated to be worth billions of dollars annually in many cases (Ribaudo, 1986; Tegtmeier and Duffy, 2004; Clark, Haverkamp, and Chapman, 1985). A recent study of the private and public benefits of soil C management in New Zealand estimated that the environmental benefits exceeded private producer benefits by 40 to 70 times (Sparling et al., 2006). The primary public benefits are related to the ability of improved soil C management to reduce wind and water erosion as well as leaching of nutrients and pesticides. Erosion and leaching can have significant effects on nutrient loading, sedimentation, wildlife habitat, water quality, human health, and many other factors (Feng, Kling, and Gassman, 2004, McCarl and Schneider, 1999; Sparling et al., 2006). Trying to value services that are provided by different ecosystems presents significant challenges. The role of and complete suite of benefits accruing from well-functioning ecosystems are not always well known. In addition, the lack of markets for ecosystem services as well as a paucity of information about how society values these services makes the translation from quantities of goods and services to economic value quite difficult (NRC, 2005). In instances where the goods and services provided by an ecosystem can be quantified it is often possible to estimate their value using economic valuation techniques.* Unfortunately, our ability to make good estimates concerning the value of ecosystem services is significantly hampered by the paucity of data available both

* In some cases, new valuation techniques are required (NRC, 2005)

nationally and locally. A comprehensive discussion of several techniques and their applicability to valuing ecosystem goods and services can be found in NRC (2005). Studies that assign economic value often focus on a subset of ecosystem services rather than provide a value for the entire ecosystem (NRC, 2005) and do not capture the full value of ecosystem services.*

Section 3 of this book contains several chapters that detail the off-site benefits of soil C management, such as provision of wildlife habitat and improvements in water quality among others. Economic estimates of the public values purely attributable to agricultural soil C management in the United States are not available, but several estimates of the public benefits from erosion reduction, improved water quality, and other benefits that are related to soil C management are available. Some of these values are summarized in Table 4.3.

Water-Induced Soil Erosion

Soil C management is one means of reducing soil erosion. Very few studies have estimated the public economic benefits from reducing water induced soil erosion. The physical and biological relationships between soil erosion and its off-site consequences are difficult to measure, and the linkages between these impacts and their economic effects are not well defined. The estimates presented next represent the value to society of avoiding damages from soil erosion and are summarized in Table 4.3.

Sedimentation

Sedimentation caused by soil erosion can create significant societal costs. For example, sedimentation can affect the enjoyment of people using waterways for recreation. Increased sedimentation at Ohio State park lakes was estimated to reduce the economic benefit of recreation to out of state boaters by an average of $0.54/tonne ($0.49/ton) of sediment (MacGregor, 1988). Bejranonda, Hitzhusen, and Hite (1999) examined the impact of sedimentation on lakeside property values at 15 Ohio State park lakes and found that homeowners would be willing to pay more for properties on lakes with less sedimentation. A broader, national estimate of the value of recreation loss attributable to siltation is between $612 million and $3,608 million (2006 dollars) (Tegtmeier and Duffy, 2004). Hansen et al. (2002) examined the costs of soil erosion and sedimentation to downstream navigation across different areas of the United States. Their results suggest that eroded soil poses no costs to navigation in areas where no downstream shipping channels or harbors are located but can create costs of up to $6.23/tonne ($5.67/ton) of soil erosion (2006 dollars) in areas where navigation is affected (Hansen et al., 2002). The total cost of inland dredging across the United States was estimated at $257 million annually ($291 million in 2006 dollars) (Hansen et al., 2002). Although Tegtmeier and Duffy (2004)

* One exception is Costanza et al. (1997). They estimated that the entire value of goods and services provided by ecosystems around the world is approximately $33 trillion. Although this study was heavily criticized, it did serve to open a lively dialogue focused on the challenges related to valuing ecosystem services.

TABLE 4.3
Examples of Societal Costs That Could Be Partially Offset by Increasing Soil C Sequestration[a]

Studies	Range of Findings
Water Erosion	
Hansen et al. (2002); Clark, Havercamp, and Chapman (1985); Ribaudo (1986); Tegtmeier and Duffy (2004).	Dredging of inland waterways estimated to cost between \$291 million and \$282 million annually. Reservoir siltation estimated between \$274 million and \$851 million annually. Total damages from water erosion estimated between \$2 billion to \$31 billion annually.
Wind Erosion	
Uri (2001); Pimentel et al. (1995); Tegtmeier and Duffy (2004)	Total damages estimated between \$13 billion and \$38 billion annually.
Wind and water erosion in total	
Uri (2001); Pimentel et al. (1995)	Total combined annual cost estimated between \$43 billion and \$59 billion annually.
Nutrient Runoff and Water Quality	
NYC watershed Tegtmeier and Duffy (2004)	Value of water improvements in the Catskills and Delaware watersheds attributable to all ecosystem services estimated to be between \$6 billion and \$8 billion. Estimated cost of nitrite and pesticide treatment is \$341 million.
Wildlife and Recreation	
Feather, Hellerstein, and Hansen (1999); Tegtmeier and Duffy (2004)	Benefits wildlife viewing from CRP lands estimated at \$674 million. Recreation loss from sedimentation estimated at between \$612 million and \$3.6 billion. Damage to fish and wildlife habitat estimated at \$1.1 billion to \$1.2 billion.

[a] Figures are reported in 2006 dollars.

estimate that the cost to shipping ranges between \$345 million and \$383 million (2006 dollars), a study by Moore and McCarl (1987) estimated the effect of sediment on municipal water treatment, road drainage system maintenance, navigation channel maintenance, reservoir capacity deterioration, and hydroelectric power plant costs within Oregon. They estimated that annual average cost of municipal water treatment in the Willamette Valley, Oregon was approximately \$1 million annually. Sediment-related ditch and culvert cleaning for the state of Oregon was estimated to be \$4.22 million annually, whereas river channel maintenance was estimated to cost \$270 million annually. The total annual average cost of erosion on navigation channel maintenance, municipal water treatment, country road maintenance, and state highway maintenance was approximately \$5.5 million per year (Moore and McCarl, 1987). Tegtmeier and Duffy (2004) updated the figures from Ribaudo (1986) and estimated that the costs of sediment removal from roadside ditches and irrigation channels throughout the United States range between \$304 million and \$895 million (2006 dollars). Soil erosion can also reduce reservoir capacity through siltation. These damages are estimated to range between \$274 million and \$6,851 million (2006 dollars), whereas flood damage attributable to erosion is estimated at between \$215 million and \$622 million (2006 dollars) annually (Tegtmeier and Duffy, 2004). These results are summarized in Table 4.3.

National Estimates of Total Damage from Water Erosion

Clark, Haverkamp, and Chapman (1985), Ribaudo (1986), and Tegtmeier and Duffy (2004) provide national estimates of the total annual damages attributable to water erosion of soils. The sectors included in these analyses are recreation, navigation, water storage facilities, municipal and industrial water users, and water conveyance systems. The estimated damages from annual soil erosion do not include all sectors of the economy or all possible activities that are affected and, as such, represent only a partial estimate of the possible benefits of reduced soil erosion of which part could be achieved by reducing soil erosion or by increasing C sequestration. Clark, Haverkamp, and Chapman (1985) estimate that the total annual off-farm costs of water erosion from agricultural lands range between \$3 billion and \$13 billion (the equivalent of between \$8 billion and \$31 billion in 2006 dollars). A similar study, by Ribaudo (1986), estimates that annual damages attributable to water erosion of soils are between \$4 billion and \$15 billion (between \$8 billion and \$30 billion in 2006 dollars). Tegtmeier and Duffy (2004) estimate that the value of damages attributable to agricultural erosion ranges between \$2 billion and \$15 billion (2006 dollars).

Since the mid-1980s, there has not been a comprehensive study to examine the damages attributable to water-based soil erosion. Water-based erosion of cropland and other agricultural soil has declined considerably over the last 20 years (NRCS, 2006) in part due to improved farming practices and government programs, such as the Conservation Reserve Program (CRP), that retired highly erodible lands. The potential damages from soil erosion have probably changed from the mid 1980s. A very rough estimate of the potential annual costs of soil erosion given 1997 erosion rates is presented in Table 4.4. Total water-based erosion in 1997 was obtained from the 1997 National Resources Inventory (revised in 2000) (www.nrcs.usda.gov/technical/NRI/1997/summary_report/table10.html). The estimated value to society of reducing soil erosion by one tonne (1.1 t) was obtained from Ribaudo (1986) Tables 4.10 and 4.11 and updated from 1993 to 2006 dollars using the consumer price index. The per-tonne value of erosion reduction calculated by Ribaudo (1986) is based on the potential annual damages from soil erosion in 1983. These values range between a low of \$1.09/tonne (\$0.98/ton) in the northern plains to \$12.44/tonne (\$11.29/ton) in the northeast. The damages created by soil erosion have probably changed over time and it is unlikely that this estimate is accurate anymore. The actual damages per tonne of soil erosion will depend on the value of downstream activities affected by soil erosion. Keeping in mind the previous caveats, the estimated societal benefits resulting from eliminating water-induced soil erosion from

TABLE 4.4
Estimated Value of Reducing Water Erosion Damages from Cropland, CRP, and Pastureland[a]

Region	Dollars in Millions[b]		
	Low Estimate	Medium Estimate	High Estimate
Pacific	110	177	358
Mountain	56	102	162
Northern Plains	95	174	805
Southern Plains	214	360	745
Lake States	296	553	954
Corn Belt	351	726	1,310
Delta States	213	347	1,247
Northeast	335	548	1,183
Appalachian	148	269	449
Southeast	96	150	219
Total	1,915	3,408	7,432

[a] Calculating the value of reducing erosion by using this approach has a number of drawbacks. A good discussion can be found in NRC, 2005.
[b] Figures are reported in 2006 dollars.

1997 levels are between $1.91 billion and $7.43 billion (Table 4.4). Tegtmeier and Duffy (2004) used a similar technique and estimated that the total costs of water induced soil erosion range between $2.2 billion and $13 billion (2006 dollars).

The additional societal value of reducing annual soil erosion has declined since the mid-1980s, in part because there has been a decline in annual soil erosion. However, the data in Table 4.4 demonstrate that a significant societal value can be obtained from reducing water-based soil erosion on agricultural soils as a result of implementing practices that enhance soil C sequestration. Unfortunately, no estimates are available regarding the degree to which additional soil C sequestration could contribute to reduced erosion. McCarl and Schneider (1999) calculate that adopting several practices to reduce GHG emissions (which include soil C sequestration) could reduce water and wind erosion by between 24% and 49.7%. The 2003 National Resource Inventory (NRCS, 2006) reports the average rate for sheet and rill erosion on cropland is 6.05 t/ha (2.6 t/acre/yr) and from wind is 4.7 t/ha (2.1 t/acre/yr). Thus, the total reduction in erosion from adopting practices to reduce atmospheric concentrations of GHGs could be in the range of 1.5 t/ha to 2.9 t/ha (0.64 t/acre to 1.3 t/acre) for water erosion and between 1.1 t/ha and 2.3 t/ha (0.5 t/acre and 1.02 t/acre) for wind erosion.

Wind-Induced Soil Erosion

Even fewer studies have examined the costs of wind erosion. Wind erosion can cause significant off-site and on-site damages (Riksen and DeGraff, 2001; Foster and Dabney, 1995; Pimentel et al., 1995). As with water erosion, many costs are associated with wind erosion, including costs associated with degraded health, reduction in agricultural production, increased cleaning, maintenance and replacement of assets, and the removal of dust from roads and ditches, among other factors (Uri, 2001; Pimentel et al., 1995; Tegtmeier and Duffy, 2004). The actual regional benefits of reducing wind erosion will depend on two main factors: the quantity of soil eroded, and the type of activities and infrastructure that are affected by the erosion. These factors can change dramatically over the landscape, thus damage estimates from one area cannot be easily used to estimate damages in another area (Uri, 2001). Huszar (1989) calculated that the off-site costs of wind erosion were approximately 45 times greater that the on-site costs across New Mexico. A study by Piper (1989) estimated that wind erosion in the *western states* costs between \$3.76 billion and \$12.08 billion annually (between \$6 billion and \$20 billion in 2006 dollars). Using values from Huszar and Piper (1986), Uri (2001), based on 1992 erosion rates, the social costs of continued wind erosion across the United States is approximately \$15.64 billion annually (\$18 billion in 2006 dollars) with a range of between \$11.30 billion and \$32.04 billion (\$13 billion to \$38 billion in 2006 dollars). Pimentel et al. (1995) estimate that wind erosion costs the United States about \$9.6 billion annually (approximately \$13 billion in 2006 dollars). Wind erosion on cropland in 2003 is about 78% of 1992 levels (NRCS, 2006). If we assume that the costs per unit of wind erosion are similar to those calculated by Uri (2001); the benefit from eradicating wind erosion from 1992 levels is approximately \$14 billion dollars annually (2006 dollars).*

Combined Benefits of Reducing Wind and Water Erosion

The combined annual costs to society of wind and water erosion were estimated to be \$37.6 billion dollars (\$43 billion 2006 dollars) annually by Uri (2001) and approximately \$44.4 billion (\$59 billion 2006 dollars) by Pimentel et al. (1995).† Although many authors agree that both water and wind erosion cause significant societal costs, the value of additional soil C sequestration, which can reduce wind and water erosion, is very difficult to quantify (Uri 2001; NRC, 1993). However, it is clear that the introduction of practices that reduce soil erosion have resulted in significant societal benefits.

Reduced Nutrient Runoff and Improved Water Quality

High water nutrient concentrations occur in agricultural basins (Ribaudo et al., 2001). Several strategies are used to reduce nutrient loads from cropland: (1) change nutrient management practices; (2) intercept runoff and filter out nutrients before they reach surface waters (Ribaudo et al., 2001); and (3) improve soils to reduce runoff.‡

* It is unlikely that the per-unit cost has remained constant across this time period because of changes in infrastructure and population density among other factors. Updated studies are required to accurately estimate the value of damages attributable to erosion today.

† Pimentel et al. (1995) used a larger classification of activities than Uri (2001).

‡ Additional information about the linkages between soil C management and water quality can be found in Chapter 12, "Environmental and Ecological Benefits of Soil Carbon Management: Surface Water Quality," by Gary Pierzynski, Dan Devlin, and Jeff Neel.

Box 4.1 Drinking Water for New York City

New York City obtains approximately 90% of its water supply from the Catskill and Delaware watersheds. This amounts to approximately 1.5 billion gallons of water per day to more that 9 million people. Up until the 1990s, the water supplied by the area contained very little contamination, however, additional development within the area threatened to decrease water quality. If quality declined sufficiently, the water would no longer meet filtration avoidance criteria set down by the Environmental Protection Agency (EPA) and New York City would be faced with building a water filtration system, effectively replacing the services that had been previously provided by the Catskill and Delaware watersheds. Building and operating the water filtration system was estimated to cost between $6 billion and $8 billion (Chichilnisky and Heal, 1998). As an alternative, the City examined the cost of protecting the upstream watersheds that had historically provided water purification processes resulting in uncontaminated drinking water that previously met EPA standards (Chichilnisky and Heal, 1998). The projected cost of protecting and restoring these watershed services was between $1 billion and $1.5 billion (significantly less than the cost of a new water filtration system) and could be achieved by offering incentives to landowners to reduce soil erosion, plant riparian buffers, land purchases and septic and sewer improvements among other measures (Chichilnisky and Heal, 1998; Ashendorff et al., 1997). In 1997, New York City crafted the Watershed Memorandum Agreement which put in place the institutional framework required to implement the watershed restoration and protection measures (Ashendorff et al., 1997). Although C sequestration was not mentioned explicitly within the agreement, activities such as tree planting result in additional C sequestration, whereas farming practices that sequester additional soil C can also reduce erosion. The value of additional soil C sequestration in increasing water quality is difficult to estimate; however, at least part of the value of water improvements in these watersheds is indirectly a result of increased C sequestration.

Several studies have estimated the physical reduction in nutrients resulting from adopting no-till practices that sequester additional soil C (examples include Feng, Kling, and Gassman (2004), and McCarl and Schneider (1999)). Very few estimates of the economic value of reducing nutrient runoff or improving water quality are available. A summary of the studies reported in this chapter is presented in Table 4.3. Tegtmeier and Duffy (2004) estimated that the costs of infrastructure for nitrite and pesticide treatment are approximately $341 million (2006 dollars). Van Vuuren, Giraldez, and Stonehouse (1997) estimated that the off-farm value of reducing nitrogen and phosphorous loading in a watershed in Ontario, Canada, by switching to no-till that sequesters additional C, range between $196,000 and $243,000 annually.

Probably the best-studied example of the value of improving water quality is the case of the New York City (NYC) watershed. The estimated benefits from ecosystem services within the Catskill and Delaware watersheds (some of which would be supplied by sequestered soil C) are $6 billion to $8 billion (Box 4.1).

No-till practices that sequester additional soil C can help lower nutrient levels in water by preventing soil erosion, reducing runoff, and reducing leaching. Several examples exist regarding how increased nutrient concentrations in U.S. waterways

Box 4.2 Hypoxia in the Gulf of Mexico

Commercial and recreational fishing within the Gulf of Mexico are thought to generate \$2.8 billion annually (CAST, 1999) in addition to revenues from fishing related tourism in the Gulf coast estimated to contribute \$20 billion annually to the economy (EPA, 1999). The Gulf is thought to support approximately one third of marine recreational fishing within the United States; in 1995, approximately 4.8 million anglers caught 42 million fish (EPA, 1999). In addition, the area is home to threatened and endangered species such as the Gulf sturgeon, brown pelican, and several types of sea turtle (EPA, 1999). Unfortunately, some of the area is increasingly suffering from hypoxia (a reduction in the oxygen concentration of the water to less than 2 mg per liter (Rowe, 2001)). Although no clear evidence exists that hypoxia has caused any economic detriment to the Gulf fisheries to date, other hypoxic zones around the world have experienced catch and recruitment failure as well as a complete loss of fisheries species creating serious economic consequences (Diaz and Solow, 1999). Hypoxia is primarily caused by organic matter settling to the bottom of the Gulf and using oxygen to decompose, creating hypoxia. The primary cause of hypoxia is thought to be sedimentation of marine organic matter from increased river nutrients flowing from the Mississippi (CAST, 1999). Since the 1950s, increased nitrogen loads from the Mississippi have contributed to hypoxia, in addition to river loading of organic C, channelization and loss of coastal wetlands, among other factors (Rabalais, Turner, and Scavia, 2002). Earlier landscape changes, such as conversion of wetlands to cropland, loss of riparian zones, and increased agricultural drainage, lowered the ability of the Mississippi basin to prevent nutrients from entering the river system and increased soil erosion (Rabalais, Turner, and Scavia 2002; Turner and Rabalais, 2003). Estimates suggest that each agricultural hectare within the basin exports 2–3 kg of nitrogen into the river system, with an estimated annual value of \$410 million annually (CAST, 1999). In addition, 31% of the phosphorous loading is attributable to commercial fertilizers and 41% from sources such as phosphorous attached to soil particles (Goolsby et al., 1999). Practices that sequester additional C can decrease nutrient transport by reducing soil erosion and potentially lead to lower nutrient levels throughout the basin (Greenhalgh and Sauer, 2003).

have contributed to the degradation of exceptional natural resources. The value of decreasing nutrient runoff has not been explicitly calculated; however, lower nutrient concentrations could partially alleviate problems of hypoxia within the Gulf of Mexico (Box 4.2) and eutrophication within the Chesapeake Bay (Box 4.3). Strategies that include soil C sequestration could reduce annual nitrogen loadings within the Gulf of Mexico by up to 9%, a little less than one half of the goal set in place by the Watershed Nutrient Task Force in 1997. In addition, activities that include soil C sequestration could significantly reduce phosphorous within the area (Pattanayak et al., 2002). Pattanayak et al. (2002) show that there could be significant improvements in U.S. water quality as a result of adopting new agricultural practices in response to a range of incentive payments tied to the production of C-credits. The study does not break out water quality improvements attributable to soil C management alone.

Box 4.3 Chesapeake Bay

The Chesapeake Bay spans part of Virginia, West Virginia, Maryland, New York, Delaware, Pennsylvania, and all of Washington, D.C. The Bay is home to about 15 million people and is an important resource for commercial and recreational fishing, camping, boating, and hunting. It also supports abundant aquatic and terrestrial wildlife (Boesch, Brinsfield, and Magnien, 2001). The area has been subject to several environmental problems including overharvesting of oysters, pollution, pesticides, and wetland loss (Davidson et al., 1997). Increases in nitrogen and phosphorous within the system are thought to contribute to eutrophication and harmful algal blooms (Tennenbaum and Costanza, 1990); 39% of nitrogen and 49% of phosphorous are attributable to agricultural sources (Magnien, Boward, and Bieber, 1995). Agriculture is the largest source of nutrients to the bay, and soil management and erosion reduction practices are encouraged under several Bay improvement plans (Boesch, Brinsfield, and Magnien, 2001). Lipton (1998) calculated that an algal bloom within the Chesapeake Bay reduced recreation fishing and resulted in an economic loss to the area of $4.3 million.

Wildlife and Recreation Benefits

Reduced erosion is one of the benefits of retiring marginal lands under the Conservation Reserve Program (CRP). Between 1982 and 2003, it was estimated that water erosion decreased by 6.5 t/ha (2.4 t/acre) — approximately 45% — whereas wind erosion dropped by 3.2 t/ha (1.2 t/acre) — approximately 47% (NRCS, 2006). Feather, Hellerstein, and Hansen (1999) calculated that the benefits from CRP enrollment in 1992 included $348 million for wildlife viewing, $80 million per year for pheasant hunting, and $36 million per year for freshwater-based recreation. The total value in 2006 dollars is approximately $674 million annually. Similarly, Van Vuuren, Giraldez, and Stonehouse (1997) estimated that the increase in recreational fishing value in an Ontario watershed ranged between $7,681 and $26,129 annually, in response to water improvements resulting from the adoption of no-till. Tegtmeier and Duffy (2004) estimated that damages to fish and wildlife habitat because of water erosion throughout the United States ranged between $1.1 billion and $1.2 billion (2006 dollars). Their results are summarized in Table 4.3.

SUMMARY OF PUBLIC BENEFITS

The preceding estimates demonstrate that the value to increased C sequestration is potentially considerable. Farm management practices that increase C sequestration can reduce sedimentation and nutrient runoff, thus improving and protecting water quality. These practices can also add value to recreation and wildlife and extend the economic life of reservoirs and shipping channels. Significant problems exist in trying to estimate the nonmarket value of increased C sequestration. First, it is not clear how much additional C will be sequestered in the future, and thus the impact of C sequestration on soil erosion, runoff, and other variables is uncertain. If soil C sequestration quantities could be estimated, tools already exist to calculate the effect on soil erosion and nutrient runoff (see McCarl and Scheider, 1999, and Feng, Kling, and Gassman, 2004). The link between physical changes in the environment and economic values is still very weak, and is an area in which considerable additional work is needed. Many of the studies cited gather their information from a few major studies completed in the 1980s. A significant need exists for more updated work as well as better regional and local level data to support this work. The studies reported previously demonstrate that the societal benefits from activities that sequester soil C are considered significant. One caution in interpreting these values is that they could include value generated by other factors in addition to soil C sequestration. Thus, the benefit of soil C sequestration is some portion of the total values reported. The considerable public benefits due to erosion reduction that could occur with the adoption of soil conservation practices to enhance C sequestration may be an important reason to provide incentives to landowners to adopt such practices.

CONCLUSIONS AND RECOMMENDATIONS

Management practices that increase the rate of soil C sequestration can partially offset increased atmospheric concentrations of GHGs and mitigate climate change. In addition to offering this benefit to society, soil C management can provide many other environmental co-benefits, such as erosion reduction, improved water quality and fish and wildlife habitats, as well as private on-farm economic benefits to some producers. Substantial research indicates that although practices that sequester additional C, such as no-till, are already adopted by many producers, others may require additional financial incentives to increase soil C sequestration because their location, management capabilities, or other factors do not make these management practices profitable to them at present. One means of encouraging greater adoption of these practices is to target future farm programs toward measures of environmental improvement, such as increases in soil C sequestration, or consider short-term incentives that could stabilize producer incomes as they switch to practices that sequester additional soil C. The societal benefits from reducing erosion and improving water quality, recreation opportunities, and wildlife habitat (which are co-benefits of C sequestration) are considered significant in the United States, perhaps saving several billion dollars annually. Unfortunately, these co-benefits are difficult to value under the best of circumstances, and data limitations at regional and local scales further exacerbate this problem. Further work in this area would help to demonstrate the benefits of future programs that encourage soil C management.

REFERENCES

Aller, T.D., Williams, J.R., Nelson, R.G., Claassen, M.M., and Rice, C.W., An economic analysis of carbon sequestration for wheat and grain sorghum production in Kansas. Paper presented at The Soil and Water Conservation Society 2001 Annual Conference, Myrtle Beach, SC, August 4–8, 2001.

American Society of Agricultural Engineers (ASAE), ASAE standard D497.4 FEBO3, St. Joseph, MI, 2003.

Antle, J., Capalbo, S., Mooney, S., Elliot, E., and Paustian, K., Spatial heterogeneity and the design of efficient carbon sequestration policies for agriculture, *J. Environ. Econ. Manage.*, 46(2), 231–250, 2003.
Ashendorff, A., Principe, M.A., Seely, A., LaDuca, J., Beckhardt, L., Farber, W., and Mantus, J., Watershed protection for New York City's supply, *J. Am. Water Works Assoc.*, 89(3), 75–88, 1997.
Bashford, L.L. and Shelton, D.P., Fuel use for field operations, *Nebraska Guide G81-578-A*. Cooperative Extension Service, University of Nebraska—Lincoln, Lincoln, NE, 1981.
Beaton, A.J., Dhuyvetter, K.C., Kastens, T.L., and Williams, J.R., Per unit costs to own and operate farm machinery, *J. Agricultural Appl. Econ.*, 37(1), 2005.
Bejranonda, S., Hitzhusen, F.J., and Hite, D., Agricultural sedimentation impacts on lakeside property values, *Agricultural Resource Econ. Review*, October, 208–218, 1999.
Boesch, D.F., Brinsfield, R.B., and Magnien, R.E., Chesapeake Bay eutrophication: scientific understanding, ecosystem restoration, and challenges for agriculture, *J. Environ. Qual.*, 30, 303–320, 2001.
Bushong, D.L. and Peeper, T.F., Potential for using no-till to increase forage and grain yields of winter wheat. Paper presented at the 26th Southern Conservation Tillage Conference, Raleigh, NC, June 8–9, 2004.
Chichilnisky, G. and Heal, G., Economic returns from the biosphere, *Nature*, 391, 629–630, 1998.
Claassen, R., Breneman, V., Bucholtz, S., Cattaneo, A., Johansson, R., and Morehart, M., Environmental compliance in U.S. agricultural policy, Agricultural Economics Report No. 832, United States Department of Agriculture, Washington, D.C., 2004.
Clark, E.H., Haverkamp, J.A., and Chapman, W., *Eroding Soils: The Off-Farm Impacts*, The Conservation Foundation, Washington, D.C., 1985.
Conservation Technology Information Center (CTIC), 2004 National crop residue management survey. Available at: www.conservationinformation.org/?action=crm. Accessed on April 6, 2005.
Costanza, R., D'Arge, R., de Groot, R., Farber, S., Grasso, M., Hannon, B., Limburg, K., Naeem, S., O'Neill, R.V., Paruelo, J., Raskins, R.G., Sutton, P., and van den Belt, M., The value of the world's ecosystem services and natural capital, *Nature*, 387, 253–260, 1997.
Council for Agricultural Science and Technology (CAST), Gulf of Mexico Hypoxia: Land and Sea Interactions, Ames, IA, 1999.
Davidson, S.G., Merwin Jr., J.G., Capper, J., Power, G., and Shivers, F.R., Jr. *Chesapeake Waters: Four Centuries of Controversy, Concern and Legislation*, Tidewater Publications, Centreville, MD, 1997.
Denton H.P. and Tyler, D.D., Making no-till conventional in Tennessee, in *Making Conservation Tillage Conventional: Building a Future on 25 Years of Research*, Evan Swanten, Ed. Proceedings of the 25th Annual Southern Conservation Tillage Conference for Sustainable Agriculture, Auburn, AL, June 24–26, 2002.
Devlin, D.L., Regehr, D.L., and Shroyer, J.P. Effects of no-till on soil and water quality, in *Kansas No-Till Handbook*, Kansas State University Agricultural Experiment Station and Cooperative Extension Service, Manhattan, KS, 10–13, 1999, chap 2.
Dhuyvetter, K.C. and Kastens, T., Yield and economic comparisons: University research trials, in *Kansas No-Till Handbook*, Kansas State University Agricultural Experiment Station and Cooperative Extension Service, Manhattan, KS, 23–32, 1999, chap. 5.
Dhuyvetter, K.C. and Kastens, T., Economics of no-till in Kansas. Available at: www.agmanager.info/dhuyvetter/presentations/default.asp. Accessed on April 11, 2005.

Dhuyvetter, K.C., Thompson, C.R., Norwood, C.A., and Halvorson, A.D., Economics of dryland cropping systems in the Great Plains: a review, *J. Prod. Agric.*, 9, 216–222, 1996.

Diaz, R.J. and Solow, A., Ecological and economic consequences of hypoxia: Topic 2 Report for the Integrated Assessment on Hypoxia in the Gulf of Mexico, National Oceanic and Atmospheric Administration (NOAA) Coastal Ocean Program, Decision Analysis Series No. 16, U.S. Department of Commerce, National Oceanic and Atmospheric Administration, National Ocean Service, Coastal Ocean Program, Silver Springs, MD, 1999.

Environmental Protection Agency (EPA), Ecological Condition in the Gulf of Mexico, EPA 620-R-98-004. U.S. Environmental Protection Agency, Office of Research and Development, National Health and Environmental Effects Research Laboratory, Gulf Ecology Division, Gulf Breeze, FL, 1999.

EPA, Greenhouse Gas Mitigation Potential in U.S. Forestry and Agriculture, EPA report 430 R-05-006, Washington, D.C., November 2005.

Epplin, F.M., Stock, C.J., Kletke, D.D., and Peeper, T.F., Economies of size for conventional tillage and no-till wheat production. Presented at the Southern Agricultural Economics Association Annual Meetings, Little Rock, AR, February 6–9, 2005.

Feather, P, Hellerstein, D., and Hansen, L., Economic Valuation of Environmental Benefits and the Targeting of Conservation Programs: The Case of the CRP. Agricultural Economics Report No. 778, Resource Economics Division, Economic Research Service, U.S. Department of Agriculture, Washington, D.C., 1999.

Feng, H., Kling, C.L., and Gassman, P.W. Carbon sequestration, co-benefits, and conservation programs, *Choices*, 19–23, Fall 2004.

Foster, G. and Dabney, S., Agricultural tillage systems: water erosion and sedimentation, in *Farming for a Better Environment*, Soil and Water Conservation Society, Ankeny, IA, 1995.

Goolsby, D.A., Battaglin, W.A., Lawrence, G.D., Artz, R.S., Aulenbach, B.T., Hooper R.P., Keeney, D.R., and Stensland G.J., Flux and sources of nutrients in the Mississippi Atchafalya River Basin: Topic 3 Report for the Integrated Assessment of Hypoxia in the Gulf of Mexico, NOAA Coastal Oceana Program Decision Analysis Series No. 17, NOAA Coastal Ocean Program, Silver Spring, MD, 1999.

Greenhalgh, S. and Sauer, A., Awakening the Dead Zone: An Investment for Agriculture, Water Quality and Climate Change, *WRI Issue Brief*, World Resources Institute, Washington, D.C., 2003.

Hansen, L.T., Breneman, V.E., Davison, C.W., and Dicken, C.W., The cost of soil erosion to downstream navigation, *J. Soil Water Conserv.*, 57(4), 205–212, 2002.

Harper, J.K., Economics of conservation tillage, in *Conservation Tillage Series No. 6*, The Pennsylvania State University, College of Agriculture Cooperative Extension Service, University Park, PA, 1996.

Heal, G., *Nature and the Marketplace: Capturing the Value of Ecosystem Services*, Island Press, Washington, D.C., 2000.

Hellwinckel, C., Larson, J.A., and De LaTorre Ugarte, D., Incentives for switching agricultural land to carbon sequestering no-tillage: What duration are incentives necessary? Presented at the Southern Agricultural Economics Association Annual Meeting, Mobile, AL, February 1–5, 2003.

Huszar, P.C., Economics of reducing off-site costs of wind erosion, *Land Econ.*, 65(4), 333–340, 1989.

Huszar, P.C. and Piper, S.L., Estimating the off-site costs of wind erosion in New Mexico, *J. Soil Water Conserv.*, November–December, 414–416, 1986.

Illinois Department of Agriculture, Illinois Launches First State Program to Offer Greenhouse Gas Emissions Credits. Press Release, January 27, 2006. Available at: www.agr.state.il.us/newsrels/r0127061.html. Accessed on February 1, 2006.

Johnson, O.S., Williams, J.R., Gwin, R.E., and Mikesell, C.L., Economic analysis of reduced tillage wheat and grain sorghum rotations in western Kansas, *Bulletin 650*, Agricultural Experiment Station, Kansas State University, Manhattan, KS, June, 1986.

Katsvairo, T.W. and Cox, W.J., Economics of cropping systems featuring different rotations, tillage, and management, *Agron. J.*, 92, 485–492, 2000.

Lal, R., Kimble, J.M., Follett, R.F., and Cole, C.V., *The Potential of U.S. Cropland to Sequester C and Mitigate the Greenhouse Effect*, Ann Arbor Press, Chelsea, MI, 1998.

Langemeier, M., Impact of adoption of less tillage practices on overall efficiency. Presented at the Southern Agricultural Economics Association Annual Meetings, Little Rock, AR, February 6–9, 2005.

Lazarus, W. and Selley, R., Farm machinery economic cost estimates for 2005, University of Minnesota Extension Service, 2005. Available at: www.apec.umn.edu/faculty/wlazarus/mf2005.pdf. Accessed on March 30, 2005.

Lipton, D.W., Pfiesteria's economic impact on seafood industry sales and recreational fishing, 1998. Available at: www.arec.umd.edu/Policycenter/Pfiesteria/lipton/lipton.htm. Accessed on July 25, 2005.

Llewelyn R.V., Williams, J.R., and Thompson, C.A., Risk analysis of conventional and no-tillage wheat and grain sorghum: Conservation tillage research, *Report of Progress 542*, Agricultural Experiment Station, Kansas State University, Manhattan, KS, 1988.

MacGregor, D.R., The value of lost boater value use and the cost of dredging: evaluating two aspects of sedimentation in Ohio's State Park lakes. Unpublished Ph.D. dissertation, The Ohio State University, Columbus, OH, 1988.

Magnien, R., Boward, D., and Bieber, S., Eds., *The State of the Chesapeake 1995*, USEPA, Annapolis, MD, 1995.

Massey, R.E., No-tillage and conservation tillage: Economic considerations, Guide G 355, University Extension, University of Missouri, Columbia, MO, December 1997.

McCarl, B. and Schneider, U.A., Greenhouse gas mitigation in U.S. agriculture and forestry, *Science*, 294, 2481–2482, 1999.

Mikesell, C.L., Williams, J.R., and Long, J.H., Evaluation of net return distributions from alternative tillage systems for grain sorghum and soybean rotations, *North Central J. Agricultural Econ.*, 10, 255–271, 1988.

Mitchell, P.D., Cost of Production System Budgets, Technical Report 97-TR37, Center for Agricultural and Rural Development, Iowa State University, Ames, IA, 1997.

Mooney, S., Antle, J.M., Capalbo, S.M., and Paustian, K., Design and costs of a measurement protocol for trades in soil carbon credits, *Canadian J. Agricultural Econ.*, 52(3), 257–287, 2004.

Moore, W.B. and McCarl, B.A., Off-site costs of soil erosion: a case study in the Willamette Valley, *Western J. Agricultural Econ.*, 12(1), 42–49, 1987.

National Research Council (NRC), *Soil and Water Quality*, The National Academies Press, Washington, D.C., 1993.

NRC, *Valuing Ecosystem Services: Toward Better Environmental Decision Making*, The National Academies Press, Washington, D.C., 2005.

Natural Resources Conservation Service (NRCS), National Resources Inventory: 2003, Annual NRI — Soil Erosion, May 2006.

Olson, K.D. and Senjem, N.B., Economic comparison of incremental changes in till systems in the Minnesota River Basin, University of Minnesota Extension Service, Minneapolis, MN, 2002. Available at: www.extension.umn.edu/distribution/naturalresources/DD6675.html. Accessed on March 30, 2005.

Parsch, L.D., Keisling, T.C., Sauer, P.A., Oliver, L.R., and Crabtree, N.S., Economic analysis of conservation and conventional tillage cropping systems on clayey soil in eastern Arkansas, *Agron. J.*, 93, 1296–1304, 2001.

Pattanayak, S. K., Sommer, A., Murray, B.C., Bondelid, T., McCarl, B.A., and Gillig, D., Water Quality Co-Benefits of Greenhouse Gas Reduction Incentives in U.S. Agriculture, U.S. Environmental Protection Agency (US EPA), Final Report, EPA Contract Number 68-C-01-142, 2002.

Pautsch, G.R., Kurkalova, L.A., Babcock, B.A., and Kling, C.L., The efficiency of sequestering carbon in agricultural soils, *Contemporary Economic Policy*, 19, 123–134, April 2001.

Pendell, D.L., Williams, J.R., Rice, C.W., Nelson, R.G., and Boyles, S.B., An economic feasibility analysis of manure applications and no-tillage for soil carbon sequestration in corn production. Presented at the 3rd U.S. Department of Agriculture (USDA) Symposium on Greenhouse Gases and Carbon Sequestration in Agriculture and Forestry, March 21–24, Baltimore, MD, 2005.

Pew Center on Global Climate Change, Climate change activities in the United States, 2002. Available at: www.pewclimate.org. Accessed on June 14, 2002.

Pew Center on Global Climate Change, Learning from state action on climate change, in *Brief 8*, 2004. Available at: www.pewclimate.org/docUploads/States%5FInBrief%2Epdf. Accessed on December 19, 2005.

Pimentel D., Harvey, C., Resosudarmo, P., Sinclair, K., Kurz, D., McNair, M., Crist, S., Shpritz, L., Fitton, L., Saffouri, R., and Blair, R., Environmental and economic costs of soil erosion and conservation benefits, *Science*, 267(24), 1117–1122, 1995.

Piper, S., Measuring particulate pollution damage from wind erosion in the western United States, *J. Soil Water Conserv.*, January–February, 70–75, 1989.

Rabalais, N.N., Turner, R.E., and Scavia, D., Beyond science into policy: Gulf of Mexico hypoxia and the Mississippi River, *BioScience*, 52(2), 129–142, 2002.

Ribaudo, M.O., Reducing soil erosion: offsite benefits. AER 561, Economic Research Service, U.S. Department of Agriculture, Washington, D.C., 1986.

Ribaudo, M.O, Heimlich, R., Classen, R., and Peters, M., Least-cost management of nonpoint source pollution: source reduction versus interception strategies for controlling nitrogen loss in the Mississippi Basin, *Ecological Econ.*, 37, 183–197, 2001.

Ribera, L.A., Hons, F.M., and Richardson, J.W., An economic comparison between conventional and no-tillage farming systems in Burleson County Texas, *Agron. J.*, 96, 415–424, 2004.

Riksen, M.J.P.M. and DeGraff, J., On-site and off-site effects of wind erosion on European Light Soils, *Land Degradation Dev.*, 12, 1–11, 2001.

Rowe, G.T., Seasonal hypoxia in the bottom water off the Mississippi River delta, *J. Environmental Qual.*, 30, 281–290, 2001.

Schnitkey, G., Lattz, D., and Siemens, J., Machine cost estimates: field operations, in *Farm Business Management Handbook FBM 0201*, Department of Agricultural and Consumer Economics, University of Illinois, Urbana-Champaign, IL, 2003.

Schrock, M.D., Kramer, J.A., and Clark, S.J., Fuel requirements for field operations in Kansas, *Trans. Am. Society Agricultural Engineers*, 28, 669–674, 1985.

Smith, E.G., Lerohl, M., Messele, T., and Janzen, H.H., Soil quality attribute time paths: optimal levels and values, *J. Agricultural Resource Econ.*, 25, 307–324, 2000.

Sparling, G.P., Wheeler, D., Vesely, E.T., and Schipper, L.A., What is soil organic matter worth? *J. Environmental Qual.*, 35, 548–557, 2006.

Stevenson, F.J., *Humus Chemistry: Genesis, Composition, Reactions*, 2nd ed., John Wiley and Sons, Inc., New York, 1994.

Tegtmeier, E.M. and Duffy, M.D., External costs of agricultural production in the United States, *Int. J. Agricultural Sustainability*, 2(1), 1–20, 2004.

Tennenbaum, S. and Costanza, R., The plight of the Chesapeake, NPG forum series, 1990. Available at: www.npg.org/forum_series/chesapeake.htm. Accessed on July 21, 2005.

Turner, R.E. and Rabalais, N.N., Linking landscape and water quality in the Mississippi River basin for 200 years, *BioScience*, 53(6), 563–572, 2003.

Uri, N., A note on soil erosion and its environmental consequences in the United States, *Water, Air, Soil Pollut.*, 129, 181–197, 2001.

USDA, Agricultural prices 2003 summary, National Agricultural Statistics Service, Washington, D.C., 2004.

U.S. Geological Survey (USGS), The quality of our nation's water's: nutrients and pesticides, Circular 1225, Reston, VA, 1999.

Van Vuuren, W., Giraldez, J.C., and Stonehouse, D.P., The social returns of agricultural practices for promoting water quality improvement, *Canadian J. Agricultural Econ.*, 45, 219–234, 1997.

Weersink, A., Walker, M., Swanton, C., and Shaw, J.E., Costs of conventional and conservation tillage systems, *J. Soil Water Conserv.*, 47, 328–334, 1992.

West, T.O. and Marland, G., A synthesis of carbon sequestration, carbon emissions, and net carbon flux in agriculture: comparing tillage practices in the United States, *Agric. Ecosystems Environ.*, 91, 217–232, 2002.

Wilhelm, W.W. and Wortmann, S., Tillage and rotation interactions for corn and soybean grain yield as affected by precipitation and air temperature, *Agron. J.*, 96, 425–432, 2004.

Williams, J.R., A stochastic dominance analysis of tillage and crop insurance practices in a semi-arid region, *Am. J. Agricultural Econ.*, 70, 112–120, 1988.

Williams, J.R., Llewelyn, R.V., Gross, L.K., and Long, S.H., Analysis of net returns to conservation tillage systems for corn and soybeans in northeast Kansas, Kansas Agricultural Experiment State Bulletin 654, Kansas State University, Manhattan, KS, 1989.

Williams, J.R., Gross, L.K., Claassen, M.M., and Llewelyn, R.V., Economic analysis of tillage for corn and soybean rotations with government commodity programs, *J. Prod. Agric.*, 3(3), 308–316, 1990.

Williams, J.R., Llewelyn, R.V., and Barnaby, G.A., Risk analysis of tillage alternatives with government programs, *Am. J. Agricultural Econ.*, 72, 172–181, 1990.

Williams, J.R. and Tanaka, D.L., Economic evaluation of topsoil loss in spring wheat production in the northern Great Plains, *Soil Tillage Res.*, 37, 95–112, 1996.

Williams, J.R., Roth, T.W., and Claassen, M.M., Profitability of alternative production and tillage strategies for dryland wheat and grain sorghum in the central Great Plains, *J. Soil Water Conserv.*, 55, 49–56, 2000.

Williams, J.R., Nelson, R.G., Claassen, M.M., and Rice, C.W., Derived carbon credit values for carbon sequestration: do CO_2 emissions from production inputs matter? Department of Agricultural Economics, Staff Paper 02-07, Manhattan, KS, 2002.

Williams, J.R., Nelson, R.G., Claassen, M.M., and Rice, C.W., Carbon sequestration in soil with consideration of CO_2 emissions from production inputs: an economic analysis, *Environmental Manage.*, 33, S264–S273, 2004.

Williams, J.R., Peterson, J.M., and Mooney, S., The value of carbon credits: is there a final answer? *J. Soil Water Conserv.*, 60(2), 36A–40A, 2005.

Yin, X. and Al-Kaisi, M.M., Periodic response of soybean yields and economic returns to long-term no-tillage, *Agron. J.*, 96, 723–733, 2004.

Section II

On-Site Benefits

5 Economic and Societal Benefits of Soil Carbon Management: Cropland and Grazing Land Systems

Ronald F. Follett

CONTENTS

INTRODUCTION

The purpose of this chapter is to provide both a historic perspective of U.S. land use as well as to address the role of agricultural technologies to enhance soil carbon

(C) management and how these are economically and environmentally beneficial. Increasingly soil C sequestration is linked to its role in helping to mitigate inputs of carbon dioxide (CO_2), a greenhouse gas (GHG), into the atmosphere, a role becoming extensively documented in the literature. The process includes the uptake of atmospheric CO_2-C by photosynthesis and the subsequent sequestration of the C contained in the plant materials into the soil as soil organic carbon (SOC). Additionally documented is that SOC is lost as a result of improper soil management or potential alternative uses of C resources that could otherwise be returned to the soil.

Soil C sequestration is worthwhile for many reasons, and in sequestering C, both improved soil productivity and environmental services can result. Soil contains the world's largest terrestrial C pool, and there is a direct link between soil and atmospheric C. Misused soils are a major source of CO_2. Practices and policies that encourage maintaining and improving soil C sequestration can consistently be associated with improved soil and water quality, reductions in silt loads and sedimentation into streams, lakes, and rivers, as well as improvements in air quality (Lal, Follett, and Kimble, 2003). Programs, such as the Conservation Reserve Program (CRP) and other incentive programs, generally improve the aesthetic and economic value of land while also enhancing the biodiversity of either the immediate land parcel or of surrounding areas.

A recent estimate of the potential for soil C sequestration in the United States is 288 million megagrams (MMg = Tg) C yr^{-1} or 1056 MMg atmospheric CO_2 yr^{-1} (Lal, Follett, and Kimble, 2003). This potential for the removal of atmospheric CO_2 and its subsequent sequestration as soil C approximates about 15% of the total U.S. emissions for 2000 of 1887 MMg of CO_2-C GHG equivalents yr^{-1} (US EPA, 2002). Our discussion here will focus on cropland and grazing land with their potential SOC sequestration ranges of 48–98 MMg C yr^{-1} and 13–70 MMg C yr^{-1}, respectively within the United States (Lal, Follett, and Kimble 2003). Together, cropland and grazing land account for approximately 40% of the U.S. potential for C sequestration. The following discussion will also include background information about how soil C stocks have been degraded, where we are now with C stocks in the United States, and how the building of C stocks needs to be addressed to approach the potential amounts that might be sequestered. Important to this discussion is how attaining such additional stocks of SOC would provide societal and economic benefits in the United States. However, SOC stocks cannot be considered alone because of the importance of inputs of crop residue-C resources to the soil, the losses of SOC through mismanagement of soils, the role of government programs, and emerging alternative uses of crop residue C (or other organic C wastes generally returned to soils). These all affect SOC sequestration and the potential of U.S. agricultural soils to sequester C. Increasing current SOC stocks by nearly approaching the potential rate of SOC sequestration cannot be realized without greater adoption of recommended management practices (RMPs). Alternatively, the removal of crop residues from agricultural lands can increase the likelihood that SOC stocks will decrease. Current estimates of rate of SOC sequestration for cropland (Intergovernmental Panel on Climate Change [IPCC] methodology; IPCC/UNEP/OECD/IEA, 1997) are that only about 20% of the potential rate of C sequestration is occurring (Sperow et al.,

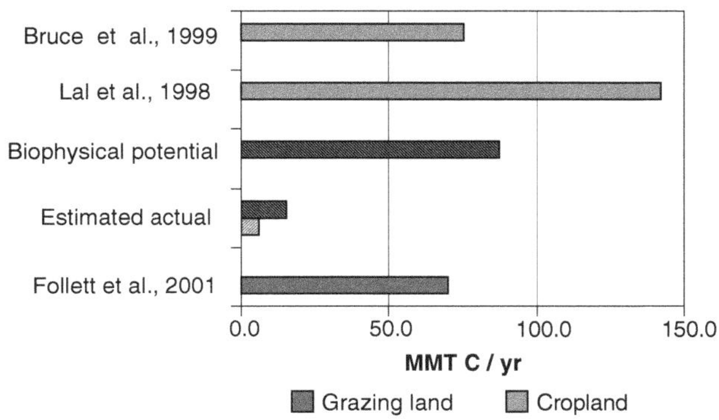

FIGURE 5.1 Estimates of SOC sequestration in U.S. croplands and grazing lands.

2003). However, when the IPCC methodology was also used to make a "biophysical" estimate, but with similar RMPs to those of Bruce et al. (1999) and Lal et al. (1998, 2003), Sperow et al. (2003) obtained estimates of potential soil C sequestration similar to those of the other authors (Figure 5.1). The IPCC methodology also results in very low estimates of SOC sequestration for grazing lands and therefore likely represents a reduced percentage of the potential soil C that can be sequestered there as well.

U.S. CROPLAND

A Brief History of U.S. Cropland Use

Following the end of the Civil War in 1865, westward population movement and population growth was increasingly dynamic. Westward movement of the U.S. population was aided by the Homestead Act of 1862 and by completion of the transcontinental railroad in 1869. Native American lands that had been seized by the government and their availability for settlement drew European immigrants whose numbers, coupled with the movement of U.S. citizen settlers increasingly opened the Great Plains and the western United States to agriculture. In 1870 and with a total population of 38.5 million people, the "center of the U.S. population" was in Highland County, Ohio, near present day Hillsboro. "Center of Population" is defined as the point at which an "imaginary, flat, weightless, and rigid map of the United States would balance perfectly" if every person – counted where they lived on the day of the census – weighed the same (U.S. Census Bureau, 2000). By 1940, with a population of 131.4 million, the "center" was in Sullivan County, Indiana, near Paxton, and by 2000, with a U.S. population of 281.4 million, it was in Phelps County, Missouri, near Edgar Springs.

Historical records also show that from the 1870s until about 1940 grain produced per unit of land area in the United States were essentially static but, depending upon the crop, yield increases after 1940 were from 75–400% (Allmaras et al., 1998; NASS, 2004). Following the shock of the 1930s "Dust Bowl," and after 1940, major subsequent technological innovations and a national response to the dust bowl occurred. As described by Power and Follett (1987), three highly important technological innovations that occurred were:

1. "Mechanization" wherein, by the end of WWII, most farms had switched to the use of gasoline tractors and there was an increase in the size of tractors and other farm machinery. There was also increased use and availability of specialized agricultural machinery, including combines, planting equipment, high-clearance tillage equipment, center-pivot irrigation rigs, and other equipment.
2. "Crop varieties," improvements in crop varieties with genetic advances in existing crops and the introduction of new crops (i.e., soybean) that pushed farmers toward "monocultures," and limited rotations, but also resulted in increased "crop-specific" knowledge by farmers about optimizing the yields that they could obtain.
3. "Agricultural chemicals," the use of "chemicals" as typified by manufactured nitrogen (N) fertilizer that was prominent after WWII and expansion of the use of agricultural pesticides.

These three technological advances after 1940 led farmers to increasingly grow a single or only a few crops and the development of increased confidence by farmers that with the use of the technologies described previously, the crops they grew would flourish.

Coupled with the preceding three technologies is a fourth technology that must also be included and that will be referred to as "improved soil management" (reduced tillage and enhanced crop residue management, etc.). Part of the national response to the Dust Bowl crisis of the 1930s, was the formation of the Soil Conservation Service (now the Natural Resources Conservation Service). The Dust Bowl provided a hard lesson about the use of large-scale, improper soil management, that when coupled with drought had devastating effects upon a nation and its society. This awakening did much for the conservation movement in the United States. As Historian Robert Worster wrote, "The ultimate meaning of the dust storms of the 1930s was that America as a whole, not just the plains, was badly out of balance with its natural environment" (PBS, 2006). The awful and worsening drought conditions, coupled with the horrible effects of the Great Depression, eventually meant that millions of U.S. citizens could no longer make a living from farming. By 1940, over 2.5 million people had moved out of the Great Plains. Life was simply unsustainable for many. Conversely, good land and resource management, and though not directly recognized at the time — good soil C management, has since been shown to restore the productivity of the land and the quality of the water and to conserve and help restore what were no doubt the previous major losses of SOC.

The lessons of the 1930s and 1940s resulted in greater awareness of the need for "soil conservation practices" and land set aside programs, which would conserve land and soil productivity, and increase soil C. Land under the "Soil Bank Act" of 1956 was set aside primarily to divert cropland from the production of major crops and thus to decrease agricultural inventories, but with a second purpose of establishing protective vegetative cover on land that needed conservation practices. Over

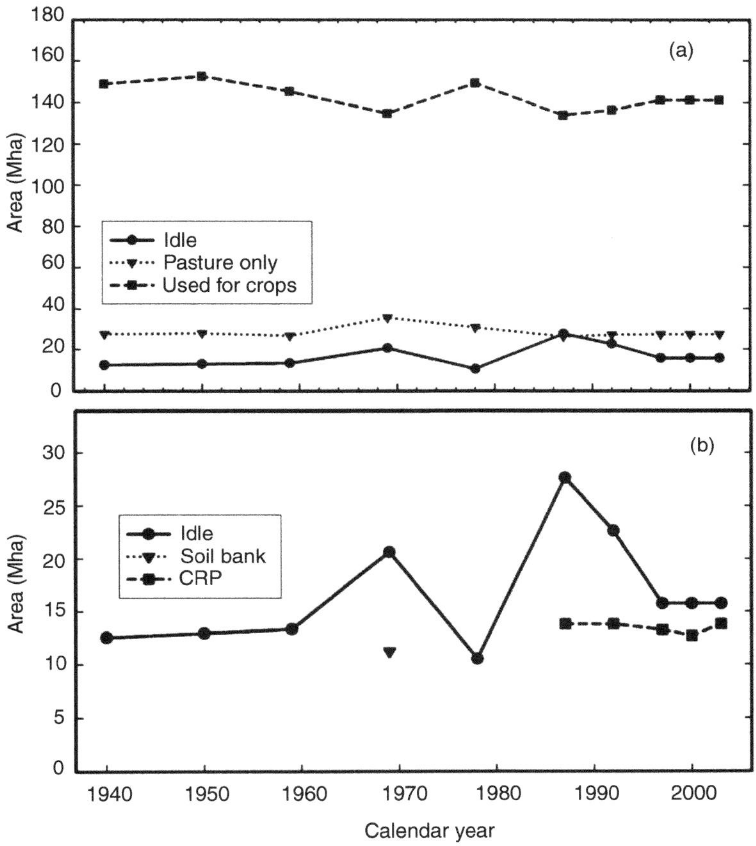

FIGURE 5.2 (a) U.S. cropland use. (b) U.S. idle and other cropland use from 1940. (From National Agricultural Statistics Service (NASS), Agricultural Statistics, NASS of USDA, U.S. Government Printing Office, Washington D.C., pp. IX–6, 2004. With permission.)

its 10-year life, the "Soil Bank Program" diverted 11.6 Mha (million hectares) or about 28.7 million acres to conservation practices on 306,000 farms. The Soil Bank Program was followed by two similar long-term contract programs: the Cropland Conservation Program, authorized in 1962, and the Cropland Adjustment Program, enacted in 1965.

In the early and mid-1970s, the prices of farm commodities rose significantly, due to diminished stocks and increased export demand. With the encouragement of the United States Department of Agriculture (USDA) policy makers U.S. producers responded by planting crops on marginal cropland and by also breaking out and planting range and pasture lands (Figure 5.2). This activity continued until the early 1980s, when overproduction and a strengthening U.S. dollar depressed prices, causing farm income to fall to its lowest level since the 1930s. However, because of unprecedented foreign grain sales, all USDA-subsidized cropland-retirement programs were suspended by 1973, and eventually few enduring resource conservation benefits remained (Berg 1994). Amazingly, the conservation lessons of the 1930s appeared to have been forgotten. Enactment of legislation creating the National Resources Inventory in 1972 and implementation of the Soil and Water Conservation Act (RCA) of 1977 resulted in the collection of data showing that soil erosion, and concurrent losses of soil organic matter, had increased to levels that were perhaps worse than during the Dust Bowl years of the 1930s. Soils once adequately treated or retired from cultivation had been tilled and planted to annual crops. Continued recognition of the adverse impacts of ongoing soil erosion resulted in key additional legislative authority being placed into the 1985 farm bill to implement a nationwide CRP. The unfortunate lesson from the 1970s is that the lack of responsible policies at the national level and the associated suspension of government programs that were supportive to conservation and reduced soil degradation resulted in the rapid loss of the economic and societal benefits of soil C that had previously been accumulated.

Cropland used for crops in 2003 accounted for approximately 140 Mha, whereas idle cropland accounts for nearly 16 Mha and cropland used as pasture accounted only for 27 Mha (Figure 5.2a). Even though land utilization statistics are only reported at about 10-year intervals in Agricultural Statistics (NASS, 2004), the influence of the 11 Mha set aside in the Soil Bank Program is apparent in both the lowered cropland area used for crops (135 Mha) and the increased amount of idle cropland (21 Mha) in the 1969 data. The subsequent suspension of cropland retirement programs became apparent in the 1978 data, wherein cropland used for crops has increased by 15 Mha–149 Mha; idle cropland had decreased by 10 Mha and land used for pastures only had decreased from 31 Mha in 1969 to 26 Mha by 1978 (Figure 5.2a and 5.2b). The influence of the 1985 farm bill that implemented the CRP has resulted in there being contracts on nearly 14 Mha by 1987 with 28 Mha of idle land being reported that same year. Initially, an administrative decision was made to limit the criterion for acceptance of land into the CRP to control soil erosion. However, because of amendments to the CRP in the 1990 farm bill, environmental benefits were expanded so that state water quality areas, conservation priority areas, and public wetland protection areas were also included. Thus, by 1992, there were just over 13.8 million ha (34 million ac) of land under CRP contract and of those 7.2 million ha (17.8 million ac) were identified as highly erodible cropland by 1982, before the CRP was implemented (Figure 5.3).

As the CRP program and other conservation programs have continued they are resulting in numerous conservation and societal benefits (ERS, 2005). As outlined in their recent Fiscal Year Summary, the Farm Service Agency (FSA) (2004) indicates that in 2003, CRP reduced soil erosion by 400 MMg, and this reduction contributes directly to:

1. Improved water quality from less movement of sediments, nutrients, and other chemicals into streams, lakes, and rivers
2. Creation of 1 Mha of conservation buffers to protect fragile resources
3. Nearly 1 Mha of restored wetlands
4. 7 Mha of vegetative cover defined as best suited for wildlife habitat
5. 1.5 Mha of sensitive wildlife ecosystem restorations

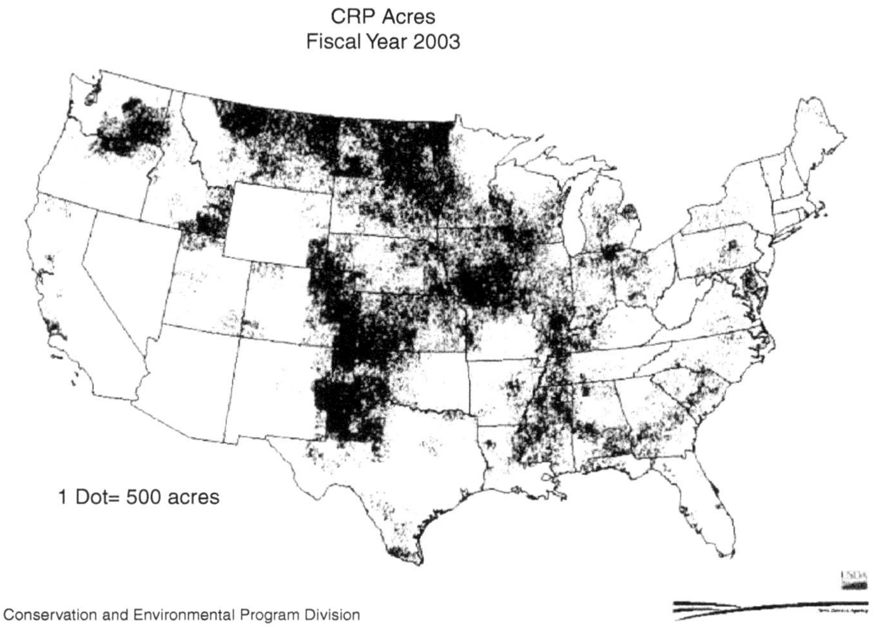

FIGURE 5.3 Land in the conservation reserve program (CRP) in 2003. Each dot on the map is equivalent to 202 ha (500 acres). (From FSA of USDA, Conservation Reserve Program, Fiscal Year Summary 1999 to 2003, 2004. With permission.)

Follett et al. (2001b), in a regional study extending across 5.6 Mha in the U.S. Great Plains and Western Corn Belt, identified that CRP had stored 5 MMg C. The FSA (2004) report estimates that 17 MMg C are stored in the soil and vegetation of U.S. land in the CRP in 2003. The CRP is an example of a government program supportive of conservation and the restoration of economic and societal benefits of soil C on the affected lands. In its first 20 years, CRP's societal contributions are substantial. It prevents 450 million tons of soil from eroding each year (Lasseter, 2006). The CRP and the related federal/state partnerships of the Conservation Reserve Enhancement Program help protect basic and vital natural resources by stopping soil and nutrients from washing into waterways or contaminating our air. It has already restored 1.8 million acres of wetlands. Drinking water is safer for millions of people, dust is reduced and air is more breathable, lakes are cleaner for swimming and fishing, and there are more fish, ducks, and other wildlife as a result of more abundant and improved habitat. As CRP contracts expire in the future, the farm and environmental communities have concerns about the implications if policy makers choose to end the program. Should the CRP program be discontinued there is need for information about the obvious impacts on soil erosion control, water quality protection, and wildlife habitat; and about benefits that may be lost, but that are not yet fully understood. The knowledge and efforts to continue with policies that have strong conservation and societal benefits must be sustained, and now extended, to recognize the role of not only the CRP or other soil conservation programs, but of the beneficial economic and societal role of agricultural lands in general. There is need also for increased recognition of how land management and use also influence the exchange of and as a sink for C containing GHGs, especially CO_2 (Lal, Follett, and Kimble, 2003).

Present Trends in U.S. Cropland Use

In 1997, there were 184.3 Mha of cropland in the United States, which included 141.2 Mha of cropland used for crops, 15.7 Mha of idle cropland, and 27.3 Mha used only for pasture (NASS, 2004). Because of its unavailability, the 2000 and 2003 U.S. cropland area will be estimated as the same as reported in 1997 (NASS, 2004). Even if the area of cropland used for crops were to increase at the expense of idle cropland and cropland used for pasture, the total cropland area in the United States will likely not expand and probably will slowly decrease. For example, irrigated land area peaked at 20.6 Mha in 1980 and has been stable at about 18.8 Mha since 1989. Even average farm size, which was 120, 151, 172, and 186 ha in 1960, 1970, 1980, and 1990 and reached a peak of 191 ha in 1996, has declined since 1996 to between 176 and 179 ha in 2003 and nearly 37% of all farms were larger than 700 ha, whereas 30% were smaller than 235 ha.

Area for other cropland used for crops has decreased from 47.5% of U.S. cropland area in 1940 to 33.4% in 2003 (Figure 5.4). Of the major crops, only the harvested area of soybean is showing a consistent increase in its fraction of U.S. cropland area from 1.3–20.7% (1.9–29.3 Mha) during 1940–2003 (Figure 5.4). The combined harvested area of rye, oats, and barley has decreased from 28.6% in 1940 to 17.2% in 2003 (21 Mha–4.3 Mha). The fraction of U.S. cropland harvested for wheat grain has remained fairly steady at 14.5% (21.6 Mha) in 1940 and 15.2% (21 Mha) in 2003. Area of harvested sorghum was 1.7% (2.6 Mha) in 1940 and 2.2% (3.2 Mha) in 2003. Area of harvested corn was 20.8% (31.0 Mha) in 1940 and 20.4% (28.8 Mha) in 2003 (NASS, 2004).

Cropland Practices to Sequester C

The net amount of SOC present is a function of the quantity and composition of crop residues, plant roots, and other organic material added or returned to the soil and the rate of SOC decomposition and loss by soil erosion and leaching of dissolved organic C from the soil. As described by Follett (2001b) plant C enters the SOC pool as plant "litter," root material, and root exudates or, if consumed by animals, as excreta. Litter-sized plant material abrades into smaller sizes (i.e., "light fraction" or "particulate organic matter"). These processes are mostly near the soil surface and therefore the C that is accumulating is easily lost as a result of soil erosion or increased tillage intensity. Some of the C and essential elements in these fractions become a source of nutrition for soil flora and fauna, including bacteria, fungi, and micro- and macro-fauna. Adsorption of organic material onto the reactive surfaces of clay minerals in soils and other clay particle surfaces is an important mechanism for binding soil particles together into soil aggregates. Among the organic materials

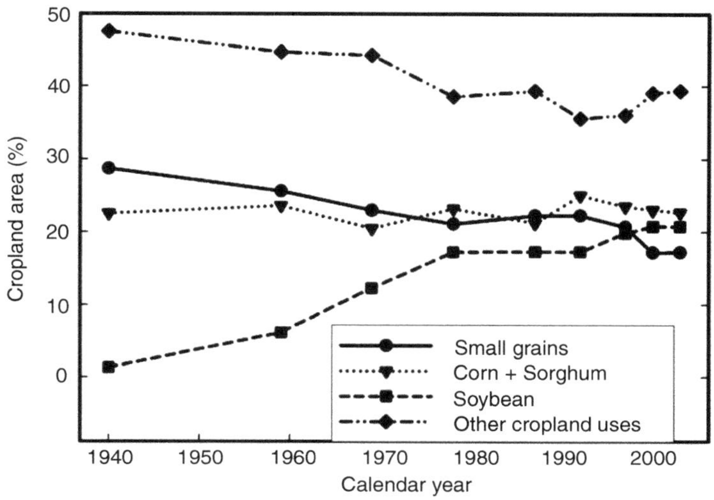

FIGURE 5.4 Fraction of U.S. cropland area on which small grain (wheat, rye, oats, and barley), corn plus sorghum, and soybean were harvested and area of other cropland used for crops from 1940–2003. Because of availability of data, 1997 values for cropland area are used as estimates for 2000 and 2003. (From National Agricultural Statistics Service (NASS), Agricultural Statistics, NASS of USDA, U.S. Government Printing Office, Washington D.C., pp. IX–6, 2004. With permission.)

important to soil aggregation are extracellular polysaccharides that soil microorganisms produce as a metabolic product while decomposing various organic substances. Polysaccharide molecules can be adsorbed to clay surfaces strongly through polyvalent bridging and because of the net negative charge that results from the presence of various organic acids (Chenu, 1995). Uncharged polysaccharides may strongly adsorb to mineral surfaces because of hydrogen bonding or van der Waals forces (Golchin, Baldock, and Oades, 1998).

Crop Yield and Residue Production

Changes in crop yields over time are highly important to the discussion of soil C because they can be related to changes in the amounts of crop residues and crop residue C available to return to the soil surface or incorporated into the soil. Recent and historical yield data are readily available through USDA-National Agricultural Statistic Service (NASS). Therefore, the C input from crop residues can be estimated using a “harvest index” (HI) equation (Donald and Hamblin, 1976; Hay, 1995) in which the crop residue (Yr) is estimated as:

$$Yr = Ygr\ [(1/HI)-1] \tag{5.1}$$

where Ygr is the harvested grain (or other biomass) and HI is the ratio of Ygr and aboveground crop residue yield. The HI, grain yield, and crop residue yields listed

TABLE 5.1
National Average Grain Yield, Aboveground Crop Residue Production, and Percent Change for Six Selected Crops in 1940 and 2000 in the United States[a]

Crop[b]	Harvest Index		Grain Yield (Mg ha^{-1})[c]		Change	Crop Residue (Mg ha^{-1})		Change
	1940[f]	2000[g]	1940[d]	2000[e]	%	1940	2000	%
Barley	0.3	0.5	1.3	3.9	200	3.5	3.9	40
Corn	0.4	0.5	1.9	8.4	340	3.5	7.5	110
Oat	0.2	0.4	1.2	2.2	80	3.9	2.8	–30
Sorghum	0.3	0.5	0.9	4.0	340	1.8	4.5	150
Soybean	0.3	0.5	1.3	2.6	100	2.9	3.0	5
Wheat	0.3	0.5	1.1	2.8	150	2.7	3.4	30

[a] Harvest index, grain yield, and crop residue production data from Johnson, J.M.F., Reicosky, D.C., Allmaras, R.R., Sauer, T.J., Venterea, R.T., and Dell, C.J., *Soil Tillage Res.*, 83, 73–94, 2005. With permission.

[b] Botanical names: Barley (*H. vulgare* L.), corn (*Z. mays* L.), oat (*A. sativa* L.), sorghum, soybean (*G. max* L. Merrill), wheat (*Triticum* ssp.).

[c] Three-year average centered on the shown year.

[d] From Cochrane, W., *The Development of American Agriculture: A Historical Analysis*, 2nd ed. University of Minnesota Press, Minneapolis, MN, 1993. With permission.

[e] From National Agricultural Statistics Service (NASS), Agricultural Statistics, NASS of USDA, U.S. Government Printing Office, Washington D.C., pp. IX–6, 2004. With permission.

[f] From Allmaras, R.R., Wilkins, D.E., Burnside, O.C., and Mulla, D.J., Agricultural technology and adoption of conservation practices, in *Advanced Soil Water Conservation*, Pierce, F.J. and Frye, W.W., eds., Sleeping Bear Press, Chelsea, MI, 1998, 99–157. With permission.

[g] From Lynch, P.J. and Frey, K.J., *Crop Science*, 33, 984–988, 1993; Prince, S.D., Haskett, J., Steiningfr, M., Strand, H., and Wright, R., *Ecological Appl.*, 11(4), 1194–1205, 2001; Halvorson, A.D., Peterson, G.A., and Reule, C.A., *Agron. J.*, 94, 1429–1436, 2002; Pedersen, P., Boote, K.J., Jones, J.W., and Lauer, J.G., *Agron. J.*, 96, 556–564, 2004; Vetsch, J.A. and Randall, G.W., *Agron. J.*, 96, 502–509, 2004; and Yang, H.S., Dohermann, A., Lindquist, J.L., Walters, D.T., Arkebauer, T.J., and Cassman, K.G., *Field Crops Res.*, 87, 131–154, 2004. With permission.

in Table 5.1 were recently reported by Johnson et al. (2005). The percent change in grain yields is attributable to genetic improvements, better fertility and pest management, and technological advances in soil management. The largest percentage change in grain yields were observed for corn and sorghum followed by barley, wheat, and soybean with oat having the least percent change. The calculated percentage increase of crop residue yields (i.e., vegetative aboveground biomass) was always less than for grain. The largest change in crop residue yield was exhibited by sorghum, followed by corn, barley, wheat, and soybean, with a decrease for oats Table 5.1. In terms of SOC contribution, aboveground vegetative biomass is only a portion of the total C input to soil by plants. To understand the soil C cycle, one must have a complete understanding of the plant C partitioned belowground that is transferred to support root growth and maintenance, contributing to root exudation and rhizodeposition, sloughed root cells, and that is released as CO_2 from the belowground respiration of plant roots and microorganisms.

Crop Residue Carbon

The annual grain yields of the crops shown in Table 5.1 are readily obtained from NASS. Use of HI allows the estimates of crop residue yields that are based upon grain yields. The following calculations were made by assuming that the HI for 2000 could be used from 1969 forward (Table 5.1). Also assumed was that crop residue contains 40% C (Johnson et al., 2005) for the calculation of the rates of aboveground crop-residue C shown in Table 5.2. The years after 1969 that were selected to show in Table 5.2 correspond to those in which NASS also reported land utilization. Though showing some increases in rates of aboveground crop residue C are small grains (wheat, rye, oats and barley) which have increased from between 0.8–2.0 Mg C ha^{-1} since 1940 to between 1.0–1.4 Mg C ha^{-1} in 2003. Soybean residue C has ranged between 0.7–1.1 Mg C ha^{-1} and since 1940, sorghum values have ranged between 1.3–1.8 Mg C ha^{-1}. The change and size of the rates of crop residue C production for corn have increased most rapidly since 1940 with corn residue C calculated to have increased from 2.1–3.6 Mg C ha^{-1}.

Multiplication of the calculated rate of crop residue C produced in the United States (Table 5.2) by area of harvested cropland area (Table 5.3) for each crop results in an estimate of total aboveground crop residue C production (Table 5.4). The last column in Table 5.3 shows how the total percentage of U.S. cropland used by these seven crops has changed between the years 1940 and 2003, ranging from 52% in 1940 to 64% in 1992. The final column in Table 5.4 shows the change over years in the calculated production of aboveground crop residue C by these seven crops. Comparison of the final columns in Table 5.3 and Table 5.4 shows that since 1987, very little change has occurred in the percent of cropland use, but that since 1992, greater than 20% more crop residue C has been produced than during or before

TABLE 5.2
Average Aboveground Residue C Produced (Mg C ha^{-1}) by Seven U.S. Crops

	Mg C ha^{-1}						
Year	Wheat	Rye	Oat	Barley	Corn	Sorghum	Soybean
1940	0.96	1.24	2.02	1.15	1.09	0.79	1.02
1969	0.82	0.88	1.15	0.96	2.10	1.36	0.74
1978	0.84	0.98	1.12	1.06	2.53	1.37	0.79
1987	1.01	1.10	1.17	1.13	3.01	1.74	0.91
1992	1.06	1.10	1.41	1.34	3.30	1.82	1.01
1997	1.06	0.97	1.28	1.25	3.18	1.74	1.05
2000	1.13	1.06	1.38	1.31	3.43	1.53	1.02
2003	1.19	1.03	1.40	1.27	3.57	1.32	0.90

TABLE 5.3
Total Area of Harvested Cropland (Mha) of Seven U.S. Crops and as a Fraction of the Total U.S. Cropland Used for Crops

	Mha							
Year	Wheat	Rye	Oat	Barley	Corn	Sorghum	Soybean	% of U.S. Cropland
1940	21.58	1.30	14.35	5.48	30.96	2.58	1.95	52.46
1969	19.27	0.55	7.26	3.86	22.11	5.48	16.60	55.70
1978	22.88	0.38	4.51	3.75	29.13	5.43	25.78	61.46
1987	22.66	0.27	2.79	4.03	24.10	4.27	23.15	60.65
1992	25.43	0.16	1.82	2.95	29.19	4.88	23.58	64.49
1997[a]	25.45	0.13	1.14	2.51	29.43	3.71	27.99	63.98
2000[a]	21.09	0.12	0.94	2.11	29.34	3.13	29.33	60.93
2003[a]	21.40	0.14	0.90	1.90	28.81	3.16	29.29	60.61

[a] The 2000 and 2003 U.S. cropland area used for crops is the same as in 1997.

TABLE 5.4
Calculated Aboveground U.S. Crop Residue C Production (MMg C) of Seven Major Crops

	MMg C							
Year	Wheat	Rye	Oat	Barley	Corn	Sorghum	Soybean	Total
1940	20.70	1.61	28.96	6.32	33.67	2.04	1.98	95.29
1969	15.85	0.48	8.39	3.71	46.54	7.46	12.22	94.66
1978	19.31	0.37	5.07	3.96	73.82	7.43	20.38	130.33
1987	22.96	0.30	3.26	4.54	72.43	7.43	21.10	132.02
1992	26.86	0.17	2.56	3.97	96.30	8.89	23.84	162.59
1997	27.02	0.12	1.46	3.14	93.55	6.44	29.27	161.00
2000	23.80	0.13	1.30	2.77	100.76	4.78	30.03	163.59
2003	25.43	0.14	1.26	2.40	102.78	4.18	26.30	162.49

1987. Although there has been a slight increase in residue C produced after 1992 and that it has been produced on a slightly smaller percent of the U.S. cropland area used for crops, the changes have been small. If these recent trends continue and alternate uses emerge to that of returning crop residues to the land then remaining crop residues must more effectively be used to sustain SOM, prevent soil degradation, and maintain environmental benefits of soil C while obtaining additional economic benefits.

Alternative uses of residues, especially on fragile lands, not only can refer to removal for biofuel, but also to such alternative uses as winter grazing of corn stocks by livestock. Recent data of the authors on a sandy soil site where, because of the fragile nature of the soil, conventional tillage treatments were being compared with strip tillage under full irrigation. The kg SOC in the top 15 cm, without winter

TABLE 5.5
The Effect of Using Cattle for Winter Grazing of Corn Residues during the Fall and Winter of 2003/2004 and the Fall and Winter of 2004/2005[a,b]

	Soil Organic Carbon in 0–5 cm			
	% C		(kg SOC ha^{-1})	
Treatment	**2003**	**2005**	**2003**	**2005**
Strip tillage	1.34a	1.23a, b	8160a	7860a
Conventional tillage	1.17a, b	1.07b	6800a, b	6050b
	Soil Organic Carbon in 5–10 cm			
	% C		(kg SOC ha^{-1})	
Strip tillage	0.66a	0.61a	7820a	7077a, b
Conventional tillage	0.60a, b	0.55b	6530b	6740b
	Soil Organic Carbon in top 10–30 cm			
	% C		(kg SOC ha^{-1})	
Strip tillage	0.66a	0.61a	17,650a	16,710a, b
Conventional tillage	0.60a, b	0.55b	16,130a, b	15,380b

[a] Cornstalks were not grazed, but remained on the field during the non-growing seasons of 2001 and 2002.
[b] Numbers followed by different letters are significantly different at the 95% confidence level.

grazing of corn stalks was stable or increasing through the spring of 2003. However, during the fall and winter of 2003/2004 and 2004/2005, a decision was made by farm management (without consultation with the researchers) to winter graze the corn stalks (at about 65 animal days/ha). As seen in Table 5.5, both the percent soil C and the weight of SOC decreased from 2003–2005. Average SOC under strip tillage had increased from 17,560 kg SOC ha^{-1} in the top 15 cm of soil to 20,390 kg SOC ha^{-1} by 2003 (or 33,630 kg SOC ha^{-1} in the top 30 cm). The corresponding values for the conventional tillage were a change from 17,600 kg SOC ha^{-1} in the top 15cm of soil to 17,540 kg SOC ha^{-1} to 17,540 by 2003 (or 29,460 kg SOC ha^{-1} in the top 30 cm). By 2005, the amounts of SOC measured in the top 30cm of soil were 31,650 kg SOC ha^{-1} and 28,170 kg SOC ha^{-1} for the strip and conventional tillage treatments, respectively. Table 5.5 shows the changes in both percent and weight of SOC for the three soil depth increments sampled and with both treatments generally being negatively affected. It should be pointed out that if residues are to be winter grazed the soils should not be fragile and susceptible to wind erosion and there may be need to improve the feed quality of the crop residues and not just the grain yields.

Crop Residues and Biofuels

Various authors are addressing issues associated with the production of ethanol from corn grain or biodiesel from soybean oil or other uses (such as the production of complex sugars for making of fabrics and biodegradable plastics) and the associated energy inputs. It is not the intent of this paper to evaluate the use of grain for the production of biofuel other uses, the interested reader is referred information included in articles by Shapouri et al. (2002, 2004), Pimental and Patzek (2005), Perlack et al. (2005), and Parfit and Leen (2005). Shapouri et al. (2002, 2004) generally indicate a net positive energy balance for the production of ethanol or biodiesel from grain whereas Pimental and Patzek (2005) contend that growing corn for ethanol results in a net negative energy balance while growing soybeans for biodiesel had a slightly negative energy balance. Interesting is that the corn grain contains 44% C and ethanol is approximately 52% C by weight. Thus, in general, the production of ethanol from grain results in a C density increase from 44–52% and the production of a C energy source that is in liquid form for use in motor vehicles. Energy inputs for the production of corn grain and its conversion to ethanol include inputs such as natural gas, diesel, electricity (including that produced from coal), and other C derived inputs. These energy inputs also express the economic and societal benefits of C; however, they include energy derived from fossil fuels that, when burned, result in an increased burden of fossil fuel derived CO_2-C being added to the atmospheric. By comparison, when CO_2-C is released from the direct burning of recently photosynthesized plant-derived biomass, it is recycled and adds little or no additional fossil C into the atmosphere.

Perlack et al. (2005) project that currently available biomass resources, using sustainable management technologies and with 40% recovery potential, is about 176 MMg dry tons annually; including 68 MMg from corn and 15 MMg from wheat and other small grains for a total of 83 MMg from these crops. Manures were estimated to provide an additional 32 MMg. If these residues are estimated to contain approximately 40% C, then they provide 27, 6, and 13 MMg residue C from corn, small grains, and manure, respectively (total 46 MMg), or about 65% of the total C from agriculture. Projected moderate (60% residue recovery)- and high (75% residue recovery)-yield increases resulting from technology with no land use change by 2014 are that 383 and 541 MMg of residues (153 and 216 MMg C) will be available, respectively. Of the residues projected to be available by 2014 for the moderate and high yield scenario, corn and small grains are projected to provide about 52 and 56%, respectively, and CRP lands are projected to provide perhaps as much as 5% of the total available residue. Perennial crops (grasses and trees) were not considered in the preceding scenarios, but when they are considered, they add large amounts of harvested biomass and C to the projections.

Serious questions were raised in a series of journal articles in 1979 about how much residue could be removed from the land without exceeding soil erosion tolerance limits. Larson (1979) summarized these data with the conclusion that crop residues could be available for energy production if the calculated erosion rates did not exceed the soil tolerance limit. With 1979 practices, there would be 49 MMg of residues available within the U.S. Corn Belt (36% of the residue produced in the region), and 65% of the residues would come from only 4 (of 14) major land resource areas that have relatively level, deep soils. In the Great Plains, about 16 MMg of residues could be harvested for energy and were produced in the more humid, eastern section. If both wind and water erosion was considered, available residue in the Great Plains was less (Skidmore, Kumar, and Larson, 1979). In Oregon, 60% of the residues could be harvested from 88% of the area planted to small grains, or about 1 MMg (Allmaras et al., 1979). In the Southeast United States, about 4 MMg of residue could be available (Campbell et al., 1979). The residues discussed in these 1979 articles are largely harvested from land planted to corn or small grains and their 1979 total was about 70 MMg. The weight of crop residue from corn and small grains projected by Perlack et al. (2005) is about 83 MMg or about 118% larger than projected in 1979. By comparison, data in Table 5.4 for 1987 and 2003 lists residue production by corn and small grains to be 103 and 132 MMg, respectively, or an increase of about 127%. The question remains whether adequate technology is available and will be used to sustain soil productivity associated with SOC in mineral soils when crop residue C is removed to produce energy. Focused research is needed on soil management, such as increased use of no-till (NT) or other practices to adequately protect SOC levels. Not addressed in this crop biomass discussion is the important role of belowground biomass C and whether, as aboveground yields increase, the belowground biomass is also increasing and how soil management may also contribute to the utilization of these residues as a resource to aid in the maintenance of SOC. Definitive studies are greatly needed about amount of C derived from root material and other below-ground inputs, their contribution to maintaining or increasing SOC, and the effects of the environment and soil management on them.

The choice of at least corn to be a potential residue crop may also be supported by the Tollenaar and Lee (2002) estimate of potential corn yields of 25 Mg ha^{-1}. From data on improved hybrids during 1930–2000 in the central United States, Duvick (1992, 2005) reported that corn grain yield and presumably increases in aboveground corn biomass yields had much higher limits than are currently observed. Corn grain and corn residue yields (Table 5.4) have increased dramatically during the past 60 years and even with continued yield increase, there is need for an increased application of advanced soil and residue management practices in the future. However, the optimistic estimates of increasingly larger corn yields must be tempered with the reality of the role of weather variability, soil, and other limitations to potential yields.

Cover Crops

"Cover crops" protect the soil from erosion and losses of nutrients via leaching and runoff. The term "winter cover crop" is used for a cover crop grown to protect the soil during the winter fallow period. Despite its acceptance, a winter cover crop does not necessarily need to be used during winter and can be used even during summer

(Delgado, Reeves, and Follett, 2004). If a legume is used, it can also potentially fix atmospheric N_2, and enhance soil N reserves (Power, Follett, and Carlson, 1983). Thus, the definition of winter cover crops can be expanded to those crops that are grown for improving soil, air and water conservation and quality; nutrient scavenging, cycling and management; increasing beneficial insects in integrated pest; and for short-term (e.g., over-winter) for animal-cropping grazing systems (Delgado, Reeves, and Follett, 2004; Reeves, 1994). Strategies that increase the cropping intensity such as the use of rotations with winter cover crops to increase the amount of biomass C returned to the soil can affect the size, turnover, and vertical distribution of both active and passive pools of SOC (Franzluebbers, Hons, and Zuberer, 1994). Winter cover crops can be an excellent source of forage for grazing for animals and can contribute to the development of sustainable animal-cropping systems. If suited to the climate and the farming operation, such cropping systems should provide an opportunity to produce more biomass C than in a monoculture system and to thus increase SOC sequestration. Winter cover crops are a viable tool for soil and water conservation. Lal et al. (1998) reviewed the literature on this topic and concluded that the potential exists for adopting winter cover-crop rotation systems on about 51 Mha, which could sequester an additional 100–300 kg C ha^{-1} yr^{-1}, thus resulting in 5.1–15.3 MMg C yr^{-1}.

Livestock Manure

Animal manure represents a valuable resource, and its potential for use as a feedstock for biofuel was briefly mentioned earlier and is discussed in more detail by Perlack et al. (2005). Additionally, manure use in many ways results in similar issues to those of other organic wastes. Consequently, though this discussion will be directed toward livestock manure the reader is asked to recognize this section in the broader context of organic wastes in general. Livestock manure and other organic wastes (assuming they do not have hazardous substances in them) have many beneficial effects, including on soil C sequestration. Organic wastes from a collectible source, such as manure from "confined" livestock operations, can be collected, stored, and made available for use as a nutrient resource, soil amendment, or for bioenergy generation. Livestock categories of manure include those from dairy (i.e., milk cows and heifers) or beef (i.e., steers, bulls, calves, and cows) cattle (*Bos taurus*); swine (i.e., growers and breeders) (*Sus scrofa domesticus*); chickens (i.e., broilers and layers) (*Gallus gallus domesticus*); and turkeys (*Meleagris gallopavo*) (NASS, 2004). Calculations were made by Follett (2001a), using procedures described by Lander, Moffitt, and Alt (1998), of the annual weight of manure N produced by confined livestock in those regions of the United States, where the manure can economically be applied to cropland. Next, an estimate was made that an average application rate equivalent to 250 kg N ha^{-1} yr^{-1} could be made to 18 Mha of cropland with the manure produced. The application of the manure, as described previously, would result in sequestration at the rate of 200–500 kg C ha^{-1} yr^{-1}, thus resulting in an estimated sequestration of 3.6–9.0 MMg C yr^{-1} (Follett, 2001a).

U.S. GRAZING LANDS

Grazing lands represent the largest and most diverse single land resource in the United States. They include the relatively undisturbed rangelands of the West (i.e., grasslands, savannas, and shrublands) and the intensively managed pastures that occur in every state, including those of the humid southeast. Studies on croplands have monitored changes in soil C contents in response to management for over a century. Yet, data on grazing land responses of soil C to use and management are sparse. Only recently have scientists begun to document soil C contents and study C dynamics on grazing lands maintained for many decades under different management regimes. According to the Natural Resources Conservation Service of USDA (NRCS), privately owned rangelands and pastures in the U.S. comprise a total of 212 Mha (USDA-NRCS, 1994). Much of the 124 Mha of publicly owned lands in the western United States properly are classified as grazing lands. Millions of additional hectares of cropland throughout the nation are planted to forage, and these forages often are grazed as well as hayed or harvested as other forms of stored forage. Grazing lands are highly important for the production of meat and other animal products. They are also important for atmospheric CO_2-C sequestration because of the magnitude of the land area involved. Even if rates of C storage are low on an aerial basis, or maximum amounts of storage are limited and slowly attained, the total mass is large. Some portion of our grazing lands, like the arid rangelands in the West, may only be capable of small additional increases of C storage because of low seasonal productivity and shallow soils. Nevertheless, other areas, even some in the West, are highly productive and have deep, well-developed soils. Annual production of plant biomass on many native perennial grasslands and most pastures exceeds that of croplands, and intensively managed pastures planted to annual grasses or legumes on deep soils might store C at rates equal to those of annual crops in many regions. Some of the highest rates of C storage found in the scientific literature were measured on perennial grass pastures, especially after establishment on former cropland, even when the stands were not managed for high productivity but were left undisturbed, as are lands enrolled in the CRP.

Historical Perspective

Numerous accounts were recorded in the journals and records of early European travelers of U.S. grazing lands before European settlement. Hart and Hart (1997) traced some of the early records that extend from early explorers, including Coronado who traveled through Texas and possibly into Colorado from 1540–1542 and Trudeau, a trader on the upper Missouri from 1803–1805, to the present day. Estimates of the numbers of bison in the Great Plains based on work by Cushman and Jones (1998) and Shaw (1995) indicate there were from 15–30 million head. By comparison, in 2002, there were over 46 million head of cattle and nearly 400,000 head of sheep in the 10 Great Plains states (NASS, 2004). Johnson (1994) recorded that trees along the Platte River near Kearney, Nebraska were rare even up until 1938, but have increased greatly since. Lewis and Clark (Moulton and Dunley, 1983) recorded that the Missouri river was bordered by timber or brush for much of its length though the surrounding prairie was treeless. W.A. Bell, an early traveler, wrote of the area west of Salina, Kansas: "... as settlers advance,... domestic herds take the place of the big game... ". Box (1979) indicates that within twenty years after the first European settlers established a livestock industry the rangelands were overgrazed both in the humid Southeast and in the arid West. Settlers and ranchers came west with an eternal optimism about the carrying capacity of the rangelands they encountered. An early quote at the time of early western rangeland settlement was,

> Men of every rank were eager to get into the cow business. In a short time, every acre of grass was stocked beyond its fullest capacity. Thousands of cattle and sheep were crowded on the ranges when half the number was too many. The grasses were entirely consumed; their very roots were trampled into the dust and destroyed." (Bentley, 1898)

Another quote from the *Desert News* is dated September 25, 1879:

> The wells are nearly all dried up and have to be dug deeper. At the present time, the prospect for next year is a gloomy one for all the farmers, and in fact all, for when the farmer is affected, all feel the effects. The stock raisers here are all preparing to drive their stock to where there is something to eat. This country, which was once one of the best ranges for stock in the Territory, is now among the poorest; the myriads of sheep that have been herded here for the past few years, have almost entirely destroyed our range.

To summarize, early accounts often conflicted. Hart and Hart (1997) from their rather extensive review of journals and other records of early travelers conclude that, much of the Great Plains before European settlement looked about like it does now. The pronghorn have returned, but cattle have replaced the buffalo. The major changes have been produced by cropland agriculture and urbanization. Finally, the vegetation on the Plains will continue to change as it has throughout the Holocene, wherein parts of the northern Great Plains were forested as little as 10,000 years ago (Dix, 1962) and warm season grasses spread from south to north during the Holocene (Follett et al., 2004; Leavitt et al., 2007). During this large and important time span of wide climatic fluctuations (summarized by Follett et al. 2004), there were droughts, widespread aeolian soil movement, and both widespread climatic warming and cooling during which the sequestration of SOC must also have changed dramatically at various times. To the degree that European settlement has resulted in the degradation or improvements to the condition of U.S. grasslands, it is from this point forward that present and future management will continue to either have beneficial or negative effects.

The Present

Following European settlement of grazing lands in the east and up until 1940, animal and cropland agriculture were strongly associated because of the need for draft animals for planting and harvesting of crops. The 1940 population was less than half of what it is today (U.S. Census Bureau, 2000) and was centered further east in more humid climates with much sparser populations further west. Consequently, much eastern grazing land use, in contrast to western grazing land use, was associated with agriculture largely for pasturage or forage harvest. Further west, the Dust Bowl not only devastated vast areas of croplands, but also grazing lands, lands broken for cropland, and reseeded grazing lands were greatly deteriorated because of the need for livestock feed and the economic needs of farmers and ranchers.

After 1940 and where suitable, more intensive management with appropriate technological inputs (e.g., improved forages, fertilizer and pesticide use, and grazing management) began to be used, resulting in greater productivity and higher potential for soil C sequestration. Box (1979) states,

> While I believe that American rangelands are in the best condition they have been in for a century, I do not intend to leave the impression that they are producing anywhere near their potential.... On the whole, the rangelands of this country have improved over the past few decades and will continue to improve in the future.

Recent national analyses of the potential for soil C sequestration on U.S. grazing lands indicate that substantial national rates exist (Follett et al., 2001a). Estimates are that grazing land sequesters nearly 37 MMg yr^{-1} and that an additional 21 MMg C yr^{-1} is sequestered through land conversion and restoration (not including CRP contributions). The roughly 330 Mha of nonintensively managed grazing land (federal and nonfederal) sequesters about 5 MMg C yr^{-1}, but 107 Mha of rangeland with improved management sequester 10 MMg C yr^{-1} and 36 Mha of improved pasture in the United States are estimated to sequester 14.8 MMg C yr^{-1}. On the negative side the combined effects of soil erosion (wind and water), acid precipitation, and fertilizer induced acidity to pastures are estimated to result in soil C losses of 16 MMg C yr^{-1} from as much as 239 Mha of grazing land soils in the United States. These data indicate that large areas of grazing lands exist that are sequestering C at very low levels and that significant soil C losses also occur, but that there is an overall net gain of 18–90 (avg. 54) MMg C yr^{-1}.

Based on 3 years (1995–1997) of CO_2 flux measurements representative of 51 Mha, Frank et al. (2001) estimated annual SOC rates at Mandan, ND, Woodward, OK, and Temple, TX, at 86, 87, and 459 kg C ha^{-1} yr^{-1}, respectively. These figures are similar to those of Schuman et al. (2001), who indicate that currently there are 107 Mha of private U.S. rangeland where improved management could sequester 50–150 kg C ha^{-1} yr^{-1} and even though the rates are low, the large area involved results in total sequestration estimates of 5.4–16.0 MMg C yr^{-1}. However, too little data on grazing land C sequestration rates have been collected in the United States and in most other regions of the World. Follett and Schuman (2005) report that the amount of meat produced from grazing animals, defined as being from beef, sheep, goat, and buffalo can be used as a metric to evaluate soil, climate, and management potential and the corresponding SOC sequestration rates across broad regions of the world. Table 5.6 shows the change in meat produced from cattle and sheep within the United States at the same time intervals as were reported in Tables 5.2, 5.3, and 5.4 (NASS various years). Based upon the Follett and Schuman (2005) report, a change in U.S. grazing animal meat production should reflect SOC sequestration on U.S. grazing lands. There is a steady increase in U.S. meat production with generally

TABLE 5.6
U.S. Production of Meat from Beef, Veal, Lamb and Mutton, and All Meats

	MMg Meat			
Year	Beef	Veal	Lamb and Mutton	All Meats
1940	3.26	0.44	0.40	8.66
1969	9.60	0.30	0.25	16.63
1978	11.00	0.29	0.14	17.51
1987	10.69	0.19	0.14	17.56
1992	10.53	0.15	0.16	18.69
1997	11.57	0.15	0.12	19.68
2000	12.21	0.10	0.11	21.02
2003	11.82	0.09	0.10	20.85

Source: From NASS 1948, 1972, 1985, and 2004.

more than half from beef but production of meat from veal and sheep has decreased during from 1940 on.

Rangeland grazing management systems have been developed to sustain efficient use of forage by livestock, but there is generally little accompanying understanding of the effects of grazing on the redistribution of and cycling of C and N within the soil plant system. The capacity of rangeland ecosystems to sequester SOC is an interaction among the spatial distribution of plant production, availability of soil and water resources, and management inputs. If management is targeted to those parts of the landscape that have higher resource availability, a high potential for enhanced productivity and C storage is more likely achieved. Management that adequately considers these factors, and if applied over a wide enough area, can potentially contribute to improved soil C sequestration, benefits to wildlife, decreased soil erosion, and to other potential social benefits associated with decreased adverse on- and off-site effects.

The majority of improved pastures in the United States are east of the Mississippi River. They occur on many soil types, with varying fertility, texture, and structure, and across a range of climatic conditions. Each of these soil properties and climatic factors has important, generally predictable effects. Overall, temperature and moisture are the most important climatic factors for determining soil C dynamics because of their effects on potential net primary productivity (NPP) and soil microbial activity. Schnabel et al. (2001) indicate that pastureland in the United States include 51 Mha of improved, native, and naturalized pastures. In the northeastern and north central United States, the pasturelands include cropland pasture, improved pasture, woodland pasture, and other. Species of grasses included in pastures are Kentucky bluegrass (*Poa pratensis* L.), orchardgrass (*Dactylis glomerata* L.), timothy (*Phleum pratense* L.), and quackgrass (*Agropyron repens* L. Beauv.). Improvements include fertilization and liming and added legumes. In the north central United States, forage species are similar but include tall fescue (*Festuca arundinacea* Schreb.) in the southern portion and smooth bromegrass (*Bromus inermus* L.) in the northern and central parts. In the southeastern United States, grass species shift to predominantly warm season species such as bermudagrass (*Cynadon dactylon*), bahaigrass (*Paspalum notatum*), and dallisgrass (*Paspalum dilatatum* Poir.) Several clovers frequently are planted as winter annual pastures or on prepared seedbeds or sod-seeded into dormant warm season grass sods. Types of management input for increased rates of SOC sequestration are improved nutrient management, plant species, and grazing systems on pasture soils. It was observed by Follett, Kimble, and Lal (2001a) that nearly 8 additional MMg C yr^{-1} could be sequestered through more intensive or improved management on 21 Mha of pasture (average 0.4 MMg C yr^{-1} more) with the improved systems. Continued improvements in forage species through plant breeding programs and the adoption of improved grazing and fertility management practices under these more humid climates should allow continued future gains in rates and amounts of SOC sequestered.

Future Considerations

Managers and policy makers need more effective information describing the potential for grazing lands to sequester C and how improvements in the science and technology can increase C sequestration through appropriate inputs and management. A policy aspect related to grazing lands are the large C stocks they contain and that if some significant part were returned to the atmosphere as CO_2, it would add considerably to the atmospheric greenhouse warming potential (GWP). Soil stocks to a 1-m depth were measured across the U.S. Great Plains and averaged from 80–90 Mg SOC ha^{-1}, not including soil inorganic C (SIC) (Follett et al., 2001b).

A concern to C sequestration on grazing land associated with global climate change is warming global temperatures. Parton et al. (2001) identify that the major factors affecting grassland soil C levels are soil texture, decomposition rates, and C input to the soil. Their modeling work to assess climate change, which included the impact of probable increases in atmospheric CO_2 on plant growth, was done for the U.S. Great Plains. Their simulated changes in soil C levels between 1980–2050 under the Hedley Centre Transient Global Circulation Model climate change scenario provides some assurance that large amounts of CO_2 will not be discharged into the atmosphere from the degradation of existing SOC and, although soil C might decrease to the North, it suggests that soil C sequestration might show an overall slight increase for the Great Plains (Parton et al., 2001).

ECONOMIC AND SOCIETAL CONSIDERATIONS

The Current Situation

As reported by Lal et al. (1998), Bruce et al. (1999), and other authors, practices that increase residue or plant growth result in enhanced SOC sequestration. Such increases also are associated with the production of C contained in food, feed, and fiber. Alternative uses of plant photosynthesized C products, such as their use for biofuels, are currently occurring or are being increasingly evaluated. Use of conservation tillage (i.e., no-till, ridge-till, and mulch-tillage), maintaining higher levels of residue cover on conventionally tilled cropland, planting cropland to permanent cover (i.e., the CRP), and improved fertility management help protect soil and increase SOC sequestration (Lal et al., 1998). Conversely, practices that remove excessive amounts of vegetation or plant residues (i.e., fallow, plow tillage, and abusive grazing practices) usually cause soil erosion and result in SOC oxidation and loss.

A primary purpose of U.S. agriculture is to produce food and agricultural products. However, agricultural practices need to be environmentally friendly. There is increasing awareness of the role of agricultural in the emission of GHGs. Energy inputs required to power farm machinery, pump water for irrigation, manufacture fertilizers and pesticides, and transport farm products to market result in GHG emissions into the atmosphere. Besides CO_2, other GHGs resulting from agriculture activities are nitrous oxide (N_2O) and methane (CH_4). A major benefit of the use of RMPs in agriculture is that they help sequester atmospheric CO_2 in the form of SOC to offset the greenhouse warming effect of GHGs. This sequestration is a contribution by farming to the mitigation of global warming, and with the use of scientific agriculture, net beneficial effects can be larger than the negative effects. Atmospheric CO_2 concentration has increased by about 32% from pre-industrial levels (280–370 ppm) and U.S. GHG emissions gases for 2000 were estimated at approximately 1887 MMg CO_2-CE yr^{-1} (US EPA 2002). Using IPCC methodology to identify changes in net GHG emissions from 1990–2001, four of the larger agricultural sinks for CO_2 were identified as being due to the conversion of cropland to hayland, conversion of cropland to grazing land, the CRP, and from land application of manure (USDA, 2004). The two largest agricultural-land sources of atmospheric CO_2 from 1990–2001 were caused by plowout of grazing lands and cultivation of organic soils. Worth noting is that even though relatively small areas of organic soils are farmed in the United States, 60% of the SOC that is sequestered by minerals soils within the rest of the United States is needed to offset their CO_2 emissions (USDA, 2004).

The length of time that SOC sequestration rates continue after adoption of improved practices is uncertain. Eventually, a new steady state is approached as the losses of C from the soil start to equal the C additions to the SOC pool. As the new steady state is approached, it tends to be maintained until management, weather patterns, or other factors cause it to change. Estimates by Lal et al. (1998) are that achieving this practical limit may require at least 50 years, and Qian and Follett (2002) reported a continued increase for up to 30 years under grass. Thus, it might be assumed that most C sequestration increases resulting from management changes approach a new equilibrium in 25 years or more. This relationship affects the estimates of the U.S. rates of increase in SOC sequestration during long-time periods and limits the time during which U.S. agriculture can help offset the C emissions from other sectors of the U.S. economy. Thus, agriculture provides a window of opportunity for the other sectors to develop alternative technologies whereby their rates of CO_2 emissions can be decreased.

Societal Expectations of Agriculture

In the 1930s, 25% of the U.S. population lived on farms, now it is less than 2%. From a peak of 6.5 million farms in 1935, the number of farms decreased to 5.5

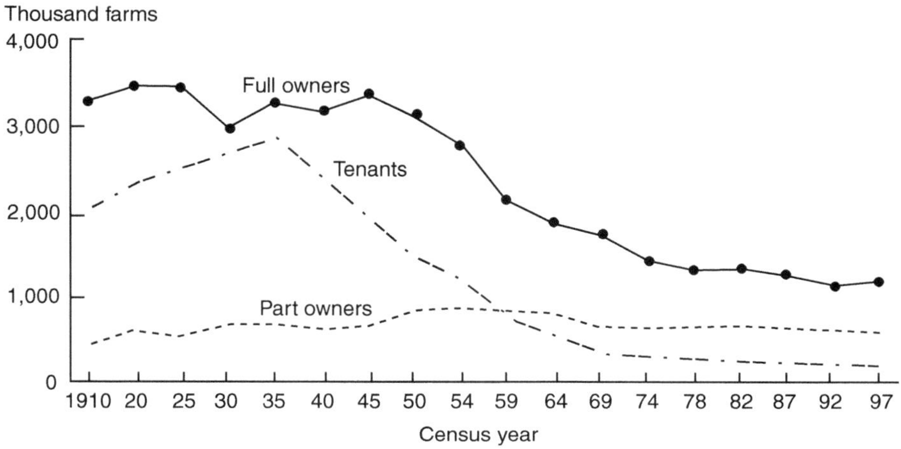

FIGURE 5.5 Number of family farms by tenure, 1910–1997. (From Hoppe, R.A., Ed., Structural and Financial Characteristics of U.S. Farms: 2001 Family Farm Report. Economic Research Service, USDA, Resource Economics Division Agriculture Information Bulletin No. 768, Washington, D.C., May 1, 2001. With permission.)

million in 1950, and by 1992 was only 1.9 million — less than in 1860 (Kime, 1996). The most rapid decline in number of farms began in approximately 1935 with most of the decline accounted for by full owners and tenants (Figure 5.5). The number of farms has since decreased, whereas the size of farms has increased (Figure 5.6). In 1996, 25% of the farms accounted for 88% of production. By another count in 1997, there were 2,190,510 farms remaining. Of these, 1,191,050 farms produced less than $10,000, 645,960 farms produced $10,000 to $99,999, and 353,500 farms produced more than $100,000. The largest operations (353,500 farms) controlled more than half of total U.S. farmland area.

Soil conservation is important for the protection of soil quality (including protection and increased sequestration of SOC). In addition, societal expectations increasingly include issues related to having clean water, high air quality (including low levels of particulates and odors), enhanced wildlife habitat, protection of or increased areas of wetlands and riparian vegetation, and finally minimal ecosystem impacts. These expectations can largely be associated with practices and approaches that protect or increase SOC in agricultural lands, or as more broadly identified to provide ecological services. Individual property rights, as currently defined, do not suffice to maintain, improve, or even provide ecological services for society's benefit. Manale (2001) identifies that many ecological services, such as floodwater retention or wildlife habitat protection, require management practices that extend beyond individual property boundaries. Intervention by agricultural programs may not necessarily be directed at the individual farm, but rather at restoring an ecological service at a broader geographic scale. Thus, the spatial scale needs to be expanded to encompass a broad range of technical and institutional options. It is not the intent here to discuss public policy options other than to point out as identified by Manale (2001) that

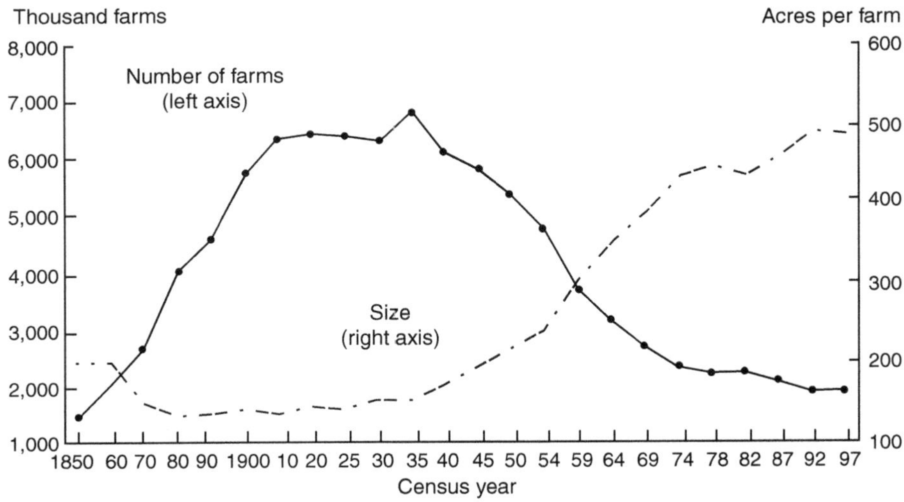

FIGURE 5.6 Number of farms and acres per farm 1850–1997. To convert acres to ha, multiply numbers on the right axis by 0.405. (From Hoppe, R.A., Ed., Structural and Financial Characteristics of U.S. Farms: 2001 Family Farm Report. Economic Research Service, USDA, Resource Economics Division Agriculture Information Bulletin No. 768, Washington, D.C., May 1, 2001. With permission.)

> Superimposed upon the effect of individual farming decisions on the environment are public policies and collective actions that exacerbate or amplify the magnitude of the impacts. In the United States and in most developed countries, government programs have historically affected how farmers farm or steward the land exacerbating agriculture's impact on the environment.... Agricultural programs that induce a bias toward intensive farming practices that boost yields, expand production onto marginal lands, and concentrate production on a small number of crops can undermine efforts to encourage adoption of conservation practices. Despite major changes in farm policy in the United States and Europe since 1989, the linkages between farm program support and production decisions remain.... Income-support programs in the United States under the 1996 farm bill, such as the commodity loan programs, influence producer decisions regarding the use of marginal lands, the intensity of land use, tillage practice, monocultural cropping practices and habitat protection...

The combined effects of items discussed previously point out how much is expected of a dwindling number of individual landowners in terms of America's expectations of them. Even though agricultural soils are important for SOC sequestration to reduce atmospheric CO_2 by improvements and by the use of RMPs, the key individuals who will make this happen will be the land managers (i.e., farmers and ranchers) as they adapt their management practices to accomplish better land stewardship. This improved management and the resulting increases in SOC is part of the emerging market in "environmental services" which supporters claim can harness market forces to provide economic incentives for environmental protection. Under the Kyoto Protocol, many developed nations have distributed tradable pollution rights to their major industries and are encouraging the trading of these rights by polluters in the electricity generation, oil, steel, cement, chemicals, pulp, and paper sectors to achieve compliance with their GHG reduction targets at lowest cost. The United States has not ratified the Protocol. At present, trading within the United States is voluntary, and regulatory measures have not been used to encourage corporations and other groups to reduce their net GHG emissions. Where trading has been more formally emplaced, it allows companies that are able to reduce their emissions beyond their stated goal to sell the extra reductions as credits on the market. Companies that cannot easily reduce their emissions can buy those credits to cover their reduction commitment. However, without a binding cap on GHG emissions, there is relatively little demand for C credits in the United States; the market for buying and selling the credits is very limited while the prices paid are low, including that for C sequestration in soils (Williams, Peterson, and Mooney, 2005). Concerns exist about the environmental effectiveness of C trading in the United States because it was established on a voluntary basis only. Still, the effort is an important step in pioneering programs to mitigate climate change. Even though considerable interest exists among agricultural producers, it is largely uneconomical for them to make major changes in their management practices to sell C credits. Management changes made by producers that are beneficial for C sequestration are most likely done to improve the efficiencies and cost savings for their operations or in response to incentives provided by government programs. Two examples where management changes have been made that encourage soil C sequestration, but yet are not specifically directed to do so include: (1) government programs that subsidize the use of certain practices such as CRP, and (2) the need for producers to improve efficiency and reduce input costs such as adoption of less intensive tillage to decrease fuel costs. Should U.S. policies toward C sequestration provide adequate incentives for producers to adopt RMPs to increase C sequestration, there is little doubt that major changes could occur. On the negative side of the soil C balance sheet may well be the diversion of crop residues to become biofuels. Little doubt exists that if the sale of crop residues and its removal from the land provide a reasonable income source for producers, then a potential loss or decreased levels of SOC sequestration could occur because of decreased return of residue C to the soil.

Numerous legislative bills, laws, policy statements, and other federal and local guidelines and rules over many years have been passed or otherwise expressed about the expectations from U.S. agriculture. What is often not identified in association with agricultural entitlement or subsidy programs is the maintenance of a cheap, plentiful, and nutritious food supply for American consumers. Certainly, the goal of maintaining low commodity prices as crops leave the farm gate has been achieved. For example, 40 years ago, a farmer purchased land for \$1235 ha^{-1} (\$500/acre), while the price of a bushel of corn was \$0.04 kg^{-1} (\$1.00/bushel). Today, the same farm ground is selling at approximately \$12,345 ha^{-1} (\$2000/acre), and the price of corn is \$0.07 kg^{-1} (\$1.80/bushel). The price for wheat that farmers are receiving today is the same price as in the year 1910. To survive, farmers have focused on the use of monocultures or simple rotations such as corn and soybeans (Power and Follett, 1987). Another result has been that because the prices that American farmers receive have not increased in the past three decades, the average annual exports usually stay between $40–50 billion. Exports of U.S. agricultural products once accounted for a huge dollar share (about 60%) of the total U.S. exports until the 1980s, but continual depressed prices have caused agricultural exports to slip in the rankings (now 5th) (NASS, 2004). Grain (and the C it contains) that was formerly exported to offset our balance of trade, and crop residue C that helps sustain and increase SOC, may now begin to help offset the cost of the large amounts of petroleum now being imported into the United States (Parfit and Leen, 2005) by their conversion to liquid fuels for motor vehicles (i.e., ethanol and biodiesel).

SUMMARY

There is a large potential for agricultural lands to produce large amounts of C based products (crops, residues, livestock, and forage). However, significant amounts of the C contained in crop residues and manures are needed to sequester SOC. In the future, there will be major challenges to obtaining the benefits of alternative uses crop residue C while also maintaining and enhancing levels of SOC and the environmental benefits associated with crop residue return to the soil to sequester SOC. In terms of privately owned lands, nearly 185 Mha of cropland and 212 Mha of grazing lands are available in the United States, which are operated by a decreasing and aging fraction of the U.S. population. A larger part of the overall U.S. production is occurring on fewer and larger farms, a widespread consolidation and specialization in agriculture is occurring, a shift away from combined crop and livestock systems is occurring, and an agricultural industry structure has developed wherein livestock feeding is in small geographic areas. The combination of these factors means that policy intervention, including economic incentives, is needed to sequester more soil C, encourage more efficient use of the nutrients in residues and manures, and capture those improvements that are consistent with economies of scale and increased efficiencies of production on large areas of U.S. agricultural land. Certainly, reduced tillage practices that require less energy inputs and seeding directly into largely undisturbed crop residues is consistent with enhancing SOC sequestration and with reducing fuel costs as well as recycling nutrients contained within the residues back into the soil for the next crop. Such practices also have the potential to conserve soil moisture. At least some, if not many, combinations of improved practices can be environmentally friendly, can contribute to SOC sequestration on agricultural lands, and can be economical and efficient enough to be suitable for larger as well as smaller operations. It will be important that practices that sustain the soil resource and efficiently use water and nutrient resources be increasingly used should U.S. agricultural lands be required to produce increasing amounts of grains and crop residues for biofuel to partially offset the large amounts of petroleum products currently imported by the United States.

REFERENCES

Allmaras, R.R., Wilkins, D.E., Burnside, O.C., and Mulla, D.J., Agricultural technology and adoption of conservation practices, in *Adv. Soil Water Conserv.*, Pierce, F.J. and Frye, W.W., Eds., Sleeping Bear Press, Chelsea, MI, 1998, 99–157.

Allmaras, R.R., Gupta, S.C., Pikul, J.L., Jr., and Johnson, C.E., Tillage and plant residue management for water erosion control on agricultural land in eastern Oregon, *J. Soil Water Conserv. Society*, 34(2), 85–90, 1979.

Bentley, H.L., Cattle range of the Southwest: a history of exhaustion of pasturages and suggestions for its restoration, *USDA Farmers' Bulletin 72.7*, Washington, D.C., 1898.

Berg, N.A., The genesis of the CRP, in *When Conservation Reserve Program Contracts Expire: The Policy Options*. Proceedings of the Soil and Water Conservation Society, February 10–11, 1994, Arlington, VA, 7–12.

Box, T.W., The American rangelands: their condition and policy implications for management, in *Proceedings of a Symposium*, January 28–31, 1979, Tucson, AZ. Forest Service publication no. GTR-WO-17, 1979, 16–22.

Bruce, J.P., Frome, M., Haites, E., Janzen, H., Lal, R., and Paustian, K., Carbon sequestration in soils, *J. Soil Water Conserv.*, 54, 382–389, 1999.

Campbell, R.B., Matheny, T.A., Hunt, P.G., and Gupta, S.C., Crop residue requirements for water erosion control in six southern states, *J. Soil Water Conserv. Society*, 34(2), 83–85, 1979.

Chenu, C., Extracellular polysaccarides: an interface between microorganisms and soil constituents, in *Environmental Impact of Soil Component Interactions, Natural and Anthropogenic Organics*, Huang, P.M., Berthelin, J., Bollag, J.M., McGill, W.B., and Page, A.L., Eds., CRS Press, Boca Raton, FL, 1995, 217–233.

Cochrane, W., *The Development of American Agriculture: A Historical Analysis*, 2nd ed. University of Minnesota Press, Minneapolis, MN, 1993.

Cushman, R.C. and Jones, S.R., *The Shortgrass Prairie*, Pruett Publishing Company, Boulder, CO, 1998.

Delgado, J.A., Reeves, D.W., and Follett, R.F., Winter cover crops, in *Encyclopedia of Soil Science*, Marcel Dekker, Inc., New York, 2004, 1–3.

Dix, R.L., A history of biotic and climatic changes within the North American grassland, in *Grazing in Terrestrial and Marine Environments*, Crisp, D.J., Ed., Blackwell Scientific Publishers, Oxford, United Kingdom, 1962, 71–89.

Donald, C.M. and Hamblin, J., The biological yield and harvest index of cereals as an agronomic and plant breeding criteria, *Adv. Agron.*, 28, 361–405, 1976.

Duvick, D.N., Genetic contributions to advances in yield of U.S. maize, *Maydica*, 37, 69–79, 1992.

Duvick, D.N., The contribution of breeding to yield advances in maize (L.), *Adv. Agron.*, 86, 83–145, 2005.

Farm Service Agency (FSA) of USDA, Conservation Reserve Program. Available at http://www.fsa.usda.gov/FSA/webapp?area=home&subject=copr&topic=crp. Search under Fiscal Year Summary. Accessed 01/17/2007.

Follett, R.F., Soil management concepts and carbon sequestration in cropland soils, ISTRO Keynote Paper, Fort Worth, TX, July 3–5, 2000, *Soil Tillage Res.*, 61, 2001a, 77–92.

Follett, R.F., Organic carbon pools in grazing land soils, in *The Potential of U.S. Grazing Lands to Sequester Carbon and Mitigate the Greenhouse Effect*, Follett, R.F., Kimble, J.M., and Lal, R., Eds., Lewis Publishers, A CRC Company, Boca Raton, FL, 2001b, 65–86.

Follett, R.F., Kimble, J.M., and Lal, R., The potential of U.S. grazing lands to sequester soil carbon, in *The Potential of U.S. Grazing Lands to Sequester Carbon and Mitigate the Greenhouse Effect*, Follett, R.F., Kimble, J.M., and Lal, R., Eds., Lewis Publishers, A CRC Company, Boca Raton, FL, 2001a, 401–430.

Follett, R.F., Samson-Liebig, S.E., Kimble, J.M., Pruessner, E.G., and Waltman, S.W., 2001b. Carbon sequestration under the Conservation Reserve Program in the historic grassland soils of the United States of America, *Soil Carbon Sequestration and the Greenhouse Effect*, Lal, R., Ed., SSSA Spec. Publ. 57, Soil Science Society of America (SSSA), Madison, WI, 2001b, 27–40.

Follett, R.F., Kimble, J.M., Leavitt, S., and Pruessner, E., The potential use of soil C isotope analyses to evaluate paleoclimate, *Soil Science*, 169, 2004, 471–488.

Follett, R.F. and Schuman, G.E., Grazing land contributions to carbon sequestration (invited Keynote Paper for the 2005 International Grassland Congress, Belfast, Ireland), in *Grazingland: A Global Resource,* Wageningen Academic Press, 266–277, 2005.

Frank, A.B., Sims, P.L., Bradford, J.A. et al., Carbon dioxide fluxes over three Great Plains grasslands, in *The Potential of U.S. Grazing Lands to Sequester Carbon and Mitigate the Greenhouse Effect,* Follett, R.F., Kimble, J.M., and Lal, R., Eds., Lewis Publishers, 167–187, 2001.

Franzluebbers, A.J, Hons, F.M., Zuberer, D.A., Long-term changes in soil carbon and nitrogen pools in wheat management systems, *Soil Science Society America J.*, 58, 1639–1645, 1994.

Golchin, A., Baldock, J.A., and Oades, J.M., A model linking organic matter decomposition, chemistry, and aggregate dynamics, in *Soil Processes and the Carbon Cycle*, Lal, R., Kimble, J.M., Follett, R.F., and Stewart, B.A., Eds., CRC Press, Boca Raton, FL, 1998, 245–266.

Halvorson, A.D., Peterson, G.A., and Reule, C.A., Tillage system and crop rotation effects on dryland crop yields and soil carbon in the Central Great Plains, *Agron. J.*, 94, 1429–1436, 2002.

Hart, R.H. and Hart, J.A., Rangelands of the Great Plains before European settlement, *Rangelands* 19, 4–11, 1997.

Hay, R.K.M., Harvest index: A review or its use in plant breeding and crop physiology, *Ann. Appl. Biol.*, 126, 197–216, 1995.

Hoppe, R.A., Ed., Structural and Financial Characteristics of U.S. Farms: 2001 Family Farm Report. Economic Research Service, USDA, Resource Economics Division Agriculture Information Bulletin No. 768, Washington, D.C., May 01, 2001. Available at: www.mindfully.org/Farm/Family-Farm-Report-USDAERSmay01.htm#4. Accessed on 01/17/07.

Intergovernmental Panel on Climate Change, United Nations Environmental Programme, Organization for Economic Co-Operation and Development, and International Energy Agency (IPCC/UNEP/OECD/IEA), *Revised IPCC Guidelines for National Greenhouse Inventories*, Intergovernmental Panel on Climate Change, United Nations Environmental Programme, Organization for Economic Co-Operation and Development, International Energy Agency, 1997.

Johnson, J.M.F., Reicosky, D.C., Allmaras, R.R., Sauer, T.J., Venterea, R.T., and Dell, C.J., Greenhouse gas contributions and mitigation potential of agriculture in the central USA, in *Greenhouse Gas Contributions and Mitigation Potential in Agricultural Regions of North America: Special North American GRACEnet Issue*, Franzluebbers, A.J. and Follett, R.F., Eds., *Soil Tillage Res.*, 83, 73–94, 2005.

Johnson, W.C., Woodland expansion in the Platte River, Nebraska: patterns and causes, *Ecol. Monogr.*, 64, 45–84, 1994.

Kime, A., *Arizona Range News,* July 1996. Available at http://matrixbookstore.biz/farm_scene.htm. Accessed on 01/17/07.

Lal, R., Kimble, J.M., Follett, R.F., and Cole, C.V, *The Potential of U.S. Cropland to Sequester Carbon and Mitigate the Greenhouse Effect*, Ann Arbor Press, Chelsea, MI, 1998.

Lal, R., Follett, R.F., and Kimble, J.M., Achieving soil carbon sequestration in the U.S.: a challenge to the policy makers, *Soil Science*, 168(12), 827–845, 2003.

Lander, C.H., Moffitt, D., and Alt, K., Nutrients Available from Livestock Manure Relative to Crop Growth Requirements, Resource Assessment and Strategic Planning Working Paper 98.1 USDA-NRCS, 1998.

Larson, W.E., Crop residue: energy production or erosion control, *J. Soil Water Conserv. Society*, 34(2), 74–76, 1979.

Lasseter, T.C., America's Conservation Program Turns Twenty, 2006. Available at: http://content.fsa.usda.gov/crp20/index.asp. Accessed on 02/13/2007.

Leavitt, S.W., Follett, R.F., Kimble, J.M., and Pruessner, E.G., Radiocarbon and δ13C depth profiles of soil organic carbon in the U.S. Great Plains: a possible spatial record of paleoenvironment and paleovegetation, *Quaternary Int. J.*, 2006, doi:10.1016/j.quaint.2006.10.033.

Lynch, P.J. and Frey, K.J., Genetic improvement in agronomic and physiological traits of oat since 1914, *Crop Science*, 33, 984–988, 1993.

Manale, A.P., New policy directions, in *Nitrogen in the Environment: Sources, Problems, and Management*, Follett, R.F. and Hatfield, J.L., Eds., Elsevier Science Publications, Amsterdam, The Netherlands, 2001, 479–503.

Moulton, G.E. and Dunley, T.W., Eds., *The Journals of the Lewis and Clark Expedition*, University of Nebraska Press, Lincoln, NE, 1983.

National Agricultural Statistics Service (NASS), NASS of USDA, U.S. Government Printing Office, Washington, D.C., 2004.

National Agricultural Statistics Service (NASS) NASS of USDA, U.S. Government Printing Office, Washington, D.C., 1985.

National Agricultural Statistics Service (NASS) NASS of USDA, U.S. Government Printing Office, Washington, D.C., 1972.

National Agricultural Statistics Service (NASS) NASS of USDA, U.S. Government Printing Office, Washington, D.C., 1948.

Parfit, M. and Leen, S., Powering the future, *J. National Geographic Society*, 208 (2), 2–31, 2005.

Parton, W.J., Morgan, J.A., Kelly, R.H., and Ojima, D., 2001. Modeling soil C responses to environmental change in grassland ecosystems, in *The Potential of U.S. Grazing Lands to Sequester Carbon and Mitigate the Greenhouse Effect*, Follett, R.F., Kimble, J.M., and Lal, R., Eds., Lewis Publishers, A CRC Company, Boca Raton, FL, 2001, 371–398.

Pedersen, P., Boote, K.J., Jones, J.W., and Lauer, J.G., Modifying the CROPGRO-soybean model to improve predictions for the upper Midwest, *Agron. J.*, 96, 556–564, 2004.

Perlack, R.D., Wright, L.L., Turhollow, A.F., Graham, R.L., Stokes, B.J., and Erback, D.C., Biomass as feedstock for a bioenergy and bioproducts industry: the technical feasibility of a billion-ton annual supply, available for a processing fee to U.S. Department of Energy and its contractors, in paper, from: U.S. Department of Energy, Office of Scientific and Technical Information, P.O. Box 62, Oak Ridge, TN 373831-0062 (email: reports@adonis.osti.gov), 2005.

Pimental, D. and Patzek, T.W., Ethanol production using corn, switchgrass, and wood: biodiesel production using soybean and sunflower, *Nat. Resour. Res.*, 14, 65–76, 2005.

Power, J.F., Follett, R.F., and Carlson, G.E., Legumes in conservation tillage systems: a research perspective, *J. Soil Water Conserv.*, 38, 217–218, 1983.

Power, J.F. and Follett, R.F., Monoculture, *Sci. Am.*, 255 (3), 78–86, 1987.

Prince, S.D., Haskett, J., Steiningfr, M., Strand, H., and Wright, R., Net primary production of U.S. Midwest croplands from agricultural harvest yield data, *Ecological Appl.*, 11(4), 1194–1205, 2001.

Public Broadcasting Service (PBS), 2006. Available at http://www.pbs.org/wgbh/amex/dustbowl/peopleevents/pandeAMEX06.html. Accessed on 01/17/07.

Qian, Y.L. and Follett, R., Assessing carbon sequestration in turfgrass soil using long-term soil testing data, *Agron. J.*, 94, 930–935, 2002.

Reeves, D.W., Cover crops and rotations, in *Crops Residue Management*, Hatfield, J.L. and Stewart, B.A., Eds., Advances in Soil Science. Lewis Publishers, Boca Raton, FL, 1994, 125–172.

Schnabel, R.R., Franzluebbers, A.J., Stout, L., Sanderson, M.A., and Studeman, J.A., The effects of pasture management practices, in *The Potential of U.S. Grazing Lands to Sequester Carbon and Mitigate the Greenhouse Effect*, R.F. Follett, Kimble, J.M., and Lal, R., Eds., Lewis Publishers, Boca Raton, FL, 2001, 291–322.

Schuman, G.E., Herrick, J.E., and Janzen, H.H., The dynamics of soil carbon in rangelands, *The Potential of U.S. Grazing Lands to Sequester Carbon and Mitigate the Greenhouse Effect,* R.F. Follett, Kimble, J.M., and Lal, R., Eds., Lewis Publishers, Boca Raton, FL, 2001, 267–290.

Shapouri, H., Duffield, J.A., and Wang, M., The energy balance of corn ethanol: an update. USDA, Office of the Chief Economist (OCE), Office of Energy Policy and New Uses, Agricultural Economic Report No. 813, Washingon, D.C., 20 pp., 2002

Shapouri, H., Duffield, J.A., McAloon, A., and Wang, M., The 2001 net energy balance of corn ethanol (preliminary), U.S. Department of Agriculture, Washington, D.C., 2004.

Shaw, J.H., How many bison originally populated western rangelands? *Rangelands*, 17, 48–150, 1995.

Skidmore, E.L., Kumar, M., and Larson, W.E., Crop residue management for wind erosion control in the Great Plains, *J. Soil Water Conserv. Society*, 34(2), 90–94, 1979.

Spero, M., Eve, M., and Paustian, K., Potential soil C sequestration on U.S. agricultural soils, *Clim. Change,* 57, 319–339, 2003.

Tollenaar, M., and Lee, A.F., Yield potential, yield stability and stress tolerance in maize, *Field Crops Res.*, 75, 161–169, 2002.

U.S. Census Bureau, Mean center for population of the United States: 1790–2000, 2000. Available at: www.census.gov/geo/www/cenpop/meanctr.pdf.

U.S. Environmental Protection Agency (US EPA), Greenhouse gases and global warming potential values, Excerpt from the inventory of U.S. greenhouse emissions and sinks: 1990–2000, 2002. Available at: www.epa.gov.

U.S. Department of Agriculture (USDA), U.S. agriculture and forestry greenhouse gas inventory: 1990–2001, Global Change Program Office, Office of the Chief Economist, USDA, Washington, D.C., *Technical Bulletin No. 1907*, 164 pp., 2004.

U.S. Department of Agriculture, Natural Resources Conservation Service (USDA-NRCS), Summary report: 1992 National Resources Inventory, USDA-NRCS, Washington, D.C., 1994.

Vetsch, J.A. and Randall, G.W., Corn production as affected by nitrogen application timing and tillage, *Agron. J.*, 96, 502–509, 2004.

Williams, J.R., Peterson, J.M., and Mooney, S., The value of carbon credits: is there a final answer? *J. Soil Water Conserv.*, 60(2), 36A–40A, 2005.

Yang, H.S., Dohermann, A., Lindquist, J.L., Walters, D.T., Arkebauer, T.J., and Cassman, K.G., Hybrid-maize: a maize accumulation model that combines two crop modeling approaches, *Field Crops Res.*, 87, 131–154, 2004.

6 Organic Farming Enhances Soil Carbon and Its Benefits

Paul Hepperly, Rita Seidel, David Pimentel, James Hanson, and David Douds, Jr.

CONTENTS

INTRODUCTION

Increasing soil carbon (C) through systematic agricultural practices provides an array of societal and farmer/producer benefits. Managing soils organically aims at optimizing C content without synthetic chemical inputs. By increasing soil C and eliminating synthetic chemical inputs, we can effectively:

1. Reduce greenhouse gases, especially carbon dioxide (CO_2) and possibly nitrogen oxides.
2. Increase soil nitrogen (N) and other key nutrients including soil aeration and water availability.
3. Reduce fossil fuel requirement for production agriculture.
4. Increase crop production potential through improved fertility and water availability.
5. Reduce production costs through reduced fertilizer and pesticide use.
6. Reduce production risks especially under drought conditions.
7. Increase water quality as soil C biologically filters surface and ground waters.
8. Increase surface and ground water supply.
9. Increase biological diversity of soil microbes, insects, and earthworms, as well as wildlife populations of birds and small mammals.

RESPONDING TO SOIL CARBON DEGRADATION

Over the last 200 to 300 years, waves of European immigrants transformed our North American virgin forests and prairie lands into a landscape dominated by tilled croplands. Plowing of native prairies and forests greatly reduced native soil C. Within a generation or two, the plowed soil lost half to three quarters of its original soil C content (Fenton, Brown, and Maubach, 1999; Hobbs and Brown, 1965; Stauffer, Muckenhim and Odell, 1940).

During the last dust bowl, the resource depleting nature of our agriculture and degradation of soil made conservation of soil a top national priority but concern about soil conservation was short-lived. With the advent of chemical fertilizers and new hybrid crops such as corn and sorghum varieties since World War II, emphasis on soil C soon faded into a background concern; soil was no longer considered a major agricultural restriction as new technologies were viewed as panaceas to agricultural constraints.

Under the current panorama of concern about environmental and fuel challenges, improving soil C content is regaining renewed prominence for its ability to improve and preserve the soil resource base. Research demonstrates that modifying existing farming practices can improve soil C to promote a better balance between conservation of natural resources and improving crop production.

This chapter reviews the specific practices that can increase soil C and its benefits. These include: (1) crop rotations; (2) cover cropping, and (3) returning organic materials back to soil as manure or compost. The chapter draws freely from the authors' 22-year review of The Rodale Institute Farming Systems Trial (Pimentel et al., 2005).

A System's Approach to Farming and Its Improvement

Farming systems refer to the combinations of crops, animals, and agricultural management practices utilized by farmers in his/her farming enterprise. Farming systems vary enormously depending on market availability, farmer know how, the geographical location of farms, capitalization for production practices, and governmental regulations and support programs. When we compare diverse farming systems, we can assess them according to their overall benefits and impacts especially when we track them over long timeframes and take quantitative measurements that can be statistically analyzed.

Organic Agriculture

According to the U.S. Department of Agriculture (USDA-AMS, 2002), organic agriculture is a type of agricultural production system that is managed according to standards of the Organic Foods Production Act and is regulated to respond to site-specific conditions through integrating cultural, biological, and mechanical practices that foster (the natural) cycling of resources that promote ecological balance and conserve biological diversity.

The National Organic Program from 2002 codifies organic production methods. Farmer and processor practices must be certified based on third-person inspections. This information in turn is certified by an independent certification agency. This system for the first time gives consumers the ability to select foods based on verified food production practices. The National Organic Program prohibits the use of synthetic chemicals, genetically modified organisms, and sewage sludge in certified organic production, and requires that organic farmers track their production practices with a formalized farm production plan. Each certified organic farmer must post a soil conservation and improvement plan that includes crop rotations coupled with other practices aimed at sustaining and improving the soil. Practices are tracked field by field and records kept on field practices and production data. This system arose in response to consumer concerns and demands for a certified organic production system. Starting from a small base, organic agriculture is now a very fast growing and profitable sector of the United States and world agricultural economy. From 1992 to 1997, organic production acreage doubled. Currently, acreage nationally exceeds 1 million acres. Double-digit growth puts organic sales at over $13 billion a year in 2005 compared with $7 billion in 2000 (Dimitri and Greene, 2002a; Dimitri and Greene, 2002b; Sloan, 2005; Odwalla, 2002).

Since 1981, the Rodale Institute Farming Systems Trial® has compared organic and conventional grain-based farming systems. The 22-year summary and update highlights the ability of organic farming to compete with conventional farming. Under organic cultivation, increases in soil C can improve soil quality, reduce energy needs for production, improve economic returns, and reduce many of the negative side effects associated with conventional agricultural systems (Pimentel et al., 2005).

Unlike conventional agriculture, organic agriculture does not depend on the use of synthetic chemical pesticides. On the other hand, no tillage systems and organic agriculture share many benefits because they both are based on increasing the level of soil C to improve productivity over time. In both approaches, soil C is accumulated in small annual increments over time to yield significant long-term improvements.

THE RODALE INSTITUTE FARMING SYSTEMS TRIAL

From 1981 through 2004, field investigations were conducted on 6.1 ha at The Rodale Institute Farming Systems Trial in Kutztown, Pennsylvania. The soil is a moderately well-drained Comly shaly silt loam. The land slopes ranged between 1% and 5%. The growing season has 180 frost-free days, average temperature is 12.4°C and average rainfall is 1105 mm per year. The experimental design included three cropping systems (main plots); each of them replicated eight times: a manure-based organic, a legume-based organic and a conventional system. The main plots were 18 × 92 m, and these were split into three 6 × 92 m subplots, which allowed for same crop comparisons in any one year. The main plots were separated with a 1.5-meter grass strip to minimize cross movement of soil, fertilizers, and pesticides. The subplots were large enough so that farm-scale equipment could be used for operations and harvesting. The rotation diagram and cultural practices for each system are illustrated in Figure 6.1 and Table 6.1. In each system, N inputs were added for the corn crop only, at equivalent available rates for the crop. These inputs included: steer manure and legume plow-down in the manure system; legume plow-down (red clover or hairy vetch) in the legume system; and ammoniated fertilizer in the conventional system.

Organic, manure-based: This "manure system" represents a mixed livestock operation in which grain crops are grown for animal feed. This rotation is typical of a dairy farming operation and includes corn, soybeans, corn silage, winter wheat and red-clover/alfalfa hay plus a rye cover crop before corn silage and soybeans. Aged cattle manure (2–3 months old) served as the N source and was applied at a rate of 5.6 t/ha (dry), for 2 out of every 5 years, immediately before plowing the soil for corn. Additional N was supplied by the plow-down of legume-hay crops. The total N applied per hectare with the combined sources was about 40 kg per year (or 198 kg ha^{-1} for any given year with a corn crop). No herbicides were applied in this system; instead, mechanical cultivation and weed-suppressing crop rotations were used for weed management.

Organic, legume-based: This "legume system" represents a mixed grain operation without livestock. It produces a cash grain crop every year, but it uses no commercial synthetic fertilizers, relying instead on N-fixing green manure crops as the primary source of N. The initial 5-year crop rotation in the legume-based system was modified twice to improve the rotation. The final rotation includes hairy vetch (i.e., winter cover crop used as a green manure), corn, rye (i.e., winter cover crop), soybeans, and winter wheat. The hairy vetch winter cover crop was incorporated before corn planting as a green manure. The total N added per hectare per year

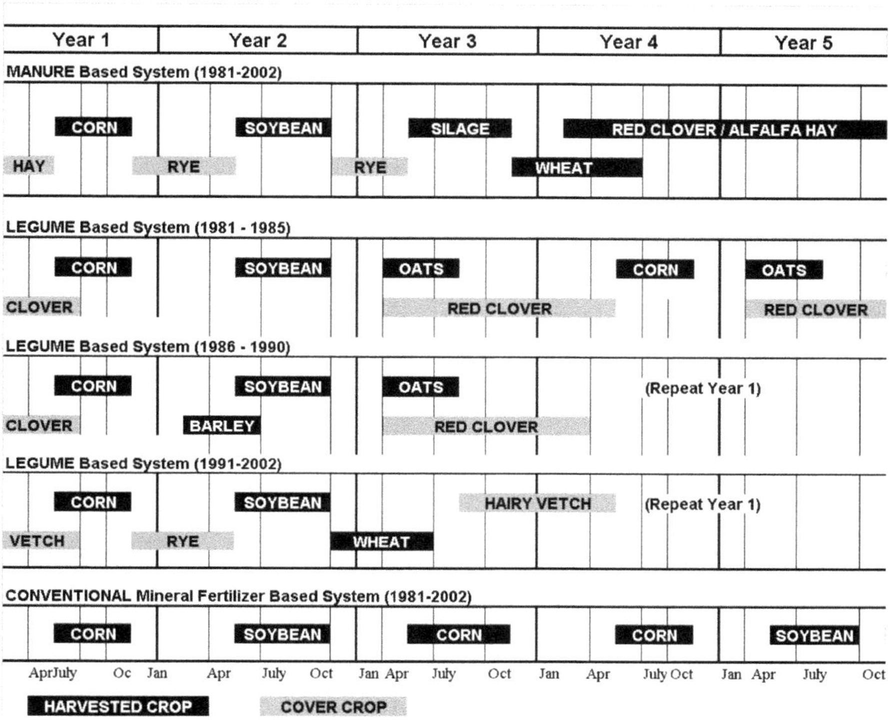

FIGURE 6.1 Rotations employed in The Rodale Institute Farming Systems Trial.

to this system averaged 49 kg (or 140 kg ha^{-1} for any given year with a corn crop). Weed control practices were the same as in the organic manure-based system.

Conventional (synthetic fertilizer and herbicide-based): This system represents a cash grain farming operation using a 5-year rotation of corn, corn, soybean, corn, and soybean, a rotation that is very common in conventional crop production in the Midwest (over 40 million hectares are in this production system in North America) (USDA, 2003). Fertilizer and pesticide applications for corn and soybeans followed Pennsylvania State University Cooperative Extension recommendations.

Data Collection and Analytical Methods

For all systems, cover crop biomass, crop biomass, weed biomass, grain yields, nitrate leaching, herbicide leaching, percolated water volumes, soil C, soil N, arbuscular mycorrhizae (AM) colonization, soil respiration and soil water content were measured. In addition, seasonal total rainfall, energy inputs and returns, and economic inputs and returns were determined.

Total soil C and N were determined by combustion using a Fisons NA1500 Elemental Analyzer by The Agricultural Analytical Services Laboratory, The Penn-

TABLE 6.1
Cultural Practices Utilized in Each System of The Rodale Institute Farming Systems Trial

Cultural Practices	Manure	Legume	Conventional
Crops	Corn, soybeans, small grains, hay Cover crop: rye	Corn, soybeans, small grains Cover crops: rye and vetch	Corn, soybeans No cover crop
Nitrogen Input	40 kg ha^{-1} yr^{-1} manure + legume hay (198 kg N ha^{-1} on corn)	49 kg ha^{-1} yr^{-1} legume cover crop (140 kg N ha^{-1} on corn)	88 kg ha^{-1} yr^{-1} mineral fertilizer (146 kg N ha^{-1} on corn)
Ground Cover	Living: 73% Dead: 20% Bare: 7%	Living: 70% Dead: 22% Bare: 8%	Living: 42% Dead: 50% Bare: 8%
Primary Tillage	Moldboard plow 0.8/yr (4 times/5 yr rotation)	Moldboard plow 1.3/yr (4 times/3 yr rotation)	Moldboard/chisel plow 1.0/yr (5 times/5 yr rotation)
Weed Control	Rotary hoeing Cultivation, rotation	Rotary hoeing Cultivation, rotation	Herbicides
Insect Control	Rotation	Rotation	Insecticides for corn only in 1986–1989, 1993

sylvania State University, University Park, PA, and soil water content was determined gravimetrically on sieved soil (2 mm).

Plant biomass was determined by taking two to five 0.5-m^2 cuts in each plot. Corn grain yields were assayed by harvesting the center four rows of each plot by combine. Soybean and wheat yields were obtained by harvesting a 2.4-meter swath in the center of each plot by combine.

A 76 cm long by 76 cm diameter steel cylinder (lysimeter) was installed during Fall 1990 in four of the eight replications in each cropping system to enable the collection of percolated water (Figure 6.2). The top of each lysimeter was approximately 36 cm below the soil surface to allow field operations to be performed in a normal fashion directly over the lysimeters. Holes were drilled in the center of the base plate to allow for unrestricted flow of percolate from the cylinder into the flexible tube leading to the collection vessel, a 20-liter polyethylene carboy. Two more tubes were connected to the carboy: the air tube, which ran from the cap of the carboy to the soil surface, and the leachate extraction tube, which ran from the base of the carboy to the soil surface. The carboy was positioned below and offset to one side of the steel cylinder to enable gravitational flow of liquid to the collection vessel. Any percolate that flowed from the cylinder into the carboy was recovered using a marine utility pump connected to the extraction tube (Moyer, Saporito, and Janke, 1996). Water could not escape from the lysimeter system. Leachate samples were collected throughout the year.

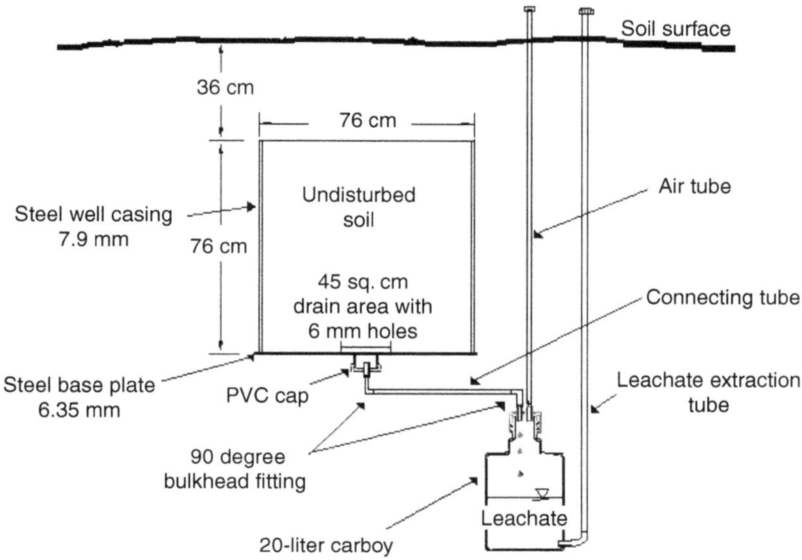

FIGURE 6.2 Lysimeter used to sample water and analyze the quantity and quality of water percolating past the crop roots in the Rodale Institute Farming Systems Trial.

Nitrate in leachate was measured using the cadmium reduction method with a Flow Injection Analysis system from Lachat Instruments; herbicides in leachate were measured by liquid–solid extraction and capillary column gas chromatography/mass spectrometry.

Statistical analyses were performed with the General Linear Model Univariate procedure and Duncan's Multiple Range Test (SPSS software version 12.0).

Results

Soil C and N

In 1981, soil C levels between the systems were not different ($p = 0.05$). However, in 2002, soil C levels in the manure (2.5%) and legume systems (2.4%) were significantly higher than in the conventional system (2.0%), resulting in soil C increases of 28, 15, and 9% for the manure, legume and conventional systems, respectively (Figure 6.3). The respective C sequestration rates for the three systems were therefore 981, 574, and 293 kg ha^{-1} year^{-1} in the manure, legume, and conventional systems (Table 6.2).

Visually, the organically managed soils were also darker and had greater soil aggregation than the conventionally managed soil (Figure 6.4).

Deep core soil samples taken in 20-cm increments from 0–80 cm in 2004 indicated higher soil C values to a depth of 60 cm for the manure and legume systems compared with the conventional system. About two-thirds of the total soil C in the 80-cm core was found in the top 20 cm in all three farming systems (Figure 6.5).

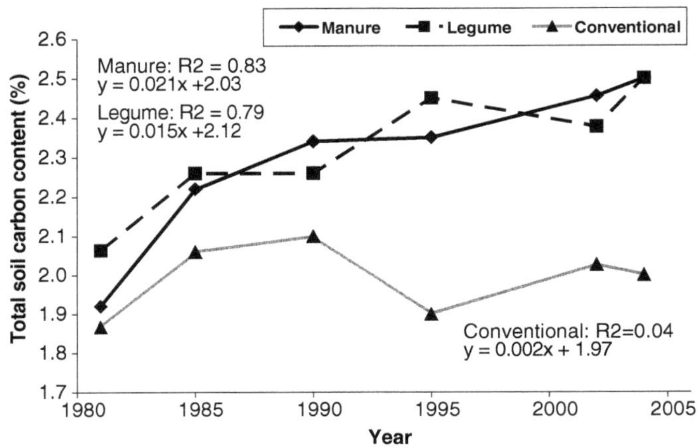

FIGURE 6.3 Linear regression of time under three farming systems of The Rodale Institute Farming Systems Trial to total soil C content.

TABLE 6.2
Soil C and N Accumulation in kg ha^{-1} year^{-1} between 1981 and 2002,[a] The Rodale Institute Farming Systems Trial[a,b]

	Carbon	Nitrogen
Manure	981 b	86 b
Legume	574 b	41 b
Conventional	293 a	−2 a

[a] Based on 4 million kg of soil per ha in the upper 30 cm.
[b] Different letters indicate statistically significant differences for that element (p = 0.05).

Soil N levels were measured in 1981 and 2002 in the manure, legume, and conventional systems. Initially, the three systems had similar percentages of soil N (approximately 0.31%). By 2002, the conventional system (0.31%) remained unchanged, whereas the manure (0.35%) and legume (0.33%) systems had increased significantly (Figure 6.6 and Table 6.2).

Soil Biology

The AM colonization of crop plant roots was determined. Soils farmed under the two organic systems had up to 30 spores per 50 cm^3 of soil versus 0–0.3 spores per 50 cm^3 of soil in the conventional system. In addition, greenhouse bioassays indicated a 2.5- to 10-fold greater AM colonization on plants grown in soil from the organic systems compared with plants grown in soil from the conventional system (Douds, Janke, and Peters, 1993). Though levels of AM fungi were greater in the

FIGURE 6.4 Side-by-side comparison of organically managed soil (left) and conventionally managed soil (right), after 24 years.

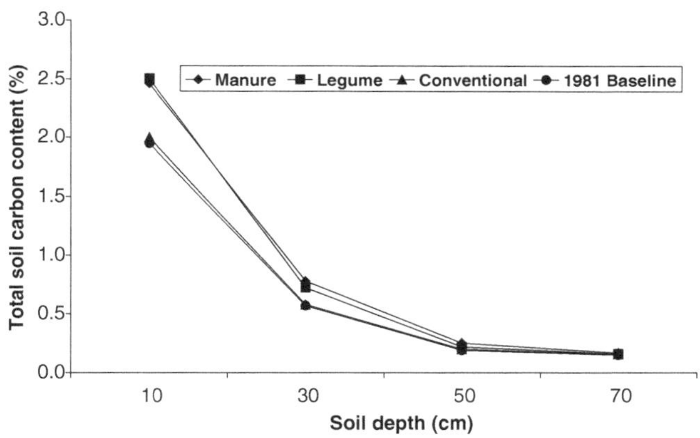

FIGURE 6.5 Total soil C content in deep core soil samples in The Rodale Institute Farming Systems Trial 2004.

organically farmed soils, ecological species diversity indices were similar in the farming systems (Franke-Snyder et al., 2001).

Wander et al. (1994) and Harris et al. (1994) demonstrated that 10 years after initiation of the trial, soil respiration was significantly higher in the two organic systems compared with the conventional system. For example, soil respiration in corn plots was 81 μg CO_2 per gram of soil in the legume system versus 34 μg CO_2 per gram of soil in the conventional system.

Nitrate Leaching

Overall, nitrate-N concentrations in leachate varied between 0 and 28 ppm throughout the year across all systems. Leachate concentrations were highest in June and

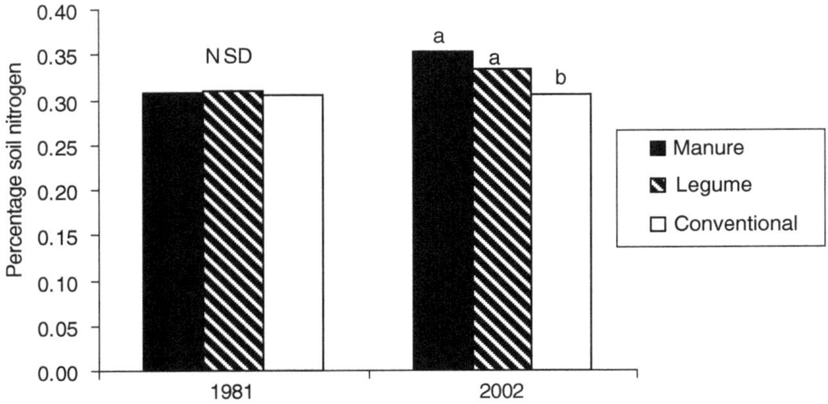

FIGURE 6.6 Total soil N content in 1981 and 2002, The Rodale Institute Farming Systems Trial. Bars with different letters indicate statistically significant differences, whereas NSD indicates no significant difference (p = 0.05).

July, shortly after fertilizer application in the conventional or plow down of the manure and legume cover crops in the organic systems. The average concentration of nitrate-N in leachate was significantly higher in the conventional system (6.7 ppm) compared with 5.7 ppm in the legume and 5.0 ppm in the manure systems. The conventional corn and soybean system also had the greatest number of leachate samples exceeding the regulatory limit of 10 ppm for nitrate-N for municipal drinking water. Twenty percent of the samples from the conventional system were above the 10-ppm limit, whereas 10% and 16% of the samples from the manure and legume systems exceeded the nitrate limit, respectively (Figure 6.7).

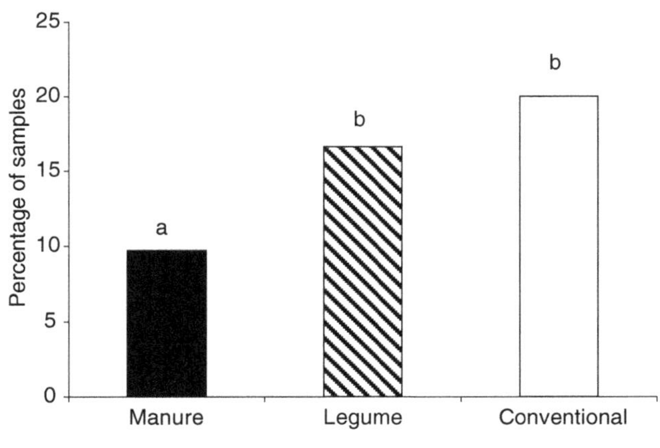

FIGURE 6.7 Percentage of lysimeter water samples that exceeded the municipal water regulatory limit of 10 ppm of nitrate-N, 1991–2002, in The Rodale Institute Farming Systems Trial. Different letters indicate statistically significant differences (p = 0.05).

Over the 12-year period of monitoring (1991–2002), all three systems leached between 16–18 kg of nitrate-N per hectare per year. When measuring these nitrate-N losses as a percentage of the N originally applied to the crops in each system, the legume, manure, and the conventional systems lost about 32%, 20%, and 20%, respectively.

Herbicide Leaching

The herbicides atrazine, metolachlor, and pendimethalin, were applied to the conventional corn and metolachlor, and metribuzin was applied to soybeans. From the conventional corn and soybean system, only atrazine and metolachlor were detected in water leachate samples collected from 2001 through 2003. Metribuzin and pendimethalin were not detected in leachate samples.

In the conventional plots where corn was planted after corn and atrazine was applied two years in a row, atrazine reached concentrations up to 4.2 ppb (Figure 6.8). These atrazine levels were higher than those in the corn-after-soybean plots, where maximum atrazine concentrations were 1.2 ppb (Figure 6.9). Metolachlor was also detected in leachate from the conventional system at average levels from 0.2–0.6 ppb. When metolachlor was applied for 2 years in a row in corn-after-corn plots, it peaked at more than 3 ppb (Figures 6.8 and 6.9).

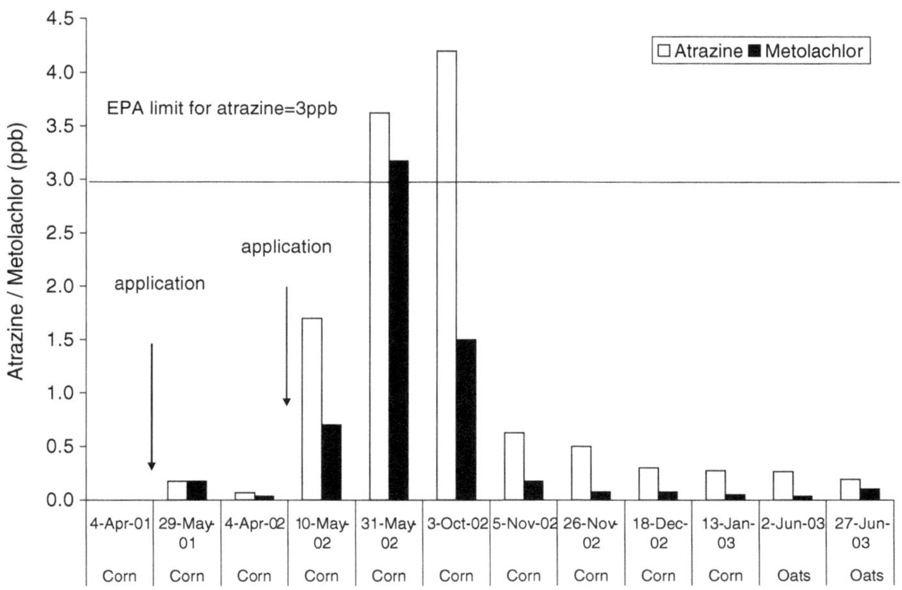

FIGURE 6.8 Herbicide (atrazine and metolachlor) concentrations in leachate from the conventional system where corn was grown for 2 years in a row, 2001–2003, The Rodale Institute Farming Systems Trial. No herbicides were detected in leachate from organic systems.

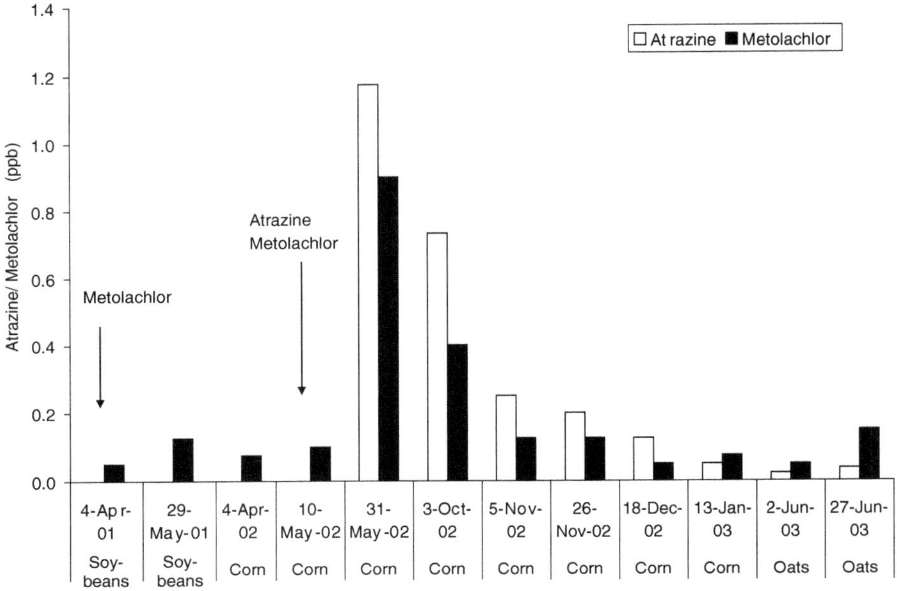

FIGURE 6.9 Herbicide (atrazine and metolachlor) concentrations in leachate found in corn after soybean plots of the conventional system, 2001–2003, The Rodale Institute Farming Systems Trial.

Leachate Volumes and Soil Water Content

Over a 12-year period, leachate volumes percolating through each system were 15% and 20% higher in the legume and manure systems, respectively, than in the conventional system (Figure 6.10). In addition, during the growing seasons of 1995, 1996, 1998, and 1999, soil water content measurements in the legume and conventional systems indicated significantly more water in the legume soil than in the conventional soil (Lotter, Seidel, and Liebhardt, 2003).

Crop Yields under Normal Rainfall

For the first 5 years of the experiment (1981–1985), corn grain yields averaged 4222, 4743, and 5903 kg ha^{-1} for the manure, legume and conventional system, respectively. The two organic systems were significantly lower than the conventional system but not different among each other. Following the transition period, corn grain yields have increased in all systems and have not been significantly different from each other. From 1986 to 2001, the manure, legume, and conventional systems produced an average of 6431, 6368, and 6553 kg ha^{-1}, respectively (Figure 6.11).

Overall, soybean yields from 1981–2001 were 2461, 2235, and 2546 kg ha^{-1} for the manure, legume, and conventional systems, respectively. The legume system was significantly lower than the other two systems. This includes, however, the crop failure in 1988 in the legume soybeans, when climate conditions were too dry to

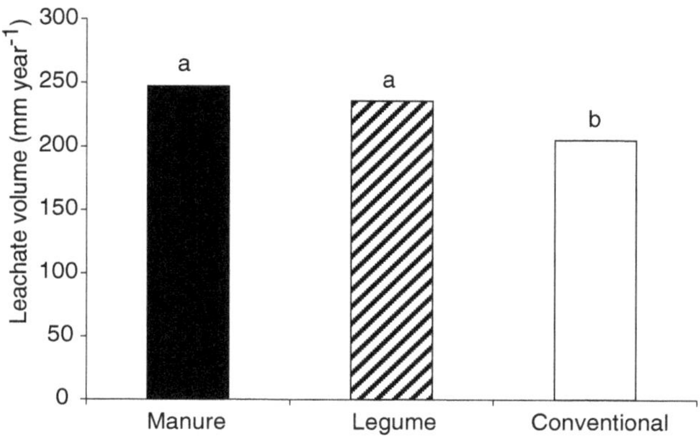

FIGURE 6.10 Average amount of lysimeter water volume per year, 1991–2002, The Rodale Institute Farming Systems Trial. Different letters above bars denote statistical differences at the p = 0.05 level.

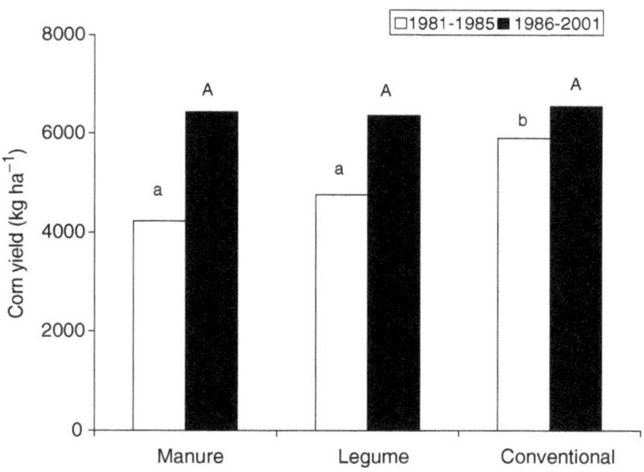

FIGURE 6.11 The Rodale Institute Farming Systems Trial long-term average corn yields, 1981–2001. Different letters above bars at each time period denote statistical differences ($p = 0.05$).

grow a relay crop of barley and soybeans. If 1988 is taken out of the analysis, soybean yields are not statistically different between the systems (Figure 6.12).

Crop Yields in Years with Extreme Conditions

The 10-year period from 1988 to 1998 included 5 years in which the total rainfall from April to August was less than 350 mm (versus 500 mm in average years). Average corn yields for the two organic systems in those 5 dry years were significantly higher compared with the conventional system but not significantly different from each other. The manure and legume systems produced 6938 and 7235 kg ha^{-1}, respectively, compared with 5333 kg ha^{-1} in the conventional system (Figures 6.13 and 6.14).

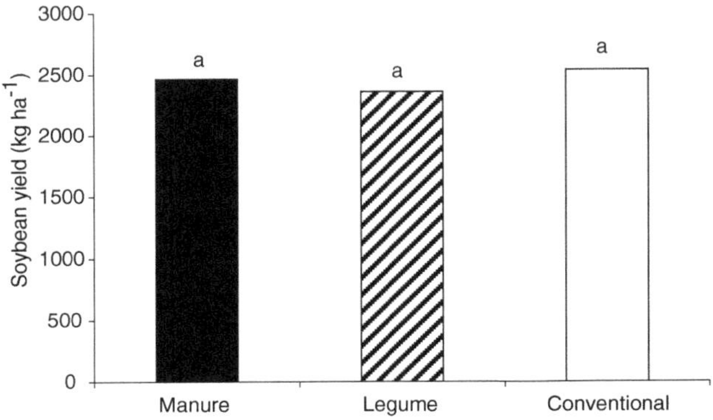

FIGURE 6.12 The Rodale Institute Farming Systems Trial long-term average soybean yields, 1981–2001, excluding 1988 (soybean crop failure). Same letters above bars denote no statistical differences (p = 0.05).

FIGURE 6.13 Side-by-side comparison of corn six weeks after planting in The Rodale Institute Farming Systems Trial during the 1995 summer drought. Organic-legume (left) and conventional (right) plots show drastically different growth responses.

During the extreme drought of 1999 (total rainfall between April and August was only 224 mm compared with the normal average of 500 mm), the manure system had significantly higher corn yields (1511 kg ha^{-1}) than both the legume (421 kg ha^{-1}) and the conventional systems (1100 kg ha^{-1}). Soybean yields responded differently than the corn during the 1999 drought. For the legume system 1400 kg ha^{-1}, for the manure system 1800 kg ha^{-1}, and for the conventional system 900 kg ha^{-1} were measured. These treatments were statistically significant ($p = 0.05$) from each other (Lotter, Seidel, and Liebhardt, 2003).

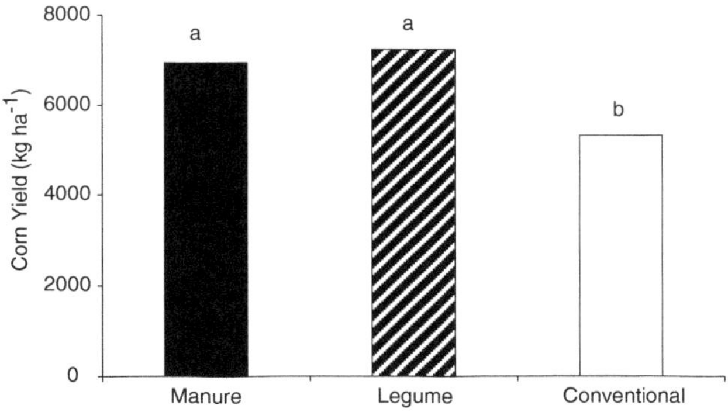

FIGURE 6.14 Mean corn yields in drought years (1988, 1994, 1995, 1997, 1998) in The Rodale Institute Farming Systems Trial. Different letters above bars denote statistical differences (p = 0.05).

TABLE 6.3
Yield, Plant Height, and Grain Protein Content for 2004 Corn Crop, The Rodale Institute Farming Systems Trial[a]

	Yield ($kg\ ha^{-1}$)	Height (cm)	Crude Protein (%)	Available Protein (%)
Organic	7902 b	242 b	8.2	7.7
Conventional	6844 a	207 a	7.2	6.7

[a] Different letters indicate statistically significant differences (p = 0.05).

In 2004, record summer rain falls (50% above normal) caused severe N deficiency in the conventional corn and resulted in differences between the organic and conventional corn in terms of plant height, yield, and grain protein content (Table 6.3 and Figure 6.15).

Economics

Economic evaluations of the Farming Systems Trial after the first 9 and the first 15 years (Hanson et al., 1990 and Hanson, Lichtenberg, and Peters, 1997, respectively) reported significant differences in net returns between the legume and the conventional systems, ranging from 2% higher to 27% lower for the legume rotation, depending on which costs (input costs, transition costs, labor costs) were taken into account (Hanson, Lichtenberg, and Peters, 1997). The third economic evaluation covered the period 1991–2001, comparing the legume system in its final rotation (10 years after the start of the trial) to the conventional system, without including the cost of transitioning to organic production (Hanson and Musser, 2003). The transition costs are the reduced income from yield losses as the organic system's soil improves during the first few years. Transition costs can be excluded for farms

FIGURE 6.15 Silk stage comparison of organic (left) and conventional (right) corn in The Rodale Institute Farming Systems Trial in 2004.

that switch to organic production by first growing crops that are less N demanding (such as legumes or small grains) and integrate N demanding corn into the rotation only after soil N is built up.

In the evaluation for 1991–2001, input costs for seed, fertilizer, pesticides, machinery, and hired labor were approximately 20% lower in the legume system compared with the conventional system (Figure 6.16), whereas labor requirements (i.e., family labor, not hired labor) were 35% higher in the legume system. Without considering price premiums for the organically produced grain, net returns were $162 per ha for the conventional system and $141 per ha for the legume system

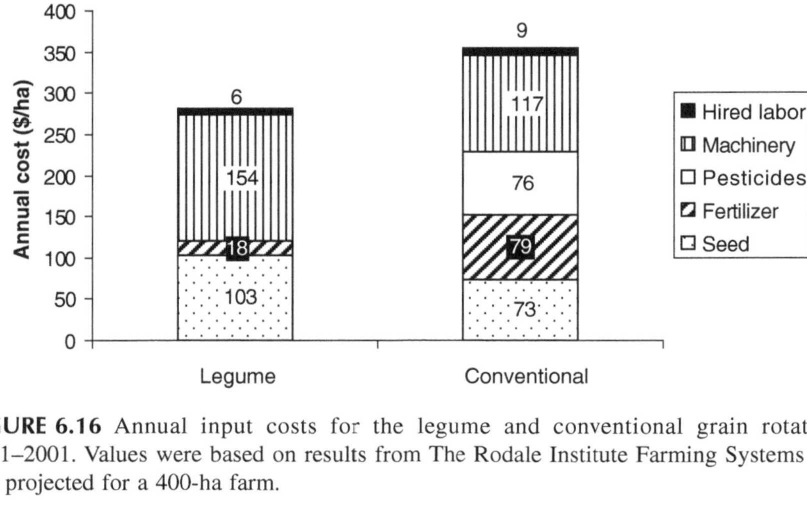

FIGURE 6.16 Annual input costs for the legume and conventional grain rotations, 1991–2001. Values were based on results from The Rodale Institute Farming Systems Trial and projected for a 400-ha farm.

TABLE 6.4
Annual Inputs and Returns for the Legume and Conventional Systems[a]

	Legume	Conventional
Input costs (\$ ha^{-1})	281	354
Labor requirements (hours ha^{-1})	3.2	2.3
Net returns after transition — including labor costs (\$ ha^{-1})	141	162

[a] Net returns did not include organic price premiums. Values were based on results from The Rodale Institute Farming Systems Trial and projected for a 400 ha farm.

(Table 6.4). This means, the organic price premium required to equalize the organic and conventional returns was only 6% above the conventional product (Hanson and Musser, 2003). Throughout the 1990s, the organic price premium for grains has exceeded this level and premiums now range between 20% and 140% (Dobbs, 1998; Bertramsen and Dobbs, 2002; NewFarm, 2003).

Energy Inputs

The energy inputs in the manure, legume, and conventional production systems were assessed. The inputs included fossil fuels for farm machinery, fuel for crop production, fertilizers, seeds, and herbicides. Commercial fertilizers for the conventional system were produced employing fossil energy, but the N nutrients for the organic systems were obtained from legumes and cattle manure. Fossil energy inputs were required to transport and apply the manure to the field. About 5.2 million kcal of energy per ha were invested in the production of corn in the conventional system. The energy inputs for the manure and legume systems were 32% and 28% less than those of the conventional system, respectively. The energy requirements for soybeans were similar for all three systems (Figure 6.17).

DISCUSSION

Soil Carbon and Biodiversity

Soil C or organic matter provides the base for productive organic and sustainable agriculture. After 22 years of differentiated management, soil C was significantly higher in both the manure and legume systems than in the conventional system. This increase in soil C resulted in C sequestration rates of 981, 574, and 293 kg ha^{-1} $year^{-1}$ in the manure, legume, and conventional systems, respectively (Table 6.2). Because the annual net aboveground C inputs (based on plant biomass and manure) were similar for all systems, this means that the two organic systems retained 2–3 times more of the aboveground C inputs in their soils than the conventional system. The increased amount of organic matter in the manure and legume systems translated

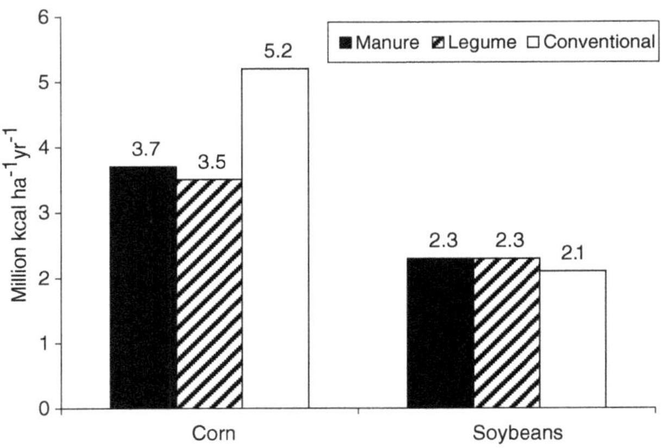

FIGURE 6.17 Average energy inputs for corn and soybeans based on analysis of The Rodale Institute Farming Systems Trial 1991–2000.

into 25% higher water holding capacities (ATTRA, 2002), helping those systems to avoid yield losses in drought years (Figure 6.14), and reduce surface runoff of water and soil erosion.

Similar C sequestration rates were found by Veenstra et al. (2006) when they evaluated the roles of tillage and cover cropping on C sequestration and other soil properties in a tomato-cotton rotation in California. Their data indicated that cover cropping was the predominant factor governing C sequestration, resulting in rates of 1100 and 1250 kg ha^{-1} year^{-1} for both standard and conservation tillage while standard and conservation tillage alone, without cover crops, decreased soil C significantly by 280 and 520 kg ha^{-1} year^{-1}. Research at Ohio State University demonstrated that in no-till agricultural systems, C sequestration rates were between 200 and 500 kg ha^{-1} yr^{-1} (Lal et al., 2004). Both these studies confirm the results from the Rodale Institute Farming Systems Trial that suggest that a diverse rotation with overwintering cover crops contributes more to soil C and organic matter increases than reduced tillage alone.

Soil organic matter is an important source of nutrients and can help increase microbial population and biodiversity (Pimentel et al., 1992; Troeh and Thompson, 1993; Mader et al, 2002; Lavelle and Spain, 2001), which was confirmed in the higher number of AM spores, greater AM colonization, and higher soil respiration rates in the two organic systems compared with the conventional system. Much of the difference in AM populations can be attributed to greater plant cover (70%) in the organic systems compared with the conventional corn-soybean rotation (40%) due to overwintering cover crops in the organic rotation (Galvez et al., 1995). In addition to fixing or retaining soil N, these cover crops provide the AM fungi with roots to colonize and maintain their viability during the interval from cash crop senescence to next year's planting. The increase in biodiversity can also provide vital ecological services including crop protection in terms of disease reduction (Cook, 1988; Hoitink, Inbar, and Boehm, 1991). In addition, not using synthetic pesticides and commercial fertilizers in organic systems minimizes the harmful effects of these chemicals upon nontarget organisms.

Overall, environmental damage from agricultural chemicals is reduced in organic systems because no commercial fertilizers or pesticides are applied. As a result, overall public health and ecological integrity should improve through the adoption of these practices (NAS, 2003).

Leachate

The average annual nitrate-N leaching rates of 16–18 kg ha in all three systems were low compared with results from other experiments with similar N inputs, where nitrate-N leaching ranged from 30–146 kg ha^{-1} $year^{-1}$ (Fox et al., 2001; Power, Wiese, and Flowerday, 2001). Nitrate-N concentrations exceeded the 10-ppm limit most often in the conventional system, which was probably due to the use of highly soluble N fertilizer in that system. On the other hand, the legume system lost the highest percentage of applied N although this loss was not steady across the entire period of the study. The surges in nitrate leaching in the legume system were sporadic and were associated with extreme weather events in certain years. In 1995 and 1999, the hairy vetch green manure supplied approximately twice as much N as needed for the corn crop that followed, contributing excess N to the soil, making it available for leaching. In 1999, the heavy N input from hairy vetch was followed by a severe drought that stunted corn growth and reduced the corn's demand for N. In both years, these N-rich soils were also subjected to unusually heavy fall and winter rains that leached the excess N into the lower soil layers.

Manure applications appeared to keep the nitrate leaching somewhat steady every year, although soil N should be monitored in all agricultural systems to avoid excessive losses.

Two of the four herbicides applied in the conventional system persisted in soil water percolating past the crop root zone, potentially allowing it to enter ground and drinking water supplies. Atrazine exceeded concentrations of 3 ppb, the maximum contaminant level (MCL) set by the Environmental Protection Agency (EPA) for drinking water, when it was applied two years in a row (in corn after corn plots). More important, however, atrazine concentrations in the conventional system usually exceeded the 0.1-ppb concentration known to produce deformities in frogs (Hayes et al., 2002). Metolachlor also reached high concentrations of more than 3 ppb, although the EPA has not yet established a MCL for metolachlor in drinking water.

Higher leachate volumes and soil water contents in the organic systems not only explain the improved yields in drought years, but also indicate an increased groundwater recharge and reduced runoff potential compared with the conventional system (Lotter, Seidel, and Liebhardt, 2003).

Crop Yields and Economics

After the first 3 years of transition to organic, N levels in the organic systems of the Rodale Institute Farming Systems Trial had improved and did not limit crop yields any more. In the short-term, there may be N shortages in organic systems that may reduce crop yields temporarily, but these can be eliminated by raising the soil N level with animal manure or legumes. Moreover, once soil C and N have improved, yields are more stable in extreme climate conditions like drought or water excess (Figures 6.14, 6.15, and 6.16).

The legume system required 35% more labor throughout the year than the conventional system; however, because the labor was spread out over the growing season, the hired labor costs per ha were about equal between the two systems. With the organic system, the farmer was busy throughout the summer with mechanical weed control, wheat harvest, and hairy vetch cover crop planting. In contrast, the conventional farmer had large labor requirements in the spring and fall — planting and harvesting — but little in the summer months.

This could have implications for the growing number of part-time farmers for whom the availability of family farm labor is severely limited. Other organic systems have also been shown to require more labor per hectare than conventional crop production. On average, organic systems require about 15% more labor (Sorby, 2002; Granatstein, 2003), but the increased labor input may range from an increase of 7% (Brumfield, Rimal, and Reiners, 2000) to a high of 75% (Nguyen and Haynes, 1995; Karlen, Duffy, and Colvin, 1995).

The legume system had lower input costs than the conventional system but required a legume cover crop before the corn, which was established after the wheat harvest. Thus, corn (a higher value crop) could only be grown 33% of the time with the organic rotation and 60% of the time in the conventional rotation. Stated in another way, the yields per ha between organic and conventional corn for grain may be similar within a given year; however, overall production of corn in this organic system was less than in the conventional system over a multiple-year period because corn was grown less frequently. On the other hand, the reduced amount of corn grown in the organic rotation was partly compensated for with the additional crop of wheat.

In the market place, prices for organic grains often range from 20 to 140% higher than conventional grains (Dobbs, 1998; Bertramsen and Dobbs, 2002; NewFarm, 2003). Thus, when the market price differential is factored in, the differences between organic and conventional would be relatively small and, in most cases, the returns on the organic grain would be higher, as in The Rodale Institute Farming Systems Trial reported here.

Energy Inputs

Significantly less fossil energy was expended for corn production in the manure and legume systems compared with the conventional system. This was due to the high-energy use for mineral fertilizer production. For soybeans, there was little difference in the energy requirements for organic and conventional systems. Other investigators have reported similar findings (Pimentel et al., 1993; Smolik, Dobbs, and Rickert, 1995; Karlen, Duffy, and Colvin, 1995; Dalgaard, Halberg, and Porter, 2001; Mader et al., 2002). In general, the utilization of less fossil energy and energy conservation by organic agriculture systems reduces the amount of CO_2

released to the atmosphere, and therefore the problem of global climate change (FAO, 2003).

CONCLUSION

For over 6000 years, organic methods have been utilized to conserve soil, water, energy, and biological resources. Many of the benefits of organic technologies identified in our long-term studies are related to soil C improvements and increased biological activity of the soil.

Specific benefits of increased soil C and biological activity include:

- Organically managed crop yields can equal those from conventional agriculture.
- During drought years, high soil organic matter under organically managed systems helps conserve soil and water resources, and proves beneficial to stabilizing yields of corn and soybean.
- During overly wet conditions, high soil C content under organic management conserves soil N, leading to higher yield and protein levels in organic systems than in conventional systems.
- The crop rotations and cover cropping typical of organic agriculture reduce soil erosion, pest problems, and the need for pesticides.
- Fossil energy inputs for organic crop production are up to 30% lower than for conventionally produced crops.
- Labor inputs average about 15% higher in organic farming systems and range from 7 to 75% higher.
- Because organic foods frequently bring higher prices in the market place, the net economic return is often equal or higher than conventionally produced crops.
- Recycling of livestock wastes can reduce pollution and at the same time benefit soil C accruals.
- Abundant biomass both above- and belowground (soil organic matter) also increases biodiversity, which helps in the biological control of pests and increases crop pollination by insects.
- The organic farming technologies leading to C sequestration include diverse rotations, cover cropping, and manure/compost utilization. Their use and benefits are not restricted to organic practitioners and may be adopted by conventional farmers to make their operations more sustainable and ecologically sound.
- Organic farming is a proven method to reduce greenhouse gases and provide multiple benefits for a wide range of other environmental concerns.

Although a centerpiece of organic farming, the organic farming practices are often utilized by sustainable and conventional farming systems to improve their efficiency and cost effectiveness.

RECOMMENDATIONS

1. Organic agriculture incentives should allow a quicker and greater expansion of this farming method and the multiple benefits associated with soil C increases. Incentives for farmers transitioning to organic should include:
   a. Financial support for the transition period
   b. Educational programs in organic transition
   c. C credits and other means
2. Cover crop incentives can be cost effective strategies to increase the multiple benefits of improved soil C for organic and conventional farmers.
3. All farmers will benefit from payments for reductions in C emissions in their farming activities, and for their ability to increase soil C.
4. Verifiable analyses are needed to quickly, accurately, and economically assay soil C dynamics. These improved assays will help speed the implementation of fair C accounting systems for all farmers.
5. Increased research and development support of innovative biological technologies will accelerate and increase total C sequestration in farming systems to advance greenhouse gas mitigation and remediate degraded soil resources.
6. Integrated use of livestock waste in farming systems will increase soil C and provide crop nutrients to crops instead of as soil, water, and air contaminants.
7. Integrating compost into farming systems will help stabilize soil C and increase soil C contents.
8. Advancing research on the relationship between mycorrhizal fungi and higher soil C accumulation rates will help optimize soil C in agricultural systems.

REFERENCES

Appropriate Technology Transfer for Rural Areas (ATTRA), Drought Resistant Soil, ATTRA, November 2002, 7 pp., National Center for Appropriate Technology, Fayetteville, AR, 2002.

Bertramsen, S.K. and Dobbs, T.L., An update on prices of organic crops in comparison to conventional crops, *Economics Commentator*, No. 426, February 22, 2002, South Dakota State University, Brookings, SD, 4 pp., 2002.

Brumfield, R.G., Rimal, A., and S. Reiners, Comparative cost analyses of conventional, integrated crop management, and organic methods, *Hort Technol.*, 10(4), 785–793, 2000.

Cook, R.J., Biological control and holistic plant-health care in agriculture, *Am. J. Alternative Agric.*, 3(2/3), 51–62, 1988.

Dalgaard, T., Halberg, N., and Porter, J.R., A model for fossil energy use in Danish agriculture used to compare organic and conventional farming, *Agric., Ecosystems, Environ.*, 87(1), 51–65, 2001.

Dimitri, C. and Greene, C., Food industry taps growing American market, Economic Research Survey, USDA, *Agricultural Outlook*, October 2002, 4–7, 2002a.

Dimitri, C. and Greene, C., Recent growth pattern in U.S. organic food market, Agriculture Information Bulletin No. 777, USDA, ERS, September, 2002b. Available at: www.ers.usda.gov/publications/aib777/. Accessed on 06/06/04.

Dobbs, T.L., Price premiums for organic crops, *Choices*, (2nd Quarter), 39–41, 1998.

Dobbs, T. and Smolik, J.D., Productivity and profitability of conventional and alternative farming systems: a long-term on-farm paired comparison, *J. Sustainable Agric.*, 9(1), 63–77, 1996.

Douds, D.D., Janke, R.R., and Peters, S.E., VAM fungus spore populations and colonization of roots of corn and soybean under conventional and low-input sustainable agriculture, *Agric., Ecosystems, & Environ.*, 43, 325–335, 1993.

Fenton, T.E., Brown, J.R., and Maubach, M.J., Effects of long-term cropping on organic matter content of soils, *Soil Water Conserv. J.*, 95–114, 1999.

Food and Agriculture Organization (FAO), Organic Agriculture and Climate Change, Environment and Natural Resources Series No. 4, Food and Agriculture Programme, United Nations, Rome, 2003. Available at: www.fao.org/docrep/005/y4137e02b.htm#P819–106360. Accessed on 02/13/07.

Fox, R.H., Zhu, Y., Toth, J.D., Jemison, J.M., and Jabro, J.D., Nitrogen fertilizer rate and crop management effects on nitrate leaching from an agricultural field in central Pennsylvania, in *Optimizing Nitrogen Management in Food and Energy Production and Environmental Protection: Proc. 2nd Int. Nitrogen Conf. on Science and Policy*, Galloway, J., Cowling, E.G., and Erisman, J.W. et al., Eds., 2001, A.A. Balkema Publishers, Rotterdam, Germany, 181–186.

Franke-Snyder, M., Douds, D.D., Galvez, L., Phillips, J.G., Wagoner, P., Drinkwater, L., and Morton, J.B., Diversity of communities of arbuscular mycorrhizal (AM) fungi present in conventional versus low-input agricultural sites in eastern Pennsylvania, USA, *Appl. Soil Ecol.*, 16, 35–48, 2001.

Galvez, L., Douds, D.D., Wagoner, P., Longnecker, L.R., Drinkwater, L.E., and Janke, R.R., An overwintering cover crop increases inoculum of VAM fungi in agricultural soil, *Am. J. Alternative Agric.*, 10, 152–156, 1995.

Granatstein, D., Tree Fruit Production with Organic Farming Methods, Center for Sustaining Agriculture and Natural Resources, Washington State University, Wenatchee, WA, 2003. Available at: organic.tfrec.wsu.edu/OrganicIFP/OrganicFruitProduction/OrganicMgt.PDF. Accessed on August 7, 2003.

Hanson, J.C., Johnson, D.M., Peters, S.E., and Janke, R.R., The profitability of sustainable agriculture on a representative grain farm in the mid-Atlantic region, 1981–1989, *Northeastern J. Agric. Resource Econ.*, 19(2), 90–98, 1990.

Hanson, J.C., Lichtenberg, E., and Peters, S.E., Organic versus conventional grain production in the mid-Atlantic: an economic and farming system overview, *Am. J. Alternative Agric.*, 12(1), 2–9, 1997.

Hanson, J.C. and Musser, W.N., An economic evaluation of an organic grain rotation with regards to profit and risk, September 2003, *Working Paper 03-10*, Department of Agricultural and Resource Economics, University of Maryland, College Park, MD, 2003.

Harris, G., Hesterman, O., Paul, E., Peters, S., and Janke, R., Fate of legume and fertilizer nitrogen-15 in a long-term cropping systems experiment, *Agron. J.*, 86, 910–915, 1994.

Hayes, T., Collins, A., Lee, M., Mendoza, M., Noriega, N., Stuart, A., and Vonk, A., Hermaphroditic, demasculinized frogs after exposure to the herbicide atrazine at low ecologically relevant doses, *Proc. Natl. Acad. Sci.*, 99(8), 5476–5480, 2002.

Hobbs, J.A., and Brown, P.A., Effects of cropping on organic matter content of a western Kansas soil, *Technical Bulletin No. 144*, Kansas Agriculture Experimental Station, Manhattan, KS, 1965.

Hoitink, H.A.J., Inbar, Y., and Boehm, M.J., Status of compost-amended potting mixes naturally suppressive to soilborne diseases of floricultural crops, *Plant Dis.*, 75(9), 869–873, 1991.

Karlen, D.L., Duffy, M.D., and Colvin, T.S., Nutrient, labor, energy, and economic evaluations of two farming systems in Iowa, *J. Prod. Agric.*, 8(4), 540–546, 1995.

Lal, R., Griffin, M., Apt, J., Lave, L., and Morgan, M., Managing soil carbon, *Science*, 304(5669), 393, 2004.

Lavelle, P. and Spain, A.V., *Soil Ecology*, Kluwer Academic Publishers, Dordrecht, 2001.

Lotter, D.W., Seidel, R., and Liebhardt, W., The performance of organic and conventional cropping systems in an extreme climate year, *Am. J. Alternative Agric.*, 18(3), 146–154, 2003.

Mader, P., Fliessbach, A., Dubois, D., Gunst, L., Fried, P., and Niggli, U., Soil fertility and biodiversity in organic farming, *Science*, 296(5573), 1694–1697, 2002.

Moyer, J., Saporito, L., and Janke, R., Design, construction, and installation of an intact soil core lysimeter, *Agron. J.*, 88(2), 253–256, 1996.

National Academy of Sciences (NAS), *Frontiers in Agricultural Research: Food, Health, Environment, and Communities*, NAS, Washington D.C., 2003.

NewFarm, *The New Farm Organic Price Index*, The Rodale Institute, Kutztown, PA, 2003. Available at: www.newfarm.org Accessed on October 22, 2003.

Nguyen, M.L. and Haynes, R.J., Energy and labour efficiency for three pairs of conventional and alternative mixed cropping (pasture-arable) farms in Canterbury (New Zealand), *Agric., Ecosystems, Environ.*, 52(2/3), 163–172, 1995.

Odwalla, *U.S. Market Trends in Organic Foods*, 2002. Available at: www.odwalla.com/enwfiles/background.html. Accessed on July 27, 2003.

Pimentel, D., Hepperly, P., Hanson, J., Douds, D., and Seidel, R., Environmental, energetic, and economic comparisons of organic and conventional farming systems, *Bioscience*, 55 (7), 573–582, 2005.

Pimentel, D., McLaughlin, L., Zepp, A., Kakitan, B., Kraus, T., Kleinman, P., Vancini, F., Roach, W.J., Graap, E., Keeton, W.S., and Selig, G., Environmental and economic effects of reducing pesticide use in agriculture, *Agric. Ecosystems, Environ.*, 46(1-4), 273–288, 1993.

Pimentel, D., Stachow, U., Takacs, D.A., Brubaker, H.W., Dumas, A.R., Meaney, J.J., O'Neil, J., Onsi, D.E., and Corzilius, D.B., Conserving biological diversity in agricultural/forestry systems, *BioScience*, 42, 354–362, 1992.

Power, J.F., Wiese, R., and Flowerday, D., Managing farming systems for nitrate control: a research review from management systems evaluation areas, *J. Environmental Qual.*, 30, 1866–1880, 2001.

Sloan, A.E., Top ten global food trends, *Food Technol.*, 59(4), 20–32, 2005.

Smolik, J.D., Dobbs, T.L., and Rickert, D.H., The relative sustainability of alternative, conventional, and reduced-till farming systems, *Am. J. Alternative Agric.*, 16, 25–35, 1995.

Sorby, K., What is sustainable coffee? Background paper, World Bank Agricultural Technology, Note 30, Washington, D.C., World Bank, 2002.

Stauffer, R.S., Muckenhim, R., and Odell, R.T., Organic carbon, pH and aggregation of the soil of the Morrow plots as affected by type of cropping and manurial addition, *J. Am. Soc. Agron.*, 32, 819–832, 1940.

Troeh, F.R. and Thompson, L.M., *Soils and Soil Fertility*, Oxford University Press, New York, 1993.

U.S. Department of Agriculture, Agriculture Marketing Service (USDA-AMS), *National Organic Program, Final Rule: 7 CFR Part 205*, USDA-AMS, Washington, D.C., 2002. Available at www.ams.usda.gov/nop. Accessed on August 1, 2003.

USDA, *Agricultural Statistics*, U.S. Department of Agriculture, Washington, D.C., 2003.

Veenstra, J., Horwath, W., Mitchell, J., and Munk, D., Conservation tillage and cover cropping influence soil properties in San Joaquin Valley cotton-tomato crop, *California Agric.*, 60(3), 146–153, 2006.

Wander, M., Traina, S., Stinner, B., and Peters, S., Organic and conventional management effects on biologically active soil organic matter pools, *Soil Science Society America J.*, 58, 1130–1139, 1994.

7 Benefits of Soil Organic Carbon to Physical, Chemical, and Biological Properties of Soil

Charles W. Rice, Karina Fabrizzi, and Paul White

CONTENTS

BENEFITS OF SOIL ORGANIC CARBON TO PHYSICAL, CHEMICAL, AND BIOLOGICAL PROPERTIES OF SOIL

Soil organic matter (SOM) is a critical component of the terrestrial biosphere that not only facilitates the production and growth of agricultural crops and other biota, but also supports a large consortium of soil microorganisms and fauna. Thus, SOM is the backbone of soil productivity and a critical link between civilization and sustainable agriculture and productivity. The importance of soil organic matter to soil and the environment led Albrecht (1938) to state, "soil organic matter is one of our most important national resources... and it must be given its proper rank in any conservation policy." Not only does SOM provide a substantial amount of plant nutrients, but it also protects water and air quality by decreasing soil erosion and sequestering greenhouse gases. Soil organic matter is comprised primarily of organic carbon (C).

The estimated amount of organic C stored in world soils is about 1100 to 1600 Petagrams (Pg), more than twice the C in living vegetation (560 Pg) or in the atmosphere (750 Pg) (Sundquist, 1993). Soil organic C (SOC) is derived mostly from dead plant residues, along with roots in the soil and root exudates, or chemicals exuded from the roots. Soil organic C is not only found in decomposing plant residues, but also soil microorganisms and fauna. The transformations and storage of this C in soil is a function of biotic, chemical, and physical controls and is related to plant residue quality, such as the carbon to nitrogen ratio (C/N), and its accessibility to organisms. Plant residues are decomposed by soil microorganisms and most of the C in the plant residue is released back into the atmosphere as carbon dioxide (CO_2). Approximately 10–20% of the C in plant residue forms SOM, sometimes referred to as "humus." Soil organic matter is composed of C and other elements, such as nitrogen (N), phosphorus (P), and sulfur (S). Some of the C incorporated into SOM can persist in soils for hundreds and even thousands of years. The storage potential of C stored in soil is a function of climate (temperature and moisture) and basic soil characteristics, including clay content and mineralogy, and crop and soil management practices, such as plant selection and tillage frequency.

Soil organic matter directly and indirectly supports the essential function of the soil and its relation to overall agricultural or natural ecosystem function. Because of the importance of SOM to soil, measurements of SOM and its fractions have been an essential component of soil science research, agricultural production, and land management. Concentrations of SOM can range from 0.2% in mineral soils to over 80% in peat soils; and the typical range for temperate mineral soils is 0.4–10% (Smith et al., 1993). Soil organic matter influences many of the chemical, physical, and biological properties of the soil. As an example of the impact SOM has on other soil properties, Larson and Pierce (1994) list SOM as an important contributor to soil cation exchange capacity, bulk density, water retention, and aeration.

Soil Chemical Properties

Although SOM is largely composed of organic C, the N, P, and S content can also positively affect soil fertility and plant growth. Approximately 90–95% of the soil N, 40% of the soil P, and 90% of the soil S is associated with SOM (Smith et al., 1993). These nutrients are recycled internally over time through plant uptake and residue deposition, microbial decomposition, and SOM formation. Within this recycling, SOM can be a source or a sink of plant nutrients. Plant productivity is directly associated with SOM content and nutrient turnover by microbial activity. For example, in agricultural soils approximately 2–4% of the organic N is made available for plant uptake on an annual basis. This N supply can be a significant portion of the plant N needs. In cropping systems, as much as 50–80% of the N can be supplied from SOM and nearly 100% of the N in native ecosystems. This percentage represents 11–300 kg N ha^{-1} (10–270 lb N ac^{-1}) for a crop (Smith et al., 1993). At \$0.50 kg^{-1} nitrogen fertilizer, this translates to a value of \$5–\$150 ha^{-1} yr^{-1} (\$2–\$61 ac^{-1} yr^{-1}) for the farmer not including any application costs.

Sulfur follows similar transformations as N. Organic S is converted to inorganic, plant-available S by soil microorganisms. Annual rates of S mineralization rates are 15–36 kg S ha^{-1} yr^{-1} (13-32 lb S ac^{-1} yr^{-1}) (Smith et al., 1993). This mineralization

TABLE 7.1
Plant Nutrients Supplied by Soil Organic Matter (SOM)

Nutrient	Value		Supplied by SOM		Savings	
	$ kg^{-1}	$ lb^{-1}	kg ha^{-1}	lb ac^{-1}	$ ha^{-1}	$ ac^{-1}
Nitrogen (N)	0.5	1.1	11–300	10–270	5–150	11–300
Phosphorus (P)	1.5	3.3	15–36	13–32	23–54	43–106
Sulfur (S)	0.4	0.9	7–40	6–36	3–16	5–32
Total	2.4	5.3			31–220	59–438

rate is sufficient to supply the majority of the annual plant S requirements for agronomic crops such as wheat (*Triticum aestivum*), alfalfa (*Medicago sativa*), and corn (*Zea maize*) (Stevenson, 1986).

Plant availability of soil P is controlled more by soil chemistry and chemical transformations instead of microbial activity. Thus, SOM is not usually a major source of P to plants, except in very high organic soils. However, organic P can represent 80% of the total P in some soils. Thus, they can supply 7–40 kg P ha^{-1} yr^{-1} (6-36 lb P ac^{-1} yr^{-1}) (Smith et al., 1993).

Significant amounts of N, P, and S are also contained within the microbial biomass. As the microbial biomass turns over a portion of the nutrients are made available to the plant. Microorganisms can also influence plant available N, S, and P levels by immobilizing the nutrients. As a rule, a C/N) ratio greater than 25 promotes N immobilization, whereas a C/N ratio less than 25 promotes N mineralization. For S and P, a C to S or C to P ratio greater than 60 promotes immobilization of S and P by the microorganisms.

If we consider the nutrient supplying capacity of just N, P, S, this would translate to 300 kg N, 40 kg P, and 36 kg of S per ha^{-1} (Table 7.1). Given current prices this would be a fertilizer value of $227 per ha^{-1} ($94 per acre), which does not include savings from application costs (Table 7.1).

An additional advantage of accounting for plant nutrients released from SOM is that the timing of the release is affected by water availability and temperature. The ideal conditions for microbial mineralization of plant nutrients from SOM often is the proper conditions for plant growth and thus nutrient needs (Rice, Havlin, and Schepers, 1995). This synchrony of mineralization and plant needs often translates to a higher use efficiency of nutrients derived from soil organic matter. Accounting for nutrients supplied from SOM reduces fertilizer applications (Rice and Havlin, 1994), fuel use, and excess nutrients lost to ground water, or surface water, or to the atmosphere in the case of N gases.

Cation Exchange Capacity

A soil's capacity to store other nutrients, such as potassium (K) and calcium (Ca), is referred to as the cation exchange capacity (CEC). These cation exchange sites, which are found on clay minerals and iron oxides, are important for retention of nutrients and making the nutrients plant available. Approximately 20–80% of the cation exchange capacity of the soil is due to SOM, and increases in SOC increases the soil CEC. It is estimated that 1 g organic C kg^{-1} soil represents approximately 4.3 mmol kg^{-1} of effective CEC in low activity clay soils (DeRidder and Van Keulen, 1990; Bationo and Mokwunye, 1991). This CEC is important in the retention of many of the basic plant nutrients, such as K, Ca, and magnesium (Mg). In some cases, soil organic compounds enhance the chelation of metals, which then increases the bioavailability of trace elements required for plant growth.

Soil organic matter also acts as a binding agent to render harmless environmental contaminants, such as petroleum products, pesticides, and heavy metals, thereby reducing toxicity and minimizing leaching (Stott and Martin, 1989). This can be accomplished through processes such as direct binding of metals by the CEC or SOM, or metal–cation bridges with soil organic C. Additionally, some organic compounds in the soil exhibit plant growth-promotion properties.

The decrease of SOM will adversely affect marginal, or easily degradable and erodible soils (Craswell and Lefroy, 2001). This loss of nutrient supply requires the application of fertilizer to compensate for the loss of nutrients because of the loss of the CEC sites from SOC. The additional fertilizer requirements are especially acute in poor or developing countries where the cost of mineral fertilizers is high. Unfortunately, these areas are often the soils with the greatest depletion of SOM and many of these areas have low-activity clays with few CEC sites.

Physical

Soil organic matter plays a key role in many soil physical properties. Soil organic matter plays a major role in the formation and stabilization of soil aggregates as binding agent. McVay et al. (2006) reported a high correlation between aggregate stability and SOC in several long-term experiments in Kansas. Six et al. (2002) reported a high correlation between SOC and aggregate stability in temperate soils, indicating the importance of the SOC found in SOM.

Increases in the SOC enhances soil structure and aggregation (Tisdall and Oades, 1982; Stengel et al., 1984; Haynes and Swift, 1990; Feller and Beare, 1997; Haynes and Naidu, 1998; Gardner et al., 1992; Hamblin and Davies, 1997; Karlen et al., 1994; Mikha and Rice, 2004). Greater aggregation makes soils less prone to crusting and compaction (Diaz-Zorita and Grosso, 2000). Good soil structure improves water infiltration resulting in less soil erosion (Schertz et al., 1994; Benito and Diaz-Fierros, 1992). Aggregation also improves aeration and root growth, which enhances microbial activity and SOM turnover, resulting in the release of nutrients.

Soil organic C directly affects soil water holding capacity (Hudson, 1994; Emerson, 1995; Gupta and Larson, 1979; Hollis, Jones, and Palmer, 1977; Salter and Williams, 1969; Salter and Haworth, 1961). Plant available soil water content may increase 1–10 g for every 1-g increase in SOM content (Emerson, 1995). This increase may be sufficient to help maintain crop growth between periods of rainfall of 5–10 days. The increase in soil water retention can be related to the hydrophilic properties of SOM and the improvement of soil structure. Higher levels of SOM reduce bulk density, thus providing an improved rooting environment. In addition, SOM holds soil water, an important attribute for plant growth in more mesic environments.

Biological

Biologically, SOM is the source of C and energy for many soil microorganisms. Soil organisms act as primary decomposers breaking down organic materials, such as plant and animal wastes, into simple compounds. As soil biological organisms decompose the plant and animal material, a portion is converted to SOM, adding to the aforementioned chemical and physical benefits to soil. Carbon also passes through the soil food web in soil organisms such as nematodes, protozoa, insects, and earthworms. These organisms aid in the mixing of soil and redistribution of nutrients. Soil fauna also promote aggregation and biopores. Chan (2004) found significantly greater numbers of earthworm burrows in no-till soils, as compared with tilled soils. Earthworm channels and other biopores have been considered beneficial to water infiltration and aeration. Increased SOC enhances the biomass and the biodiversity of the microbial community of the soil. Generally, a diverse microbial community leads to greater biological control of plant diseases and pest. Soil organic C also drives many of the previously mentioned biological and chemical transformations in soil, such as increasing or decreasing the flow of N, P, and S to plants through mineralization and immobilization, respectively.

SUMMARY

Soil organic C is the central component of SOM that drives many other functions of soil that relate to plant production and ecosystem services. These relationships are illustrated in Figures 7.1 and 7.2. Figure 7.1 illustrates the relationships to other soil functions. Figure 7.2 illustrates the direct and indirect relationships to ecosystem services. The direct effect on nutrient supply can be converted to a monetary value. The integrated effect on crop yield can also be converted into a monetary value.

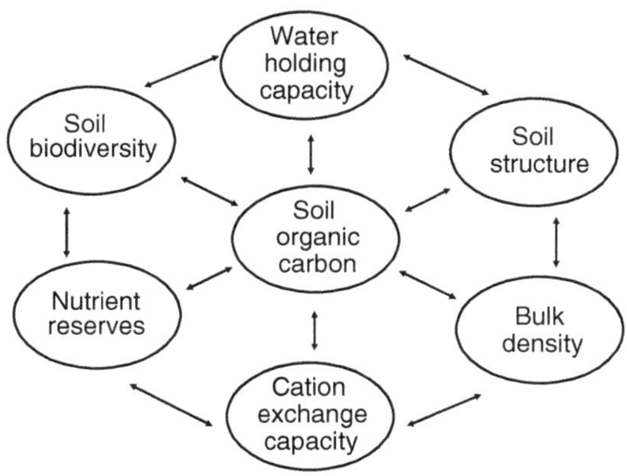

FIGURE 7.1 This figure illustrates the relationships to other soil functions.

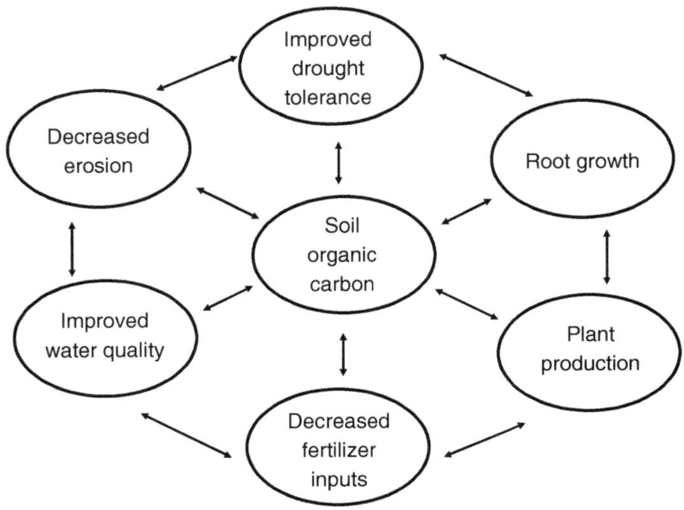

FIGURE 7.2 This figure illustrates the direct and indirect relationships to ecosystem services.

However, the value of other ecosystem services has not been defined quantitatively and needs to be developed for a full assessment of the value of soil C.

REFERENCES

Albrecht, W.A., Loss of soil organic matter and its restoration, *Soils and Men*, USDA Yearbook of Agriculture, USDA, Washington, DC, 347–360, 1938.

Bationo, A. and Mokwunye, A.U., Role of manures and crop reissued in alleviating soil fertility constraints to crop production with special reference to Sahelian zones of West Africa, *Fertilizer Res.*, 29, 125–177, 1991.

Benito, E. and Diaz-Fierros, F., Effects of cropping on the structural stability of soils rich in organic matter, *Soil and Tillage Res.*, 23, 153–161, 1992.

Chan, Y.K., Impact of tillage practices and burrows of a native Australian anecic earthworm on soil hydrology, *Appl. Soil Ecol.*, 27, 89–96, 2004.

Craswell, E. T. and Lefroy, R.D.B., The role and function of organic matter in tropical soils, *Nutrient Cycling Agroecosystems*, 61, 7–18, 2001.

DeRidder, N., and Van Keulen, H., Some aspects of the role of organic carbon in sustainable intensified arable farming systems in West Africa semi-arid tropics, *Fertilizer Res.*, 26, 299–310, 1990.

Diaz-Zorita, M. and Grosso, G.A., Effect of soil texture, organic carbon and water retention on the compatibility of soils from the Argentinean Pampas, *Soil Tillage Res.*, 54, 121–126, 2000.

Emerson, W.W., Water retention, organic carbon and soil texture, *Australian J. Soil Res.*, 33, 241–251, 1995.

Feller, C. and Beare, M.H., Physical control of soil organic matter dynamics in tropical land-use systems, *Geoderma*, 79, 49–67, 1997.

Gardner, W.K., Fawcett, R.G., Steed, G.R., Pratley, J.E., Whitfield, D.M., and Rees, H., Crop production on duplex soils in south-eastern Australia, *Australian J. Experimental Agric.*, 32, 915–927, 1992.

Gupta, S.C. and Larson, W.E., Estimating soil water retention characteristics from particle size distributions, organic matter content and bulk density, *Water Resour. Res.*, 15, 1633–1635, 1979.

Hamblin, A.P. and Davies, D.B., Influence of organic matter on the physical properties of some East Anglian soils of high silt content, *J. Soil Science*, 28, 11–22, 1997.

Haynes, R.J. and Swift, R.S., Stability of soil aggregates in relation to organic constituents and soil water content, *J. Soil Science*, 41, 73–83, 1990.

Haynes, R.J. and Naidu, R., Influence of lime, fertilizer and manure applications on soil organic matter content and soil physical conditions: A review, *Nutrient Cycling Agroecosystems*, 51, 123–137, 1998.

Hollis, J.M., Jones, R.J.A., and Palmer, R.C., The effects of organic matter and particle size on the water retention properties of some soils in the west midlands of England, *Geoderma*, 17, 225–231, 1977.

Hudson, B.D., Soil organic matter and available water capacity, *J. Soil Water Conserv.*, 48, 188–193, 1994.

Karlen, D.L., Wollenhaupt, N.C., Erbach, D.C., Berry, N.C., Swain, J.B., Eash, N.S., and Jordahl, J.L., Crop residue effects on soil quality following 10 years of no-till corn, Soil and Tillage Research, 31, 149–167, 1994.

Larson and Pierce, The dynamics of soil quality as a measure of sustainable management, *SSSA Special Publication*, 35, 37 pp., 1994.

McVay, K.A., Budde, J.A., Fabrizzi, K., Mikha, M.M., Rice, C.W., Schlegel, A.J., Peterson, D.E., Sweeney, D.W., and Thompson, C., Management effects on soil physical properties in long-term tillage studies in Kansas, *Soil Science Society Am. J.*, 70, 434–438, 2006.

Mikha, M. and Rice, C.W., Tillage and manure effects on soil and aggregate-associated carbon and nitrogen, *Soil Science Society of America J.*, 68, 809–816, 2004.

Rice, C.W. and Havlin, J.L., Integrating mineralizable N indices into fertilizer N recommendations, in *New Directions in Soil Testing for Nitrogen, Phosphorus, and Potassium, Special Publication No. 40*, Havlin, J.L., et al., Eds., Soil Science Society of America, Madison, WI, 1994, 1–13.

Rice, C.W., Havlin, J.L., and Schepers, J.S., Rational nitrogen fertilization in intensive cropping systems, *Fert. Res.*, 42, 89–97, 1995.

Salter, P.J. and Haworth, F., The available water capacity of a sandy loam soil II, The effect of farmyard manure and different primary cultivations, *J. Soil Science*, 12, 335–342, 1961.

Salter, P.J. and Williams, J.B., The moisture characteristics of some Rothamsted, Woburn and Saxmundham soils, *J. Agric. Science*, (Cambridge) 73, 155–158, 1969.

Schertz, D.L., Moldenhauer, W.C., Livingston, S.J., Weeisies, G.A., and Hintz, A.E., Effect of past soil erosion on crop productivity in Indiana, *J. Soil Water Conserv.*, 44, 604–608, 1994.

Six, J., Feller, C., Denef, K., Ogle, S., Moraes Sa, J., and Albrecht, A., Soil organic matter, biota and aggregation in temperate and tropical soils: Effects of no-tillage, *Agronomie*, 22, 755–775, 2002.

Smith, J.L., Papendick, R.I., Bezdicek, D.F., and Lynch, J.M., Soil organic matter dynamics and crop residue management, in *Soil Microbial Ecology*, Metting, F.B., Jr., Ed., Marcel Dekker, Inc., New York, 1993, 65–94.

Stengel, P., Douglas, J.T., Guerif, J., Goss, M.J., Monnier, G., and Cannell, R.Q., Factors influencing the variation of some properties of soils in relation to their suitability for direct drilling, *Soil Tillage Res.*, 4, 35–53, 1984.

Stevenson, F.J., *Cycles of Soil: Carbon, Nitrogen, Phosphorus, Sulfur, Micronutrients*, John Wiley & Sons, New York, 1986.

Stott, D.E. and Martin, J.P., Organic matter decomposition and retention in arid soils, *Arid Soil Res. Rehabil.*, 3, 115–148, 1989.

Sundquist, E.T., The global carbon dioxide budget, *Science*, 259, 934–941, 1993.

Tisdall, J. and Oades, J., Organic matter and water stable aggregates in soils, *J. Soil Science*, 33, 141–163, 1982.

Section III

Off-Site Benefits

8 Soil Physical Properties and Erosion

Rattan Lal

CONTENTS

INTRODUCTION

Soil physical properties are strong determinants of soil's susceptibility to erosion. Important among these are structural stability, water infiltration rate, available water capacity, and soil organic carbon pool. Soil detachability increases with decrease in aggregation amount and strength. Soil erodibility, or susceptibility of soil to erosion, comprises detachability and transportability and increases with decrease in structural stability. Structurally unstable soils, such as those containing predominantly kaolinitic and illitic clay minerals and low soil organic carbon (SOC) concentration in conjunction with high silt content, are prone to slaking of aggregates, formation of surface seals or crust, decline in the infiltration rate, and high soil erosion risks. Two opposing hypotheses exist regarding the impact of soil erosion on the atmospheric carbon (C) pool. Sedimentologists and geomorphologists argue that soil erosion and C-laden sediments are a net C sink of about one Pg C yr^{-1}. This hypothesis is based on the assumption that SOC depleted in eroding soils is dynamically replaced by photosynthesis and it is protected against mineralization in depositional sites by reduced microbial activity. In contrast, soil scientists and agronomists hypothesize that 0.4–0.6 Pg C yr^{-1} transported through erosional processes into aquatic ecosystems and depressional sites is sequestered. However, an additional 0.8–1.2 Pg C yr^{-1} displaced by erosional processes is mineralized and emitted into the atmosphere. The mineralization is increased because of breakdown of aggregates by erosional processes. Further, decline in quality and fertility of eroding soil reduces agronomic/biomass productivity, which hinders the process of restoration of the depleted SOC pool. Depletion of the SOC pool has strong adverse impacts on agronomic productivity and economic returns, especially in rainfed/dry farming where fertilizers and other amendments are not available or not affordable. Because of numerous adverse off-site effects of erosion, adoption of conservation effective measures for controlling soil erosion is essential to sustainable management of soil and water resources.

Even though an objective resolution of opposing hypotheses concerning the role of soil erosion on greenhouse gas (GHG) emissions to the atmosphere is not yet resolved, the rationale is strong for policies that result in the adoption of effective erosion control measures because of numerous adverse on-site and off-site effects of accelerated erosion. On-site effects, caused by decline in the effective rooting depth and loss of plant-available water and nutrients, lead to a decline in agronomic yield over the short-term and a reduction in soil quality over the long-term. Off-site effects lead to sedimentation of water reservoirs and waterways, eutrophication, and emission of greenhouse gases (GHGs). The nature and magnitude of emission of GHGs varies among soil types, landscape positions, drainage characteristics, and the magnitude and characteristics of SOC as well as soil organic nitrogen (SON) pools. It is the fate of the transformations of SOC and SON pools, which are transported and redistributed over the landscape and acted upon by physical and microbial processes, that needs to be addressed because of its possible impacts on the global climate through emission of GHGs. Depletion of the SOC pool has strong and long-term adverse impacts on agronomic productivity (Lal, 2006) and economic returns (Sparling et al., 2006), and these effects are over and above those caused by accelerated erosion (Lal, 1987; 1998; Lal et al., 1998; den Biggelaar et al., 2003).

Change in land use and land cover and the attendant alternations in surface features are important (Ruddiman, 2003), but poorly recognized drivers of long-term climate change (Pielke, 2005). Effects of land use, land use change and land cover on microclimate, weather patterns, and climate may be as important as those associated with increase in atmospheric concentration of GHGs (Feddema et al., 2005). Land use affects climate patterns directly and indirectly. Directly, it alters water and energy balance along with temperature and moisture regimes, microclimate and weather patterns on local and regional scales. Indirectly, land use and land use change affects ecosystem C and N budgets thereby affecting the magnitude and direction of the GHG fluxes between the atmosphere and Earth's ecosystems. In general, conversion of natural to managed ecosystems leads to emission of carbon dioxide (CO_2) from soil into the atmosphere.

Natural or geologic erosion is a slow process, and an important soil-forming factor. Some of the most fertile soils of the world were formed through geologic erosion, including those in the valleys of rivers such as the Nile, Indus, Ganges, Yangtze, Tigris, and Euphrates. It was in these fertile soils, with regular soil fertility restoration through periodic flooding, where settled agriculture began about 10 millennia ago. In contrast to the slow and constructive geologic erosion, the accelerated erosion caused by anthropogenic perturbations is a rapid and destructive process. It has numerous adverse on-site and off-site effects. On-site, it reduces top soil depth, decreases soil fertility, depletes SOC and SON pools, reduces degree and strength of aggregation leading to crusting, decreases the water infiltration rate, and reduces agronomic yield/net primary productivity (NPP) depending on the antecedent soil properties and management (e.g., fertilizer rate, irrigation, tillage, and cropping systems). Off-site, it leads to run-on and sedimentation with attendant flooding and deposition. In some cases, agronomic yield/NPP may increase because of the increase in rooting depth, soil fertility, and available water capacity. A prolonged inundation and anaerobiosis, however, may reduce agronomic yield of upland crops, increase methanogenesis and methane (CH_4) emission, and enhance nitrous oxide (N_2O) emissions due both to nitrification and denitrification. The depositional sites usually are enriched in SOC and SON pools.

In view of the rapid increase in atmospheric concentrations of CO_2 and other GHGs due to anthropogenic activities, emphasis on the identification of sources and sinks of GHGs is strong. It is in this regard that the importance of soil erosion and physical properties needs to be critically addressed. Two diametrically opposed hypotheses have been proposed. The first, proposed by soil scientists and agronomists, states that accelerated soil erosion is a source of atmospheric CO_2 (Lal, 2003). The second, proposed by sedimentologists and geomorphologists, states that soil erosion (i.e., geologic or accelerated) represents a net sink of atmospheric CO_2 (Stallard, 1998; Smith et al., 2001).

The objective of this chapter is to review the role of soil physical properties and processes on the SOC budget at different spatial (i.e., soilscape, landscape, watershed, regional, and global) and temporal (i.e., human and geologic) scales, and outline conditions/assumptions under which erosional processes can be a source or sink for atmospheric CO_2.

SOIL PHYSICAL PROPERTIES AND CARBON SEQUESTRATION

Edaphologically important soil physical properties are: clay content and mineralogical composition, aggregation and aggregate stability, bulk density, total porosity and pore size distribution, p^F curves and the available water capacity, effective rooting depth, infiltration rate, and hydraulic conductivity. Some key soil physical properties that integrate several properties and processes include: bulk density, water stable aggregation, available water capacity, and infiltration rate. The optimum range of these properties, in relation to sustainable management of soil and water resources for agricultural land use, are listed in Table 8.1. An important strategy of soil surface management is to maintain physical properties within the optimum range for optimizing crop growth and returning an adequate amount of aboveground and belowground biomass to the soil. Management practices that may optimize soil physical properties include conservation tillage with crop residue mulch, conservation-effective measures for soil and water conservation, use of integrated nutrient management (INM) practices to maintain an optimum elemental/nutrient balance in the root zone, and diverse cropping systems to enhance soil biodiversity.

TABLE 8.1
Optimum Range of Key Soil Physical Properties for Sustainable Use of Soil and Water Resources

Soil Property	Units	Optimum Range
1. Effective rooting depth	cm	50–100
2. Bulk density	Mg m^{-3}	1.1–1.4
3. Aggregation	%	25–75
4. Drainable porosity at 60-cm suction	%	15–20
5. Available water capacity	cm	8–30
6. Infiltration rate	cm h^{-1}	1–5
7. Saturated hydraulic conductivity	cm h^{-1}	0.2–2
8. Soil organic carbon concentration	%	2–5

Source: From Lal, R., *Methods and Guidelines for Assessing Sustainable Use of Soil and Water Resources in the Tropics*, SMSS/USDA-SCS, Washington, D.C., 1994. With permission.

A judicious management of soil physical properties is crucial to maintaining/enhancing SOC pool, which is enhanced if the amount of biomass returned to the soil exceeds the losses. The biomass C is returned to the soil through crop residue return as mulch, as input of root biomass (especially in the subsoil), and use of organic amendments including manure, compost, and organic material from agricultural/industrial byproducts. The SOC pool is depleted through erosion, leaching, and mineralization. Thus, at any given point in time, the SOC pool is influenced by its gains and losses and increases if gains exceed losses and vice versa.

An important prerequisite for maintenance of soil structure is minimal soil disturbance. Two activities that cause drastic soil disturbance include soil tillage and vehicular traffic. Three types of soil tillage are used: primary, secondary, and tertiary. Primary tillage involves soil inversion with a moldboard plow. Secondary tillage involves breakdown of soil clods with disc plow, and tertiary tillage comprises mixing and pulverization with cultivators and harrows. Soil tillage breaks down aggregates, exposes soil to raindrop impact, releases soil organic matter (SOM), which was hitherto protected against microbial processes through encapsulation, and increases mineralization of SOM and release of CO_2 to the atmosphere. The principal objectives of mechanical tillage (weed control) can be achieved with chemical growth regulators (herbicides) in combination with growing cover crops in the rotation cycle. Therefore, presowing seedbed preparation through tillage operations must be minimized (minimum tillage) completely eliminated (no-till or zero till), or reduced by combination of minimum tillage with other conservation effective measures (conservation tillage). Conversion of plow tillage to conservation tillage or no-till farming reduces soil erosion, decreases mineralization and increase SOC concentration in the surface layer. The rate of SOC sequestration through conversion to no-till farming ranges form 100–1000 kg ha^{-1} yr^{-1} depending upon soil type and climate (Lal et al., 2004; Lal et al., 1998; Lal, Follett, and Kimble, 2003).

TABLE 8.2
Soil and Crop Management Practices to Optimize Soil Physical Properties and Enhance the SOC Pool

Property	Objective	Management Options
1. Bulk density	Alleviate soil compaction	No-till farming, crop residue mulch, guided traffic, low axle load, dual tires, periodic chiseling
2. Aggregation	Improve soil structure	Mulch farming, including cover crops, promoting earthworm activity through manuring/compost use, use of soil amendments/conditioners
3. Infiltration rate	Increasing infiltration rate	No-till farming with crop residue mulch, reducing vehicular traffic, cover cropping
4. SOC pool	Increasing SOC pool	Increasing input of biomass C through residue mulch, compost, cover crop, manuring
5. Available water capacity	Enhancing available H_2O capacity in coarse-textured soils	No-till farming, manuring, crop residue, mulch

Vehicular traffic is another cause of decline in soil structure. Harvest traffic, with grain carts of 10- or 20-Mg load on a single axle, causes severe soil compaction. The problem of soil compaction or densification, leading to decrease in porosity and reduction in proportion of macropores, is typically more on clayey soils than on loamy or coarse-textured soils and in snow-free belt (i.e., tropical and subtropical regions) than in soils of temperate climates that undergo several freeze-thaw cycles during winter.

Soil and crop management practices for optimization of physical properties and enhancement of SOC pool are outlined in Table 8.2. However, soil specific practices vary among soil types and cropping/farming systems.

Soil Erosion as a Sink for Atmospheric CO_2

The hypothesis that soil erosion is a sink for atmospheric CO_2 is proposed by several researchers investigating C budget in sediments deposited by erosional processes (Stallard, 1998; Smith et al., 2001; Van Oost et al., 2005a, 2005b). This hypothesis is based on the following assumptions:

1. World rivers transport a large quantity of C-enriched sediments (Meybeck, 2005).
2. Depletion of SOC pool on-site by erosional processes (i.e., water and wind) is replenished by photosynthesis and addition of new biomass carbon.
3. No loss of SOC occurs by mineralization from eroded sediments either during the transport or during redistribution over the landscape in depres-

sional/aquatic sites when deposited. This assumption implies that no soil respiration occurs by microbial processes at any stage of the erosional process. This hypothesis is supported by Ritchie (1989).
4. No adverse impact of erosion occurs on NPP on site, and photosynthesis and productivity are continuously maintained.

With these assumptions, Stallard (1998) estimated that sedimentation on land could bury about 1 Pg Cyr^{-1} (0.6-1.5 Pg C yr^{-1}). This sink presumably reaches a maximum strength between 30° and 50° N. This hypothesis postulates that eroding soils have the potential for replacing most of the SOC removed by erosion, and the depositional/burial sites have no losses due to mineralization or respiration. Further, the replacement of SOC on eroding sites virtually occurs simultaneously, as does depletion by erosion.

Experimental testing of this hypothesis requires full accounting of the eroded C at different scales: eroding sites, landscape over which the sediments are transported including the aquatic ecosystems, and depositional sites. The knowledge about the fate of eroded C is very important. Equally important are the rate and magnitude of replacement of SOC on eroding soils in relation to land use and management history.

Soil Erosion as a Source for Atmospheric CO_2

The hypothesis that accelerated erosion is a source of atmospheric CO_2 is proposed by soil scientists (Lal, 1995b, 2003, 2005; Lal et al., 2004). This hypothesis is based on the following assumptions:

1. Aggregate disruption by raindrop impact releases soil organic matter and exposes it to microbial activity (Figure 8.1).
2. The SOC pool is depleted on eroding soils due to its preferential removal as a light fraction of low density (1.0–1.3 Mg m^{-3} vs. 2.5–2.65 Mg m^{-3} for the mineral fraction).
3. The loss of SOC and the nutrient-rich surface layer depletes soil fertility, reduces agronomic productivity of and the quantity and quality of biomass C returned to the eroding soil. There is a long time lag in the restoration of eroding soils, depending on the land use and management.
4. Erosional processes lead to breakdown of aggregates, which increase soil respiration rate during the transport (Jacinthe and Lal, 2001).
5. Depositional sites may have lower respiration/mineralization due to burial and re-aggregation. However, depositional sites may have higher CH_4 emission due to inundation/anaerobiosis leading to methanogenesis, and more N_2O emission caused by both nitrification/denitrification processes.

With the assumptions outlined previously, Lal (2003) estimated that erosion leads to emission of 0.8–1.2 Pg C yr^{-1}. The assumption of the dynamic replacement of the SOC pool on-site is not proven by the act that eroded soils have much more SOC pool than the uneroded places (Gregorich, Drury, and Baldock, 2001; Monreal, Zentner, and Robertson, 1997). Once again, experimental testing of this hypothesis

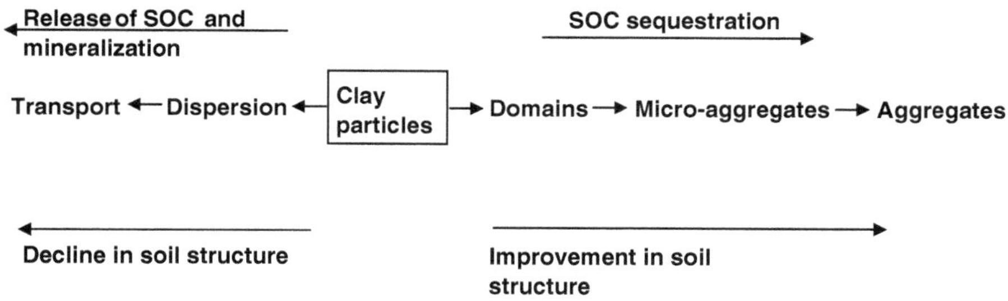

FIGURE 8.1 Soil aggregation through formation of organo-mineral complexes encapsulates soil organic carbon (SOC) and protects it against microbial action. In contrast, soil dispersion by raindrop impact and flowing runoff causes erosion, release of SOC, increase in mineralization of organic matter, and emission of CO_2.

involves monitoring: (1) the fate of eroded SOC during its transport and deposition phases, and (2) the fate of eroding soils under the current land use and soil/vegetation management with regard to the rate of soil erosion and the status of SOC pool (i.e., whether it is depleting or restoring).

The Basis of Apparent Controversy

Several factors have caused an apparent controversy, as outlined previously. Important among these are:

1. *Definition of Soil Erosion*: Soil erosion is a four-stage process: aggregate disruption, or slaking particle detachment, particle/aggregate transport, and sediment deposition. The controversy can be resolved if the SOC pool and gaseous fluxes (CO_2, CH_4) are measured during each phase, especially with regard to the fate of SOC removed, transported and deposited in depressional sites. Modeling studies, although important, must be validated against experimental data obtained from field studies conducted at a range of spatial scales (e.g., soilscape to river watershed scale). Assessment of the SOC pool at the depositional sites or the eroding soils is not adequate to resolve the controversy.
2. *Temporal Scale*: Assessment of the ecosystem C budget on eroding soils must be done over a long period of time (i.e., 2–3 decades) and related to land use and management history. The rate of erosion, both of soil and of C, must be measured/assessed for a long period. Measurements at a specific time (snapshot) are not adequate.
3. *Net Primary Productivity*: Research data on NPP of eroding soils, above- and belowground and in relation to land use and management, is essential to understanding whether these soils are degrading or aggrading. These experimental data must be obtained over a long period of 2–3 decades to establish definite temporal trends with management and land use.
4. *Methodology*: Use of standard methodology, for field and laboratory analyses, is essential to ascertaining that results are comparable and conclu-

sions valid and objective. Methodological protocol must be scale neutral or effectively address the scaling issues.

Commonalities in Two Opposing Views

Both hypotheses, which emphasize that accelerated soil erosion is a "source" and "sink" of atmospheric CO_2, state that the SOC transported to and buried in depressional sites and aquatic ecosystems are sequestered. The estimates of the amount of C being sequestered, however, vary. Stallard (1998) estimated that roughly 1 Pg (0.6 to 1.5 Pg C yr^{-1}) might be buried/sequestered by erosional/depositional processes. In comparison, Lal (2003) estimated that 0.4–0.6 Pg C yr^{-1} is being transported and buried into the aquatic ecosystems.

Differences in Two Opposing Views

1. Mineralization: The "sink" hypothesis assumes that SOC being transported or deposited has minimal (zero) mineralization. The "source" hypothesis argues that disruption/breakdown of aggregates increases mineralization, and 20 to 30% of erosion-displaced soil organic matter may be mineralized (Jacinthe and Lal, 2001). The "source" hypothesis further assumes that periodic inundation and anaerobiosis leads to emissions of CH_4 due to methanogenesis and N_2O due to nitrification/denitrification processes.
2. SOC Pool of the Eroding Soil: The "sink" hypothesis assumes that the SOC pool depleted in eroding soils is replenished (dynamic replacement) immediately by photosynthesis and the attendant biomass returned. The "source" hypothesis argues that erosion causes decline in soil quality and reduction in biomass/agronomic yield and NPP. Unless the erosion is effectively controlled, the SOC pool of eroding soils is progressively depleted. Restoration of the depleted SOC pool depends on conversion to a restorative land use and availability of biomass-C and of N, P, S, and other elements, which are essential to conversion of biomass-C into stable humus and humic substances.
3. Time Lag: The "sink" hypothesis argues that SOC pool in eroding soil undergoes "dynamic replacement and loss," with no time lag between loss and replacement. In contrast, the "source" hypothesis argues that restoration of the depleted pool takes a long period of time, and there is no instantaneous or dynamic replacement of the loss. The recovery period may be as long as 125 years (Sparling et al., 2006).
4. Ecological Impact of Erosion: The "sink" hypothesis assumes that even accelerated soil erosion is ecologically beneficial, and has a moderating/beneficial impact on the atmospheric concentration of CO_2. In contrast, the "source" hypothesis argues that anthropogenically accelerated erosion is a destructive process with strong adverse ecological impacts.

The schematics in Figure 8.2 summarize the adverse on-site and off-site impacts of accelerated erosion, which are not addressed by the "sink" hypothesis. On-site

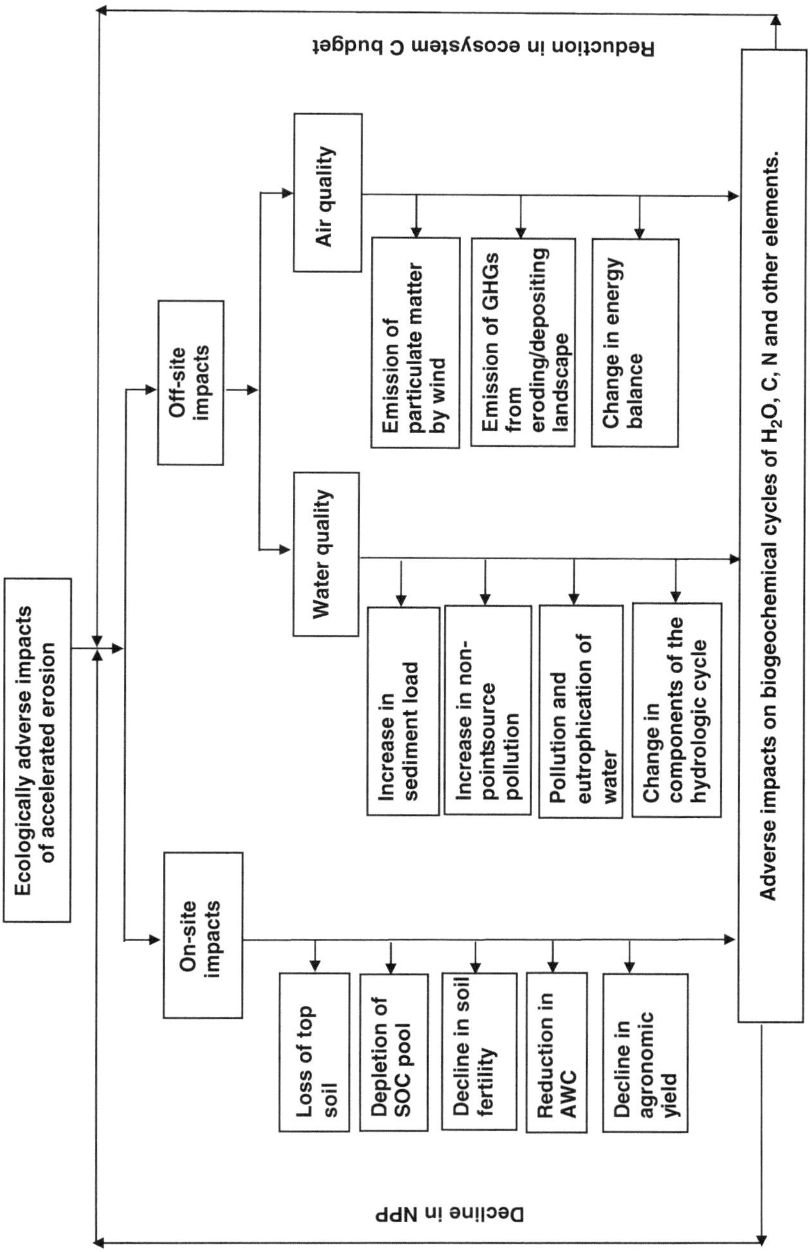

FIGURE 8.2 Adverse ecological impacts of anthropogenically accelerated soil erosion (AWC = available water capacity; GHGs = greenhouse gases).

impacts cause decline in agronomic yield over a short-time horizon and reduction in soil quality and its capacity to perform ecosystem services over a long-time period. Short-term adverse impacts on yield are caused by loss of water as runoff and plant nutrients as dissolved load or as sediment-entrained suspended load. Long-term adverse impact on soil quality is caused by truncation of the topsoil depth, reduction in plant-available water and nutrient retention capacities, and decline in soil structure. The long-term adverse impacts on soil structure are primarily caused by depletion of the SOC pool (Figure 8.2).

Off-site, erosion adversely affects quality of water and air (Figure 8.2). Decline in water quality is caused by transport of agricultural chemicals and sediments, which are also the cause of anoxia of coastal ecosystems. In addition to the ejection of particulate material, impact on air quality of relevance to the risks of global warming is due to emission of GHGs exacerbated by erosional processes at all four stages: detachment, transport, redistribution, and deposition (Figure 8.2).

The net impact of accelerated erosion is: (1) reduction in NPP and in the amount of biomass-C returned to the soil, (2) depletion of the terrestrial C pool and the attendant increase in the atmospheric C pool, and (3) pollution of water resources (Figure 8.2). Thus, there are at least three reasons for adopting conservation-effective measures to curtail this menace. Restoration of SOC pool is an important strategy of increasing crop yields and advancing food security in developing countries (Lal, 2006).

Interaction Between Soil Physical Properties and Erosion

Soil erosion is a physical process, a work function for which the required energy is supplied by the impacting raindrops and blowing/flowing wind/water. ($E = 1/2\ mV^2$, where m is the mass of the drop/water, and V is the impact/flow velocity of water or wind.) The first two of the four erosional processes (e.g., aggregate disruption and detachment) are set in motion through the kinetic energy of raindrops and flowing water. Soil's detachability (i.e., its ability to resist detaching force of impacting raindrops and blowing/flowing wind/water) depends on structural stability. Soils with strong/stable structure (e.g., clayey soils containing a high SOC pool) are less easily detached than those with weak structure (silt loam soils containing illite/kaolinite minerals and low SOC pool). Similarly, soil's erodibility also depends on its structural stability. Erodibility has two distinct but related components: detachability and transportability, and is inversely proportional to structural stability.

In addition to soil structure, infiltration rate is another important soil physical property. Both instantaneous and equilibrium infiltration rates affect runoff rate and amount. Infiltration rate, which is the rate of water entry into the soil through the soil-air interphase, also depends on soil structure. Structurally stable soils are not prone to crusting or formation of surface seal. Furthermore, structurally stable soils have a large faction of macropores or transmission pores, which can rapidly conduct water from surface into the subsoil. Soils with high infiltration rates have a low runoff rate/amount and are less prone to erosion than those with lower infiltration rates.

Horizonation or layering is another physical property with a strong impact on erosion. Soil layers/horizons that create hydrologic discontinuity exacerbate the soil

TABLE 8.3
Rationale for Erosion Control and Technological Options to Achieve It

Rational	Technological Options
1. Agronomic productivity	No-till farming, residue mulching, contour hedges, diverse cropping system
2. Water quality	Combination of biological/agronomic measures (e.g., no-till with residue mulching) with engineering techniques of installing terraces, waterways, and reservoirs
3. Sedimentation control	Wetland restoration, constructing water reservoirs, contour hedgerows, mulching, cover crops, and no-till farming
4. Soil carbon sequestration	Integrated nutrient management (INM), manuring, cover cropping, no-till farming, use of biosolids and organic amendments, establishing deep-rooted crops/plants
5. Improving use efficiency of input	Precision/soil specific farming, conservation tillage, INM, guided traffic

erosion hazard. Soils with relatively uniform deep profiles are less prone to erosion than those with distinct layers of contrasting texture, bulk density, and hydraulic conductivity.

Rationale for Soil Erosion Control

Because of its numerous adverse on-site and off-site impacts (Figure 8.2), it is important to adopt conservation-effective measures to ascertain sustainable use of soil and water resources. Reasons for soil erosion control and technological options to achieve it are outlined in Table 8.3. Conservation-effective technologies are those that can reduce the rate of erosion to the tolerable level or "A" value. Land use and land cover are important factors that affect the global C budget with potential strong impacts on the climate change (Feddema et al., 2005; Pielke, 2005).

The most important rationale for erosion control is the need for sustainable use of soil and water resources. This is especially true for fragile soils in ecologically sensitive ecoregions, such as Alfisols in the West African and South Asian tropical regions, where resource-poor, small-size landholders are unable to use the recommended off-farm inputs. These subsistence farmers use extractive farming practices based on mining the soil fertility that depletes the SOC pool. The mineralization of the SOC pool leads to a release of plant nutrients (e.g., N, P, K, S, Ca Mg, Zn, and Cu) and emission of CO_2 into the atmosphere.

Sustainable use of such soils demands the losses of SOC and fertility by erosion and mineralization be minimized, nutrients removed by plants and animals replaced, and the biomass C added (as mulch, compost, roots) to maintain a positive ecosystem C budget.

Impact on climate change aside, increase in SOC pool is essential to enhancing crop yields and increasing use efficiency of input (Entry et al., 1996; Francis, Tabley, and White, 2001; Lal, 2006). The soil C pool, enhanced through deliberate change in land use and management practices, can also be traded nationally (e.g., on the Chicago Climate Exchange) or internationally (Point Carbon, 2004; Newsweek, 2006; Sparling et al., 2006). The SOC pool may be of limited value in irrigated soils. If fertilizers are available economically and readily (Sojka and Upchurch, 1999; Letey et al., 2003), it is crucial to enhancing productivity of degraded/desertified soils managed under uncertain rainfed/dry farming conditions and where resource-poor farmers can neither readily access nor economically afford fertilizers and other essential input. Thus, the negative economic significance of declining a SOC pool, which has been strongly depleted on-site by accelerated erosion, cannot be overemphasized.

CONCLUSIONS

Soil physical properties are strong determinants of soil erosion. Important among these are structural stability and infiltration rate. Soil erosion is a four-stage process, and its effect on the fate of organic matter transported by erosional processes must be assessed at all stages and over 20–30 years. Two opposing hypotheses exist concerning erosional/depositional effects on atmospheric CO_2. The "sink" hypothesis states that erosion is a sink for atmospheric CO_2 with a capacity of 0.6–1.5 Pg C yr^{-1}. This hypothesis is based on the assumptions that the SOC pool in eroded soils is replaced by photosynthesis and that organic matter in sediment is not lost through respiration. The "source" hypothesis states that accelerated soil erosion is a net source of atmospheric CO_2 with a magnitude of 0.8–1.2 Pg C yr^{-1}. This hypothesis is based on the assumption that 20–30% of the soil organic matter transported by erosion is mineralized, the eroding soils are prone to decline in fertility and reduction in NPP, and a time lag of 2–3 decades is required for the restoration of the depleted SOC pool depending on land use and management history. An objective resolution of these opposing hypotheses requires research data from long-term field experiments conducted to assess the fate of erosion-displaced soil organic matter at all four stages of erosion performed at different scales. In the meantime, the rationale is strong for adopting effective erosion control measures because of numerous adverse on-site and off-site effects of accelerated erosion.

REFERENCES

Breslau, K., It can pay to be green: clean air means profits at the Climate Exchange, *Newsweek*, 45, 2006.

den Biggelaar, C., Lal, R., Wiebe, K., and Breneman, V., The global impact of soil erosion on productivity, I. Absolute and relative erosion-induced yield losses, *Adv. Agron.*, 81: 1–48, 2003.

Entry, J.A., Mitchell, C.C., and Beckman, C.B., Influence of management practices on soil organic matter, microbial biomass and cotton yield in Alabama's old rotation, *Biol. Fertil. Soils*, 23, 353–358, 1996.

Feddema, J.J., Oleson, K.W., Bonan, G.B., Mearns, L.O., Buja, L.E., Meehl, G.A., and Washington, W.M., The importance of land-cover change in simulating future climates, *Science*, 310, 1674–1678, 2005.

Francis, G.S., Tabley, F.J., and White, K.M., Soil degradation under cropping and its influence on wheat yield on a weakly structured New Zealand silt loam, *Australian J. Soil Res.*, 39, 291–305, 2001.
Gregorich, E.G., Drury, C.F., and Baldock, J.A., Changes in soil carbon under long-term maize in monoculture and legume-based rotation, *Canadian J. Soil Science*, 81, 21–31, 2001.
Jacinthe, P.A. and Lal, R., A mass balance approach to assess carbon dioxide evolution during erosional events, *Land Degradation Dev.*, 12, 329–339, 2001.
Lal, R., Effects of soil erosion on crop productivity, *Crit. Rev. Plant Science*, 5, 303–368, 1987.
Lal, R., *Methods and Guidelines for Assessing Sustainable Use of Soil and Water Resources in the Tropics*, SMSS/USDA-SCS, Washington, D.C., 1994.
Lal, R., Erosion-crop productivity relationship for soils of Africa, *Soil Science Society America J.*, 59, 661–667, 1995a.
Lal, R., Global soil erosion by water and carbon dynamics, in *Soils and Global Change*, Lal, R., Kimble, J., E. Levine, E., and Stewart, B.A., Eds., Lewis Publishers, Chelsea, MI, 1995b, 131–142.
Lal, R., Soil erosion impact on agronomic productivity and environment quality, *Crit. Rev. Plant Science*, 17, 318–464, 1998.
Lal, R., Soil erosion and the global carbon budget, *Environ. Int.*, 29, 437–450, 2003.
Lal, R., Influence of soil erosion on carbon dynamics in the world, in *Soil Erosion and Carbon Dynamics*, Roose, E.J., Lal, R., Feller, C., Barthes, B., and Stewart, B.A., Eds., Taylor & Francis/CRC, Boca Raton, FL, 2005, 23–35.
Lal, R., Enhancing crop yields in developing countries through restoration of the soil organic carbon pool in agricultural lands, *Land Degradation Dev.*, 17, 197–209, 2006.
Lal, R., Kimble, J.M., Follett, R.F., and Cole, C.V., *The Potential of U.S. Cropland to Sequester Carbon and Mitigate the Greenhouse Effect*, Lewis Publishers, Chelsea, MI, 1998.
Lal, R., Follett, R.F., and Kimble, J.M., Achieving soil carbon sequestration in the United States: a challenge to the policy makers, *Soil Science*, 168, 827–845, 2003.
Lal, R., Griffin, M., Apt, J., Lave, L., and Morgan, M.G., Managing soil carbon, *Science,* 304, 393, 2004.
Letey, J., Sojka, R.E., Upchurch, D.R., Casel, D.K., Olson, K.R., Payne, W.A., Petrie, S.E., Price, G.H., Reginato, R.J., Scott, H.D., Smethurst, P.J., and Triplette, G.G., Deficiencies in soil quality concept and its application, *J. Soil Water Conserv.*, 58, 180–187, 2003.
Meybeck, M., Origins and behavior of carbon species in world rivers, in *Soil Erosion and Carbon Dynamics*, Roose, E.J., Lal, R., Feller, C., Barthe's, B., and Stewart, B.A., Eds., Taylor & Francis/CRC, Boca Raton, FL, 2005, 209–237.
Monreal, C.M., Zentner, R.P., and Robertson, J.A., An analysis of soil organic matter dynamics in relation to management erosion and yield and yield of wheat in long-term crop rotation plots, *Canadian J. Soil Science*, 77, 553–563, 1997.
Pielke, R.A., Land use and climate change, *Science*, 310, 1624–1626, 2005.
Point Carbon, Carbon Market Europe, Point Carbon, Oslo, 2004. Available at: http://www.pointcarbon.com/. Accessed December 13, 2005.
Ritchie, J.C., Carbon content of sediments of small reservoirs, *Water Resour. Bull.*, 25, 301–308, 1989.
Ruddiman, W., The anthropogenic greenhouse era began thousands of years ago, *Climatic Change*, 61, 261–293, 2003.
Smith, S.V., Renwick, W.H., Buddenmeier, R.W., and Crossland, C.J., Budgets of soil erosion and deposition for sediments and sedimentary organic carbon across the contiguous United States, *Global Biogeochemical Cycles*, 15, 697–707, 2001.

Sojka, R.E. and Upchurch, D.R., Reservations regarding the soil quality concept, *Soil Science Society America J.*, 63, 1039–1054, 1999.

Sparling, G.P., Wheeler, D., Vesely, E.-T., and Schipper, L.A., What is soil organic matter worth? *J. Environ. Qual.*, 35, 548–557, 2006.

Stallard, R.F., Terrestrial sedimentation and the carbon cycle: coupling weathering and erosion to carbon burial, *Global Biogeochemical Cycles*, 12, 231–257, 1998.

Van Oost, K., Govers, G., Quine, T.A., Heckrath, G., Olesen, J.E., DeGryze, S., and Merckx, R., Landscape scale modeling of carbon cycling under the impact of soil redistribution: the role of tillage erosion, *Global Biogeochemical Cycles*, 19, 1–13, 2005a.

Van Oost, K.T., Quine, A., Govers, G., and Heckrath, G., Modeling soil erosion-induced carbon fluxes between soil and atmosphere on agricultural land using SPEROS-C, in *Soil Erosion and Carbon Dynamics*, Roose, E.J., Lal, R., Feller, C., Barthes, B., and Stewart, B.A., Eds., Taylor & Francis/CRC, Boca Raton, FL, 2005b, 37–53.

9 Wetlands and Global Carbon

Steven I. Apfelbaum

CONTENTS

OVERVIEW

THE CHALLENGES

The conversion of wetlands to agricultural production lands is the primary cause of wetland losses in the United States. Urbanization has also contributed to substantial wetland declines in some parts of the United States. These former carbon (C) sinks, ecological settings that incorporate atmospheric C extracted from the air into plant tissue in the basic process of building soil, have been replaced with a prevailing agricultural land use that has shifted them to become net contributors of atmospheric greenhouse gases (GHGs). Modern agriculture depends largely on the deep fertile soils built by historic wetlands, but conventional row crop tillage and high-impact agricultural uses continues to degrade these historic soils, exacerbates contributions of carbon dioxide (CO_2) and other GHGs into the earth's atmosphere, and compromises the sustainability of the modern agricultural landscape. Despite the contentious history of wetlands and agriculture, the management of wetlands to enhance soil C stocks is complementary to sustainable agriculture. Restoring wetlands also is a cost-effective way to sequester atmospheric CO_2, mitigate GHGs generated by modern agriculture, and enhance organic stocks of soils to insure the sustainability of food production for future generations of Americans.

Background

The United States has spent considerable time in debate over wetland regulations in the past thirty years. Although most debate has focused on definitions and regulatory constructs for technically identifying and regulating wetlands, the largest impact on wetlands, on an acreage basis, the agricultural community, has largely been exempt from regulations. At this time, the largest wetland restoration benefits for mitigating global climate change now may rest with the agricultural community and soil management and restoration practices on private lands. Restoration is also being implemented on public lands, such as wildlife refuges, national monuments, and parks. However, it is on the vast private land holdings where wetland restoration and wet site soil C management can be most successful for global GHG management. As such, farm policy may provide beneficial leverage in fostering this restoration on the land.

The loss of wetlands is associated with declines in the quality of watersheds, water quality in streams, rivers, lakes and estuarine environments, and substantially increased amounts and rates of water flow (Demissie et al, 1991). These changes have contributed to significant and costly floods, human hazard and safety risk, erosion and sedimentation in these waterways, and the loss of valuable topsoil from croplands into the sediments in lakes, rivers, and wetlands. In sediments, under water, the organic C decomposes and liberates methane, whereas nitrous oxides exacerbated from agricultural fertilizers are emitted along wetland margins; both gases are GHGs considerably more potent that CO_2. Preventing this loss of topsoil in the first place is a first line of defense against the decomposition and enrichment of organic matter in these sediments. The retention of topsoil in agricultural fields will prevent the loss of productivity in upland cropland areas while simultaneously benefiting critical ecological processes in these aquatic and wetland habitats (Gleason and Euliss, 1998) and represents a major benefit for the nation's agricultural, wetland and aquatic resources.

Strong relationships exist between increased peak stream flow and watershed percentage of wetland areas. Demissie et al. (1991) identified that as the ratio of wetland to watershed area increases, peak flows decreased, and the ratio of precipitation to runoff decreased.

Decreases in these ratios have contributed to significant and costly floods, human hazard and safety risk, erosion and sedimentation in these waterways, and the loss of valuable topsoil from croplands into the sediments in lakes, rivers, and wetlands. The underwater deposition of sediments results in organic C decomposition and methane liberation, while nitrous oxides exacerbated from agricultural fertilizers are emitted along wetland margins; both gases are GHGs considerably more potent than CO_2. Prevention of topsoil loss is a first line of defense against the decomposition and enrichment of organic matter in these sediments. The retention of topsoil in agricultural fields prevents the loss of productivity in upland cropland areas while simultaneously benefiting critical ecological processes in down slope aquatic and wetland habitats (Gleason and Euliss, 1998) and represents a major benefit for the nation's agricultural, wetland, and aquatic resources.

Watersheds with a larger percentages of extant wetlands have the lowest flood risk, higher quality water resources, and higher levels of biodiversity (Demissie et al., 1991). Similar trends have been documented for damages and costs associated with flooding impacts in river floodplains, such as the Mississippi (National Research Council, 2002) and Missouri Rivers (National Research Council, 2004). These and others (e.g., Miller and Nudds, 1996) suggest a strong relationship between agricultural drainage of wetlands and flood frequency in rivers. The National Research Council Recovery plans call for the restoration of significant acreages of wetland to reduce flooding, improve water quality, and enhance the biodiversity of rivers, and as essential for the recovery of such endangered species as the pallid sturgeon and the piping plovers (National Research Council, 2004). The sturgeon requires deeper faster moving waters over gravel, and connections to floodplain wetlands during juvenile life stages, and the piping plover nests on sandbars created from normal energy dissipation from flood events. Both habitats have declined in rivers managed by man (National Research Council, 2004).

Significant changes in the ecology of wetlands have occurred, including sedimentation, changes in hydrology, chemical inputs, and invasive species gaining a foothold in our nation's wetlands (Figure 9.1). As wetland decline has occurred, so too have declines of wetland dependant wildlife and former economies including wild rice harvests, duck and waterfowl subsistence and waterfowl sport hunting. A large percentage of the nation's threatened and endangered plants, birds, fishes, and other groups of organisms are wetland dependant (USFWS, Endangered Species Act). This agency documents that over 80% of the listed special concern species (plants and animals) are wetland dependent at some point in their life cycle, and that the number of fisheries, mollusks and many important aquatic insects are declining throughout the United States (USFWS Website). Wildlife declines and changes associated with poor soil C management are addressed further in Chapter 10, "Wildlife Benefits."

INTRODUCTION

Wetlands are the most productive ecosystems in the world (Whittaker and Likens, 1973) and they have played a critical role in the history and evolution of man. In the Carboniferous period, extensive wetlands sequestered the fossil fuels that man depends upon today and the rich biotic communities of wetlands provided sustenance for man for centuries. With an average net primary productivity of 1180 g C m^{-2} yr^{-1} (Pant, Rechcigl, and Adjei, 2003) and a surface area of 7×10^6 to 9×10^6 (Mitsch, Mitsch, and Turner, 1994), it is not surprising that wetlands comprise the largest pool of stored C in the world (Eswaran, Van den Berg, and Reich, 1993). Wetlands also provide habitats critical to the maintenance of global biodiversity, retention of surface water to mitigate flooding, improvement of water quality in streams and rivers, and maintenance of intricate geochemical characteristics within complex groundwater pathways (Euliss and Laubhan, 2005).

The value of wetlands to man has not always been recognized as positive and large areas of wetlands have been drained or converted to alternate land uses. Although it is difficult to quantify the loss of wetlands globally (Mitsch and Gosselink, 2000), total global losses in excess of 50% are likely (Dugan, 1993). Reasons for wetland loss are numerous and include such things as drainage for agricultural production, urban expansion, and mosquito control. In the coterminous United

FIGURE 9.1 The degradation of wetlands from receiving sedimentation, additional or less hydrology (including wetland drainage), and declining biodiversity have occurred widely in the Midwestern United States. The conversion from diverse healthy wetlands to simple polycultures and monocultures of invasive plant species (e.g., hybrid cattail — *Typha glauca* — and Reed Canary grass — *Phalaris arundinacea*) is illustrated in these images. *Continued.*

States, overall wetland loss was estimated at 53% in the 1980s (Dahl, 1990) with losses from individual states varying from a 9% loss in New Hampshire to over a 90% loss in California. With continuing wetland loss, coupled with increasing knowledge of the important functions and values they provide to society, there is much public support to restore previously drained wetlands, especially for conservation purposes (Knutsen and Euliss, 2001).

In the United States, many previously farmed wetlands have been restored within perennial grasslands in response to incentives provided under various conservation programs of the U.S. Farm Bill (e.g., The Conservation Reserve Program), and in other federal, state, and private programs. Because wetlands naturally sequester atmospheric C, these restorations store atmospheric C as well as perform other ecosystem functions. The latest inventory of GHG emissions and sinks for the United States listed prairie wetlands in Conservation and Wetland Reserve Program lands as C sinks

FIGURE 9.1 *Continued.*

(US EPA, 2003). The size of the historic wetland C sink in the prairie pothole region of North America was recently estimated by Euliss et al. (in review) at 378 Tg, with over half (197 Tg) of the total C stores lost to the atmosphere from cultivation of farmed or drained wetlands. In other areas of the United States, the wetland losses and C sinks were significantly larger than in the pothole region of the Great Plains. For example, instead of isolated potholes, much of the northern half of Illinois, parts of northern Indiana, and large areas of southeastern Wisconsin were historic wetland. In addition, large floodplain areas of the Mississippi and Missouri, Illinois, and many other states included vast wetlands (Mitsch and Gosselink, 2000.). Now drained for agricultural use, these historic C sinks are being depleted through erosion and organic matter decomposition (i.e., C) loss to the atmosphere.

Wetlands are important ecotones between terrestrial and aquatic habitats, providing many services to society, including C storage, mitigation of other GHGs, improving water quality, attenuating floodwater, reducing topsoil loss into our river system, and by providing critical habitat to support biodiversity. These services result from managing the wetlands to increase C storage. Thus, managing wetlands for soil C storage provides multiple benefits, both direct and indirect. Over 90% of the nation's streams, groundwater, stream and estuarine sediments, and freshwater fish have at least one contaminant at detectable levels (Heinz Report, 2002), mainly from excess sediment, nutrients, and pathogens (US EPA, 2003). Excluding marine species, about a third of all plants and animals whose status is known are listed in various reports, including the Heinz Report (2002) and the National Audubon Society (2004), as being "at risk;" about a third of all North American bird species are declining significantly due to unnatural factors (e.g., loss of habitat, invasive plants, pollution, and poor land use decisions). A great number of wildlife species are wetland dependent or utilize wetlands seasonally, including species of considerable economic importance such as waterfowl, white-tailed deer, and ring-necked pheasants. Moreover, it has been estimated that from over a third to nearly 50% of the plant and animal species listed as threatened or endangered in the United States are wetland dependent or use wetlands at some point in their lives (Boylan and MacLean 1997; US EPA, 1995).

A recent report estimated the damage from agricultural production to wildlife and ecosystem biodiversity in the United States exceeded $3.200 billion dollars annually (Tegtmeier and Duffy, 2004). This damage estimate suggests a loss of productivity and sustainability of agricultural land because much of the damage their estimate was based on was due to surface runoff exacerbated by traditional land practices. Managing agricultural production fields for soil C (e.g., no-till and minimum tillage) has been demonstrated to reduce surface runoff and increase crop yields due to better topsoil moisture and nutrient retention (Kimble, Follett, and Lal, 2002). Agricultural lands managed for soil C are expected to benefit all wetland functions because sediment, nutrient, agrichemical, and excess surface water inputs are reduced.

FACTORS INFLUENCING CARBON LOSS IN WETLANDS

Existing C sinks in wetlands have and continue to be lost through three primary mechanisms:

1. *Drainage of Wetlands*: Wetland drainage dewaters and exposes to decomposition the historic organic matter in wetlands. As in a garden compost pile, the presence of oxygen hastens the decomposition of this organic matter, releasing C, among other materials, and sediments to the environment. As drainage infrastructure systems age and fail, which is hastened in drained wetlands, the farmer has the choice to put another layer of farm field tiles into the ground and lower the elevation of the ditch systems. On a typical farm of 100 acres with this problem, the added operational costs are generally in the range of $100,000 every decade or so. Applied to the larger agricultural community across the United States, the continued use and agricultural operations in drained wetlands and updating their infrastructure, if they even can do so, translates into millions and perhaps billions of dollars; costs incurred through farm subsidy or personally financed by farmers.

Some types of wetlands display seasonal hydroperiods, containing water only for a portion of the year. This typically means the wetlands do not have standing water, although the substrates may remain saturated and protected from oxidative processes that hasten decomposition compared with the intentional dewatering effects associated with crop production needs. In the Florida Everglades and in the Sacramento-San Joaquin Delta of California, wetland drainage to facilitate production of agricultural crops has led to rapid land subsidence (1 to 3 inches per year) as sequestered C in the form of dead wetland vegetation was exposed to oxygen and allowed to decompose (USGS, 2000). Although there may be some minor decomposition of organic matter and release of C during normal "dry" periods, one of the real benefits is typically a resurgent growth of plant seeds and a subsequent increase in plant growth that contributes to the stored C stocks. Studies have reported that prairie wetlands, which go dry periodically due to normal climate variation, do not lose significant quantities of the sequestered C unless the land is cultivated, which exposes the buildup of organic matter to excess oxygen (Euliss et al., 2006). This finding is consistent with studies that have demonstrated that cultivation is the primary mechanism that releases soil C in agricultural fields (Lal et al., 1999).

Wetland soils vary from deep and long-lived C storage types that include muck and peat deposits, to wetlands located in sand and gravel deposits in locations with high ground water tables, such as along river and lake margins. In addition, most notably in the Midwestern U.S. wetlands are formed in surficial depressions in glacially deposited clay substrates that hold water; the prairie pothole region of the United States and Corn Belt states are replete with wetlands formed in clay and silt clay soils.

Wetlands in sandy and graveling soils, such as in riverine floodplains, dry down as water levels decline. Wetlands formed in muck and peat deposits very seldom dewater. Wetlands in clay and silt soils, unless they are deeper depressions, typically experience seasonal dry down.

2. *Farming of Wetlands*: Farming of the historic wetland soils created using annual row crop tillage hastens the soil decomposition process or "composting effect," by mixing and remixing these susceptible soils and exposing them to high concentrations of oxygen. In many parts of the United States, it is clear that farming wetlands is unpredictable and the failures and added costs can lead to significant requirements for financial subsidies. Grave economic consequences for farmers and the public, who often pay and repay for the failed crops, must be considered because both new drainage infrastructure and more costly types of equipment (to operate in wetter ground) may be required and unpredictable crop failures may occur annually to every other year.
3. *Decomposition of Historic Soil Organic Materials*: A typical urban backyard compost pile reduces in volume by sometimes > 60% through the process of decomposition. Dewatering and farming of historic wetlands contributes similarly to the decreased volume and subsidence of the soil

surface in farmed wetlands and can directly lead to significantly higher operational costs for maintaining drainage infrastructure and farming costs as the soil surface subsides and approaches the shallow ground water table in these soils. No financial impact studies have summarized the overall impact of soil depletion to the agricultural community, but crop production costs and yields have been documented to decline precipitously as soil organic matter levels decline. Declines of over 50–75% have been documented in depleted soils, which when projected over the acreage of the United States where this depletion is occurring may translate to billions of dollars in additional annual operational costs for the farming community.

The drainage of organic soils, developed beneath wetlands, is a leading factor of land subsidence in the United States. Although seemingly a small area, the total area of organic soils in the United States is approximately the size of Minnesota (USGS, 2000). In the conterminous United States, the two best examples of major land subsidence are in the Sacramento/San Joaquin River Delta in California and the Everglades in Florida. Typically, these rich organic soils are drained for agricultural purposes and the oxidation of organic matter in reclaimed soils has resulted in over 25 feet of soil subsidence in certain areas. The loss of organic matter in such areas represents an enormous pool of formerly stored C that was released into the atmosphere as CO_2.

4. *Application of Fertilizers Increases the Emission Rates of Trace Greenhouse Gases*: Applying fertilizers to lands that drain into existing wetlands leads to nutrient enrichment and eutrophication. The deteriorated water quality and increased decomposition process that occurs with this enrichment can generate significant quantities of GHGs that are 21 to 310 times more potent in contributing to global warming than CO_2. Nitrous oxide and methane are two such potent constituents that are released when the decomposition of wetland soil organic matter is hastened. Thus, wetlands are important for C storage, but there are valid concerns over the release of GHGs such as methane (CH_4) and nitrous oxide (N_2O). Wetlands are the largest global source of CH_4, emitting approximately 20% of the annual global emissions to the atmosphere (Wang, Dong, and Patrick, 1996). Methane is an important GHG with a global warming potential (GWP) of 21 (IPCC, 1996). Nitrous oxide is an even more potent GHG with a GWP of 310 (Lal et al., 1999) and it persists in the atmosphere about as long as CO_2, about 120 years whereas CH_4 only lasts 12–15 years (US EPA, 2003). Although nitrogen enrichment of ombrotrophic mires apparently has little or no effect on CH_4 emission, data from a glaciated region in northeastern Germany suggests that enrichment from nitrogen fertilizers and accelerated mineralization of soil organic matter significantly elevates the emission of CH_4 and N_2O (Merbach et al., 2002). These findings are consistent with ongoing studies in the United States (Euliss, unpublished data) and with other studies that demonstrate that nitrogen fertilizers enhances the emission of N_2O (Thornton and Valente,

1996; Davidson et al., 2000). Restored wetlands in North America are generally sited within marginal farmlands that have been voluntarily idled by private landowners (e.g., Conservation Reserve Program of the U.S. Farm Bill) and reseeded to perennial grasses; thus, they receive little or no enrichment from agricultural fertilizers. Consequently, converting cultivated cropland to perennial vegetation within restored wetlands should reduce nutrient enrichment in wetlands and lower emissions of N_2O and possibly CH_4.

In short, agricultural operations both hasten the release of C emissions from dewatered and farmed wetlands, and from wetland receiving nutrient enrichments in runoff waters from cropped lands that receive fertilizers. Losses of soil organic matter, burning off the historic C contained in organic matter, and emissions of potent GHGs are well-documented problems worldwide.

FACTORS INFLUENCING SEQUESTRATION OF CARBON IN WETLANDS

1. *Ceasing Artificial Drainage of Wetlands and Hydric Soils*: Dewatering historic wetland soils and exposing deeper soil strata to the decomposition process can be avoided by only farming historic wetlands soils when they seasonally dry down, and confining the agricultural uses to only those areas that dry down adequately to support agricultural uses. Farming these wetlands will release the stored C they accumulated in prior wet periods when they were not farmed. In addition, unless protected with grassy buffers they have the potential to emit large quantities of nitrous oxide from their margins and methane from the anoxic portions of the flooded basin (Merbach et al., 2002). For GHG management, these systems may fare best if not farmed.
2. *Reduced Mechanical Disturbance of Soils*: In coordination, and separate from the act of artificially draining historic wetland (i.e., hydric soils) by reducing the mechanical disturbance of these seasonally drained wetland soils through annual tillage (e.g., disking, plowing, and rototilling), existing C stocks found in the soil can be best protected from decomposition and the release of C to the atmosphere. The use of no-till seeding techniques may, for example, allow for the continued use of seasonally dry wetland soils while minimizing C loss.
3. *Restoring Hydrology of Wetlands*: Restoring degraded wetlands, particularly wetland hydrology, may be the most effective way to influence C sequestration in wetlands. In fact, ongoing studies indicate a general reduction in GHG emissions from most wetland restorations (Merbach et al., 2002) and suggest wetland restoration to be a most important strategy for enhancing C sequestration in wetlands (Euliss et al., 2006).

   Once hydrology is restored, the water saturated and waterlogged environment reduces the decomposition potentials and rates annually as plant matter dies back. A very rapid accumulation of plant matter (sequestered C) can build up and over a decade or so, in the least impacted currently drained wetlands, significant levels of lost C may be replaceable by hydrological restoration.

4. *Buffering Wetlands from Fertilizer and Nutrients*: The use of wetland buffers can be utilized to mitigate sediment and nutrient import into wetlands and to curtail the subsequent wetland enrichment process and the release of trace gas emission. The buffer idea can be extended into the upland drainage areas to retain topsoil, reduce runoff to wetlands in high gradient areas, and thus reduce sediment and nutrient import into wetlands.
5. *Restoring the Biodiversity of Wetlands*: Basic ecology courses teach the interdependency of plants and animals, nutrients, water, and all life, in healthy ecological systems. Relationships between biotic and hydrogeochemical attributes of wetlands have long existed in healthy wetlands that contribute especially to a reduced emissions of trace GHGs. The soil bacteria and fungi and some plants have the capacity to utilize nutrients so effectively that they historically were less vulnerable to release as trace gas emissions than at present. Disturbances including nutrient enrichment from farm fertilizers entering wetlands overwhelm these important organisms and their efficiency at removing available nitrogen declines. In general, this decline can be associated with an increase in nitrous oxides and methane gas and other traces gases that contain nitrogen and other excess nutrients.

PROGRAMS AND INCENTIVES FOR WETLAND RESTORATION IN THE UNITED STATES

The Solution(s)

Conclusive evidence documents that an unprecedented opportunity exists for strategically restoring wetlands to return the lost and declining functions, and to help halt the existing problematic trends in flooding, water quality deterioration, and wildlife decline (NRC, 2002, 2004). Moreover, these investigations suggest clearly that one of the lowest costs and biggest bang for the dollar strategies is to restore wetlands.

The United States has programs such as the Wetland Reserve Program of the NRCS, which is the U.S. Fish and Wildlife Services program for wetland protection and restoration. Supporting the numerous private organizations (e.g., Ducks Unlimited, The Nature Conservancy, Prairie Enthusiasts, LandKeeper's, and farm organizations) in this endeavor can have the largest and immediate benefits for reducing global C loading.

Recommendations

1. Develop a National Strategic Vision and Implementation Plan for Watershed and Wetland Restoration and focus financial and technical expertise from complimentary programs to undertake and implement such a pro-

gram. The plan should map, watershed by watershed the wetland restoration opportunities and develop a desired general performance-based watershed condition model that can then be applied to each watershed to determine the acreage of wetland restoration that would be targeted to achieve the performance outcomes that are desired.
2. Create an Agricultural Carbon Management Reserve program with agricultural land management (see Chapter 5), wetland restoration and management, urban landscape management (see Chapter 13), and grassland and forest management (see Chapter 14) as cornerstones. This Reserve Program will need a substantial vision and marketing presence across all federal agencies, state agencies, and will need to be focused in creating real changes in the land that result in wetland restoration and other C program benefits.
3. Encourage and incentivize landowners through new and creative strategies, to undertake wetland restorations on their land. Perhaps establish, in association with the Agricultural Carbon Management Reserve Program, a fundamental Carbon Utility. This would be similar to a potable water supply utility, whereby all landowners who provide wetland restoration on their lands are given annual payments for supplying the C management services to the public benefit and at much lower cost than, for instance, an industrial strategy to remove C from the atmosphere.
4. Require, in urban areas and other areas where wetlands are subject to development pressure, the continued use and expansion of wetland mitigation banking as the primary compensation tool used by the Army Corps of Engineers and U.S. Environmental Protection Agency in their processing of Clean Water Act, section 404, regulatory permits. Wetland banks are restorations that are constructed to compensate for the losses of small often isolated wetlands directly through filling or dredging and indirectly effectively through hydrological and water quality changes and sedimentation from development activities. Wetland banks are constructed for the public good in strategic locations where they will be most successful. Privately constructed wetland banks typically become public parks and open spaces as well. This ensures their protection and use for the public good.
5. Establish a program to unite the various existing government farm, water, and air quality programs together into a synthesis to store C, enhance biodiversity, and improve water quality. New market-based opportunities could help achieve these outcomes through C markets. A unified program could be essential to develop standards to allow wetlands to enter the marketplace.

CONCLUSIONS

More than most other nations on earth, the United States has taken proactive steps to protect existing wetlands. However, although these efforts are still under way, the United States has experienced a 95% loss of wetlands on hundreds of millions of acres, many of which have been converted for agricultural land uses, among other uses. It is both the converted wetlands, which are annually cropped and the soils being depleted annually through land tillage that is a primary global C and GHG emitter. In addition, the existing wetlands that occur within agricultural landscapes where they receive nutrient enrichment from fertilizers and the nitrogen and phosphorous released from other drained and farmed wetlands that are also contributors of C and potential trace elements.

To curb the release of these emissions, the nation needs to recognize the C and GHG storage potential of historic wetlands and encourage their restoration, and to buffer against nutrient and fertilizer influxes into the existing wetlands where trace gas emissions are occurring. In addition, converting the farm program subsidies to a "carbon farming" and "water quality and quantity farming" enterprise may be a better long term strategy than focusing on crop failures and subsidies that encourage farming soils that are delivering crops and commodities at great "downside" costs to the global atmosphere. In other words, if every dollar spent in subsidy of crop failures in former wetland soils that encourages those farming historic wetlands soils and contributing to a very expensive clean-up cost by elevating the global GHG levels were reduced, and the money spent in wetland restoration instead, the public would realize at least two benefits: reduced GHG emissions and more predictable pricing on commodities that could be grown in upland soil systems.

REFERENCES

Boylan, K.D., and MacLean, D.R., Linking species loss with wetland loss, Environmental Law Institute, *National Wetlands Newsletter*, 19, 12–17, 1997.

Dahl, T.E., *Wetland Losses in the United States, 1780s to 1980s*, U.S. Department of Interior, Fish and Wildlife Service, Washington, D.C., 1990.

Davidson E.A, Keller, M., Erickson, H.E., Verchot, L.V., and Veldkamp, E., Testing a conceptual model of soil emissions of nitrous and nitric oxides, *Bioscience*, 50, 667–680, 2000.

Demissie, M., Khan, A., and Mubarak, A., Influence of wetlands on stream flow in Illinois, *National Conference on Hydraulic Engineering*, Nashville, TN, 949–954, 1991.

Dugan, P., *Wetlands in Danger*, Michael Beasley, Reed International Books, London, 1993.

Eswaran, H., Van den Berg, E., and Reich, P., Organic carbon in soils of the world, *Soil Sci. Soc. Am. J.*, 57, 192–194, 1993.

Euliss, N.H. Jr., and Laubhan, M.K., Quantifying the environmental benefits of the conservation reserve program on prairie wetlands: separating acts of nature from acts of Congress, in *The Conservation Reserve Program — Planting for the Future,* A.W. Allen and M.W. Vandever, Eds., U.S. Geological Survey, Biological Resources Division, Biological Science Report, 2005, 11–15.

Euliss, N.H. Jr., Gleason, R.A., Olness, A., McDougal, R.L., Murkin, H.R., Robarts, R.D., Bourbonniere, R.A., and Warner, B.G., North American prairie wetlands are important non-forested carbon storage sites, *Science of the Total Environment*, 2006.

Gleason, R.A., and Euliss, N.H. Jr., Sedimentation of prairie wetlands, *Great Plains Res.*, 8, 97–112, 1998.

Heinz Report, *The State of the Nation's Ecosystems: Measuring the Lands, Waters, and Living Resources of the United States,* The H. John Heinz III Center for Science, Economics and the Environment, Cambridge University Press, Cambridge, 2002.

Hutchin, P.R., Press, M.C., Lee, J.A., and Ashenden, T.W., Methane emission rates from an ombrotrophic mire show marked seasonality which is independent of nitrogen supply and soil temperature, *Atmos. Environ.*, 30, 3011–3015, 1996.

Intergovernmental Panel on Climate Change (IPCC), Houghton, J.T., Meira Filho, L.G., Callander, B.A., Harris, N., Kattenberg, A., and Maskell, K., Eds., *Climate Change the Science of Climate Change*, Cambridge University Press, Cambridge, 1996.

Kimble, J., Follett, R., and Lal, R., Eds., *Agricultural Practices and Polices for Carbon Sequestration in Soil*, Lewis Publishers, Boca Raton, FL, 2002.

Knutsen, G.A. and Euliss, Jr., N.H., Wetland restoration in the prairie pothole region of North America: a literature review, *Biological Science Report, USGS/BRD/BSR--2001-0006*, U.S. Geological Survey, Biological Resources Division, 54 pp., 2001.

Lal, R., Kimble, J.M., Follett, R.F., and Cole, C.V., *The Potential of U.S. Cropland to Sequester Carbon and Mitigate the Greenhouse Effect*, Ann Arbor Press, Chelsea, MI, 1999.

Merbach, W., Kalettka, T., Rudat, C., and Augustin, J., Trace gas emissions from riparian areas of small eutrophic inland waters in Northeast-Germany, in *Wetlands of Central Europe, Soil Organisms, Soil Ecological Processes, and Trace Gas Emissions*, Broll, G., Merbach, W., and Pfeiffer, E.-M., Eds., Springer, Berlin, 2002, 235–244.

Miller, M.W. and Nudds, T.D., Prairie landscape change and flooding in the Mississippi River Valley, *Conservation Biology*, 10, 847–853, 1996.

Mitsch, W.J. and Gosselink, J.G., *Wetlands*, 3rd ed., Van Nostrand Reinhold, New York, 2000.

Mitsch, W.J., Mitsch, R.H., and Turner, R.E., Wetlands of the old and new worlds — ecology and management, in *Global Wetlands: Old World and New*, Mitsch, W.J., Ed., Elsevier, Amsterdam, 1994, 3–56.

Pant, H.K., Rechcigl, J.E., and Adjei, M.B., Carbon sequestration in wetlands: concept and estimation, *Food, Agric. Environ.*, 1, 308–313, 2003.

National Audubon Society, 2004.

National Research Council (NRC), The Missouri River Ecosystem: Exploring the Prospects for Recovery, Committee on Missouri River Ecosystem Science, 188 pp., 2002.

NRC, Review of the U.S. Army Corps of Engineers Restructured Upper Mississippi River-Illinois Waterway Feasibility Study, Committee to Review the Corps of Engineers Restructured Upper Mississippi River-Illinois Waterway Feasibility Study, National Research Council, 66pp., 2004.

Tegtmeier, E.M. and Duffy, M.D., External costs of agricultural production in the United States, *Int. J. Agric. Sustainability*, 2, 1–20, 2004.

Thornton, F.C. and Valente, R.J., Soil emissions of nitrous oxide from no-till corn, *Soil Sci. Soc. Am. J.*, 60, 1127–1133, 1996.

U.S. Environmental Protection Agency (US EPA), *America's Wetlands: Our Vital Link Between Land and Water*, Report No. EPA843-K-95-001, Office of Water, Office of Wetlands, Oceans and Watersheds, Washington, D.C., 1995.

US EPA, *Inventory of U.S. Greenhouse Gas Emissions and Sinks: 1990–2001*, US EPA, Washington, D.C., 596 pp., 2003.

U.S. Fish and Wildlife Service (USFWS), Available at www.fws.gov.

U.S. Geological Survey (USGS), Land subsidence in the United States, in *USGS Fact Sheet 165-00*, USGS, 6 pp., 2000.

Wang, Z., Dong, Z., and Patrick, W.H. Jr., Methane emissions from natural wetlands, *Environ. Monit. Assess.*, 42, 143–161, 1996.

Whittaker, R.H. and Likens, G.E., Primary production: the biosphere and man, *Hum. Ecol.*, 1, 357–369, 1973.

10 Wildlife Benefits

N. H. Euliss, Jr., M. K. Laubhan, and R. A. Gleason

CONTENTS

OVERVIEW

Farmers, ranchers, and other private landowners manage two-thirds of the land base within the United States; therefore, it is essential to involve these individuals in strategies to increase carbon (C) storage. Similarly, the long-term sustainability of our nation's biological diversity also requires proper stewardship of private lands (Gilbert and Dodds, 1992). These endeavors are not mutually exclusive given that proper C management can also improve wildlife habitat quality. Wildlife provides both marketable goods and recreation value in excess of $125 billion dollars annually within the United States (U.S. Department of Interior, U.S. Fish and Wildlife Service and U.S. Department of Commerce, and U.S. Census Bureau, 2002). To date, some existing Farm Bill programs have been instrumental in providing these benefits simultaneously, helping to conserve our nation's natural resources. Given the diverse environmental concerns of modern society, future Farm Bill programs may need to be targeted at providing a full range of ecosystem goods and services.

Despite the large societal value attributed to wildlife, many populations continue to decline. The response of society has varied, but environmental and conservation goals, including those related to wildlife, have increasingly become key factors in the formulation of the U.S. Department of Agriculture (USDA) policies that influence management of private land. Most experts agree that conserving wildlife resources over the long-term will require federal, state, and private conservation entities to cooperate with private landowners to increase and preserve habitat on agricultural lands. The most successful and influential legislation affecting the quantity and quality of wildlife habitat on private lands is the Farm Bill. For example, wetlands and grasslands restored as part of the Conservation Reserve and Wetland

Reserve Programs are recognized for their importance to wildlife, but also have been recognized recently for their importance in the sequestration (storage) of atmospheric carbon (CO_2-C) that has the potential to reduce atmospheric concentrations of these greenhouse gases. Thus, existing Farm Bill programs provide multiple ecosystem services that are not mutually exclusive, suggesting that proper management of agricultural lands to sequester C is also conducive to enhancing wildlife habitat.

INTRODUCTION

Environmental and conservation goals have increasingly become key factors in the formulation of USDA policies. Many of the programs administered by the USDA have been designed to have direct impacts on specific ecosystem processes, including the conservation of soils, improvement of water quality, and increasing the quality and quantity of wildlife habitat. In addition to these direct benefits, many indirect benefits also have been realized. For example, wetlands and grasslands restored as part of the Conservation Reserve Program (CRP) and Wetland Reserve Program (WRP) as established in the Farm Bill have recently been recognized for their importance in the sequestration of CO_2-C in soils (Lal et al., 1998). Thus, it should be no surprise that existing Farm Bill programs and policies have the potential to influence emerging environmental concerns. For example, the ability of agricultural lands to sequester C has recently been recognized as important because of concerns of global climate change. Less obvious, but not mutually exclusive, are the positive influences that sequestering C can exert on wildlife populations.

Economic Importance of Wildlife

The diverse vegetation communities or habitats in the United States have historically supported an abundant and diverse wildlife community that pioneers and other early Americans relied on for food and clothing. Following settlement of the United States, wildlife products, such as meat and skins, played an important role in daily life. However, overexploitation resulted in the extirpation or reduced populations of many species, which was further exacerbated by habitat loss as human populations expanded, and the United States developed a larger agricultural sector. Although human dependence on wildlife for daily subsistence has declined, society still depends directly and indirectly on this resource. For example, myriad wildlife species provide marketable goods that are commercially important; the U.S. commercial fisheries harvest in 1991 was valued at more than $3.3 billion annually and formed the basis of a $26.8 billion fishery processing and sales industry (US EPA, 1995). Further, recreational pursuits that involve wildlife continue to be important with more than 82 million U.S. residents spending $108 billion dollars annually to fish, hunt, or observe wildlife (U.S. Department of Interior, U.S. Fish and Wildlife Service, U.S. Department of Commerce, and U.S. Census Bureau, 2002; Cordell and Herbert, 2002). Additionally, the indirect benefits provided by wildlife and their associated habitats also are important to society because the biological, chemical, and physical processes that function to sustain ecosystems provide important services to humans, including improvement of water quality, reduction in flood risk, and enhancement of C sequestration (NAS, 2004).

Future of Wildlife

Despite the recognized economic value of wildlife, many populations continue to decline as further habitat loss and degradation occurs to support human population growth. The widespread loss of wildlife habitat due to the conversion of native lands to agriculture and urban areas has been well documented. Also well documented, but lesser known, are the modifications that these changes have wrought on remaining wildlife habitats (NAS, 2004). For example, more than 90% of the nation's streams, groundwater, stream and estuarine sediments, and freshwater fish have at least one contaminant at detectable levels (Heinz Report, 2002); the primary source of contaminants is excess sediment, nutrients, and pathogens (US EPA, 2000). Land uses that alter fundamental processes (e.g., erosion, nutrient cycling) are leading contributors of degradation. In combination, the loss and modification of natural landscapes has had devastating effects on the diversity and abundance of the nation's wildlife resources. For example, 389 animals are listed as endangered and 129 animals are listed as threatened in the United States (U.S. Fish and Wildlife Service, 2007), and, excluding marine species, about a third of all plants and animals whose status is known are considered "at risk" (Heinz Report, 2002).

The response of society to declining wildlife populations has varied depending on many factors, including the types of ecosystem services deemed important and the potential economic returns of land uses other than wildlife habitat. Initially, the U.S. strategy to conserve wildlife focused on purchasing land holdings to be managed specifically for wildlife. However, less than 6% of the coterminous United States is in nature reserves and many are located at higher elevations on less productive soils that support fewer wildlife species (Scott et al., 2001). Preliminary assessments indicate that more than 90% of threatened and endangered species occur on private lands (U.S. Government Accounting Office, 1995; Groves et al., 2000). It is now recognized that land held in ownership primarily for wildlife is being increasingly impacted by surrounding land uses that, in many cases, can negatively impact habitat quality by altering erosion or sedimentation rates, hydrology, and nutrient concentrations. For example, the estimated damage from agricultural production to wildlife and ecosystem biodiversity in the United States exceeds $320 million annually (Tegtmeier and Duffy, 2004). To some extent, this has prompted the development of federal, state, and local regulations that are intended to protect, conserve, and restore wildlife populations. Although these actions have helped sustain wildlife populations, most experts agree that conserving wildlife resources over the long-term also will require cooperating with private landowners to implement management activities to increase the area and quality of wildlife habitat.

ROLE OF THE FARM BILL

Numerous government programs have been established to provide incentives to manage privately owned lands for the benefit of wildlife. It was not until the 1985

Farm Bill, however, that increasing concerns about the environmental effects of agricultural activities resulted in the enactment of a conservation title that brought about a merging of commodity support policy and resource conservation policy (Congressional Research Service, 1996). Even with this legislation, however, activities on private lands originally were implemented to benefit only one or a small suite of environmental services and it was not until later that ancillary benefits were recognized. Perhaps the most recent recognition of ancillary benefits of the Farm Bill is the continued positive relationship between C sequestration and improvement of wildlife habitat. Increases in the atmospheric concentration of CO_2 from a pre-industrial level of 280 parts per million (ppm) to a current level of 375 ppm have resulted in growing concern regarding methods to effectively manage C. In addition to using fossil fuels more efficiently and increasing the use of low-carbon and carbon-free fuels and technologies, the newest method of lowering atmospheric concentrations of greenhouse gases is sequestering C by maintaining or enhancing natural processes (U.S. Department of Energy, 2004). This latter method has great potential on private lands because activities already encouraged by current Farm Bill programs to enhance other ecosystem services, including wildlife, also have great potential to store C. Representative examples include programs designed to increase the quantity (e.g., CRP, WRP, Grassland Reserve Program, and the Healthy Forests Reserve Program) and quality (e.g., Environmental Quality Incentives Program, Wildlife Habitat Incentives Program, Conservation Security Program, and Conservation Technical Assistance) of wildlife habitats. These programs are sufficiently diverse to provide incentives to private landowners, regardless of location and the vegetation biome (e.g., grassland, wetland, and forest) being considered. For example, CRP enrollment exceeds 34 million acres and includes lands in all 50 states and Puerto Rico (Allen and Vandever, 2003).

The value of these programs for wildlife has been extensively documented in the scientific literature with benefits encompassing multiple species in all major wildlife taxa (Hall and Willig, 1994; Heard et al., 2000). From a biological perspective, these programs provide direct benefits by increasing the area of suitable habitat and diversifying the foods and cover available for wildlife to meet annual life cycle requirements. These programs also provide indirect wildlife benefits (e.g., reduced exposure to agrichemicals) and simultaneously can improve soil C stores. For example, cropping practices, such as no-till and minimum tillage, improve soil tilth while simultaneously reducing surface runoff and exports of sediments, nutrients, and agrichemicals. Consequently, agricultural lands become more productive and sustainable, while the quality of critical off-site wildlife habitats, such as wetlands, rivers, and estuaries, are maintained or improved.

The total economic value of all direct and indirect benefits is difficult to calculate. However, simulation models suggest that termination of CRP contracts in 2001 would have caused spending on outdoor recreation to decrease as much as $300 million per year in rural areas (Sullivan et al., 2004). In addition, the economic value of C sequestration is yet to be realized, but it has been estimated that 5.14 million metric tons of C have been sequestered on 5.6 million ha of CRP lands in a 13-state region in the Central United States (Follett et al., 2001). Finally, economic benefits to our nation's waters are likely to be significant. For example, over $63 billion per year is spent in the United States to control water pollution (US EPA, 1995). Sediment is the number one pollutant in the United States and the greatest source of sediment is erosion of agricultural lands (Robinson, 1971; Wayland, 1993). Consequently, the implementation of agricultural practices that reduce surface runoff would represent a significant and positive step to mitigate habitat degradation and help ensure the sustainability of the nation's priceless natural ecosystems and the renewable wildlife populations they sustain. For example, specific USDA conservation practices reduce runoff from large blocks of actively farmed land (e.g., no-till and various residue management practices), and other practices target specific areas within agricultural fields susceptible to soil loss (e.g., Conservation Cover, Filter Strips, Grass Waterways). In addition, USDA has a number of programs (e.g., CRP, WRP, Conservation Reserve Enhancement Program [CREP], Environmental Quality Incentives Program [EQIP]) that retire highly erodable or environmentally sensitive lands from production agriculture to perennial grassland for extended periods.

CONCLUSIONS

Given the current climate, future Farm Bill programs may place more emphasis on C sequestration benefits. However, doing so will require a broad ecological approach to optimize ancillary benefits for all ecological goods and services, including wildlife. Although it is possible to develop programs that specifically target C storage on agricultural lands, the challenge will be to implement practices that do not simultaneously compromise the value of other ecosystem goods and services.

REFERENCES

Allen, A.W. and Vandever, M.W., A national survey of Conservation Reserve Program (CRP) participants on environmental effects, wildlife issues, and vegetation management on program lands, *Biological Science Report, USGS/BRD/BSR-2003-0001*, 51 pp., 2003.

Congressional Research Service, *Farm Bill Issues: Overview*, June 6, 1996. Available at: http://cnie.org/NLE/CRSreports?agriculture/ag-8.cfm. Accessed on 01/11/07.

Cordell, H.K. and Herbert, N.G., The popularity of birding is growing, *Birding*, 34, 54–61, 2002.

Follett, R.F., Pruessner, E.G., Samson-Liebig, S.E., Kimble, J.M., and Waltman, S.W., Carbon sequestration under the Conservation Reserve Program in the historic grassland soils of the United States of America, *Soil Science Society America J.*, 57, 27–40, 2001.

Gilbert, F.F. and Dodds, D.G., *The Philosophy and Practice of Wildlife Management*, 2nd ed., Krieger Publishing Company, Malabar, FL, 1992.

Groves, C.R., Kutner, R.L.S., Stoms, D.M., Murray, M.P., Scott, J.M., Schafale, M., Weakly, A.S., and Pressey, R.L., Owning up to our responsibilities: who owns land important for biodiversity, in *Precious Heritage: The Status of Biodiversity of the United States*, Steine, B., Kutner, L.S., and Adams, J.S., Eds., Oxford University Press, New York, 2000, 275–300.

Hall, D.L. and Willig, M.R., Mammalian species composition, diversity, and succession in Conservation Reserve Program grasslands, *Southwestern Naturalist*, 39, 1–10, 1994.

Heard, L.P., Allen, A.W., Best, L.B., Berger, W., Lesser, A.J., Hackett, E., Johnson, D.H., Pederson, R.L., Reynolds, R.E., Rewa, C., Ryan, M.R., Molleur, R.T., and Buck, P., A comprehensive review of Farm Bill contributions to wildlife conservation, 1985–2000, *Wildlife Habitat Management Institute Technical Report, USDA/NRCS/WMI-2000*, U.S. Department of Agriculture, Natural Resources Conservation Service, Washington, D.C., 208 pp., 2000.

Heinz Report, *The State of the Nation's Ecosystems: Measuring the Lands, Waters, and Living Resources of the United States,* Heinz, John H. III, Ed., Center for Science, Economics, and the Environment, Cambridge University Press, Cambridge, 288 pp., 2002.

Lal, R., Kimble, J.M., Follett, R.F., and Stewart, B.A., Eds., *Management of Carbon Sequestration in Soil*, CRC Press, Boca Raton, FL, 1998.

National Academy of Sciences, *Valuing Ecosystem Services: Toward Better Environmental Decision Making,* The National Academies Press, Washington, D.C., 2004.

Robinson, A.R., Sediment, *J. Soil Water Conserv.*, 26, 61–62, 1971.

Scott, J.M., Davis, F.W., McGhie, R.G., Wright, R.G., Groves, C., and Estes, J., Nature reserves: do they capture the full range of America's biological diversity?, *Ecological Appl.*, 11, 999–1007, 2001.

Sullivan, P., Hellerstein, D., Hansen, L., Johansson, R., Koenig, S., Lubowski, R., McBride, W., McGranahan, D., Roberts, M., Vogel, S., and Bucholtz, S., The Conservation Reserve Program: Economic Implications for Rural America, U.S. Department of Agriculture, Economic Research Service, Agricultural Economic Report 834, Washington, D.C. 112 pp., 2004.

Tegtmeier, E.M. and Duffy, M.D., External costs of agricultural production in the United States, *Int. J. Agric. Sustain.*, 2, 1–20, 2004.

U.S. Department of Energy. Carbon sequestration. Available at http://cdiac2.esd.ornl.gov. Accessed on 01/11/07.

U.S. Department of Interior, U.S. Fish and Wildlife Service, U.S. Department of Commerce, and U.S. Census Bureau, 2001 National Survey of Fishing, Hunting, and Wildlife-Associated Recreation, U.S. Government Printing Office, Washington, D.C., 2002.

U.S. EPA, National water quality inventory: 1994 report to Congress, *U.S. Environmental Protection Agency, EPA841-R-95-005*, Washington, D.C., 1995.

U.S. EPA, National water quality inventory: 2000 report to Congress, *U.S. Environmental Protection Agency, EPA841-R-02-001*, Washington, D.C., 2000.

U.S. Fish and Wildlife Service, USFWS threatened and endangered species system (TESS), January 9, 2007. Available at http://ecos.fws.gov/tess_public/SpeciesCountForm.do. Accessed on 01/07.

U.S. Government Accounting Office, Species protection on nonfederal lands, Government Accounting Office, Washington, D.C., 1995.

Wayland, R., What progress in improving water quality? *J. Soil Water Conserv.*, 48, 261–266, 1993.

11 Soil Carbon and the Mitigation of the Risks of Flooding

*Andrew P. Manale**

CONTENTS

INTRODUCTION

What does soil carbon have in common with Hurricane Katrina, the monster 2005 hurricane that flooded New Orleans, the Great Upper Mississippi Flood of 1993, and the Red River Flood of 1997? For each of these extreme weather and flooding events, substantial investments in soil carbon (C), and thus soil organic matter in upland and coastal soils, could have saved the public both trouble and money.

Floods cause billions of dollars in damages in the United States every year. Between 1965 and 2003, over two-fifths of all presidential disaster declarations involved floods. A preponderance of these declarations occurred in a minute fraction of the nation's watersheds and counties (Figure 11.1). We cannot yet prevent extreme weather events from occurring, but we can mitigate their social and environmental impacts by increasing the C content of agricultural soils.

Policies at all levels of government must recognize the importance of managing land, especially agricultural and wetlands, to maintain and restore soil C. The benefits are many and extend well beyond temporary water storage for flood mitigation to water quality, improved wildlife habitat, and air quality. The savings from averted damages from floods or excessive runoff or for outlays for expensive structural

* The views expressed in this chapter are those of the author and do not necessarily reflect the policy of the U.S. Environmental Protection Agency or the federal government.

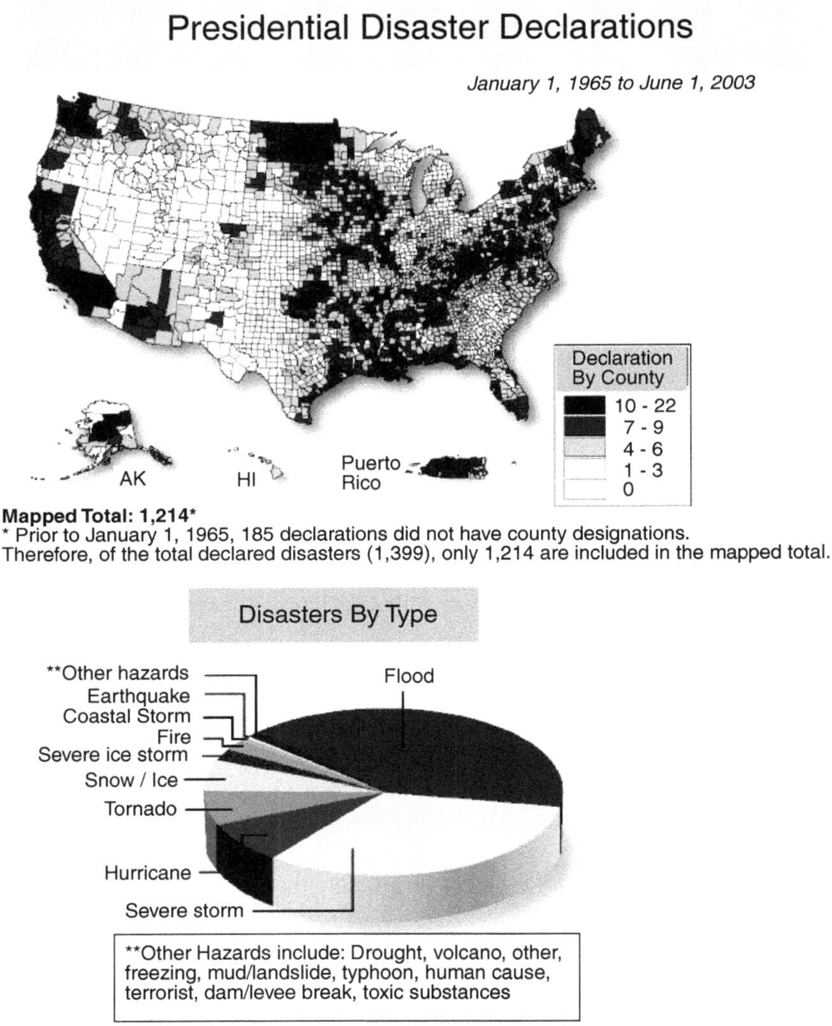

FIGURE 11.1 Presidential disaster declarations and floods. (Source: FEMA.)

measures compensate for the cost of practices that contribute to restoration of soil C. Consideration of the other tangible, let alone intangible benefits of soil C that continue far into the future, will show benefits that far exceed the upfront cost of these public investments in security against uncertain catastrophic events. The technologies and systems for protecting and enhancing soil C are known and feasible. What is needed are the policies that lead to their adoption.

HOW DOES SOIL C AFFECT RUNOFF?

Soil C increases the moisture content of soils (Rawls et al., 2003; Huntington, 2003), improves permeability and water infiltration, and reduces compaction, thereby decreasing the amount of water available to leave the land. Less available water means less runoff or stormwater that can cause downstream or lowland flooding. Carbon in the form of highly organic and vegetation of coastal wetlands attenuates storm surge (Mitch and Gosselink, 2000). Less water that comes ashore means less water that can inundate low-lying inland areas.

Organic matter acts like a sponge, absorbing up to six times its weight in water (UMN Extension Service, 2002; Olness and Archer, 2005). Functioning like glue that causes soil to clump and form aggregates, soil C improves soil structure. Better soil structure means greater permeability. With higher organic matter, a greater percentage of the precipitation that falls on the land partitions into the ground or is retained by the soil. There is less runoff that leaves the land contributing to downstream flow. Greater water-holding capacity also means more water is available to growing plants. The U.S. Global Climate Change Research Program estimates that enhancing the percentage of SOM from current baseline levels of 1–2%, can, in some circumstances, increase the amount of water that soils retain four- to sixfold (USGCRP, 2001).

LANDSCAPE MODIFICATION AND THE LOSS OF SOIL C

Humans have extensively modified the natural landscape, often causing large losses of soil C. Land in areas with recurrent flooding often has lost its natural tendency to slow down, store, or dissipate runoff from extreme weather events—an important ecosystem service (natural amenity that benefits society). The consequence has been amplified social and environmental impacts of extreme weather events on communities and the environment (IFMRC, 1994). A decline in the natural resilience of the land and its ability to recover from extreme events can lead to greater (i.e., more costly) future impacts to society. Should global warming continue and extreme weather events occur more frequently (whether or not anthropogenic activities are the cause), the social cost of flooding can be expected to increase absent actions to mitigate or otherwise reduce their impact.

Scientists and authorities concerned with flooding and its associated social, environmental, and economic costs have long known that managing landscapes, particularly agricultural, and implementing certain agricultural practices that retard water flow or retain water on the land contribute to effective flood management (Leopold and Maddock, 1954; IFMRC, 1994). The role that organic matter and, in particular, soil C plays in this process has only recently been scientifically elucidated, however (Patriquin, 2006). Watersheds containing large portions of land with C-rich soils, such as wetlands, have the lowest flood risk, higher quality water resources, and higher levels of biodiversity (NRC, 2005).

Up until the 1990s, the clay content of soils was believed to be the key component influencing its capacity to store water, largely through role in soil texture. Olness and Archer (2005) explain that soil C was believed to contribute to only a minor extent. They point out that more recent research, especially on the relationship between soils and greenhouse gas emissions and C sequestration, has led to the shift to the current scientific consensus on the major role played by soil C and thus soil organic matter. It is no wonder that organic matter now plays a central role in the U.S. Department of Agriculture's recently developed Soil Quality Index (USDA-NRCS, 2006).

RED RIVER EXAMPLE OF THE RELATIONSHIP BETWEEN SOIL C AND FLOODING

Recurrent flooding in the Red River Basin of North Dakota and Minnesota illustrates how soil C can mitigate flood risks. The communities in the Red River Basin of North Dakota, Minnesota, and Canada suffer from frequent floods. The spring flood of 1997 caused an estimated $5 billion in damages and left the City of Grand Forks inundated for weeks (IJC, 2000).

Most land in the Red River Basin is devoted to agriculture. From the air, the landscape resembles a waffle with a north-south, east-west crisscrossing of elevated roads that are often paralleled by drainage ditches. Over 95% of the wetlands have been drained. Half the preagricultural levels of soil C, the component that is integral to soil structure and water retention, let alone the productivity of soils, has been lost (Blann et al., 2004).

On average, a flood event occurs every 5 years in the Red River Basin. Indeed, a 95% chance exists that a flood will occur within 12 years of the last flood event. In recent years, disastrous floods have occurred with greater frequency and nearly all have historically occurred in the early spring. To prevent a 1997-magnitude event, Wang et al. (2003) estimate that 1.3 million acre-feet (1.6 billion cubic meters) of floodwater would have to be stored in the upland areas of the basin.

Managing soils in the Red River Basin, and other primarily agricultural watersheds prone to flooding to increase soil C, and thus the soil organic matter (SOM),* would retain enough flood water to mitigate downstream flooding. How much so depends upon the nature of the soils and their location within the watershed. The following calculation illustrates this relationship between soil C and flood mitigation.

For the loamy, relatively poorly drained soils of the Red River Basin, a 1% increase in SOM would, generally upon the condition of the soils and other factors, such as the clay content, result in a 2.5% increase in the volume of water retained

* Soil C constitutes 57% of SOM by weight.

TABLE 11.1
Soil C and Water Storage

Additional Soil Organic Matter	Additional Soil C	Cubic Meters of Water Per Hectare (acre-ft per acre)
1%	0.57%	37.5 (0.03)
2%	1.14%	75 (0.06)
3%	1.71%	112.5 (0.09)

in the soil (Saxton, 2005). This additional organic matter per hectare (2.47 acres) of soil measured to a depth of 15 centimeters (roughly 6 inches) translates into 37.5 cubic meters of water (see Table 11.1).

Roughly 14,803,260 acres (6 million hectares) of cultivated soils are available in the North Dakota and Minnesota portions of the watershed. Increasing the soil organic content of roughly half these acres by 1%, as listed in Table 11.2, could meet 7% of the water storage requirement.* Affecting the change on all of the cultivated land could meet 14%. Increasing the soil organic content by 2% in absolute terms (soil C by 1.14%) over 5 million hectares could fulfill over 23% of the flood storage requirement† In combination with aboveground temporary storage practices on agricultural lands, even greater mitigation can be achieved (Manale, 2006).

How much would it cost to mitigate flood risks through managing soil C? To increase the soil organic matter content by 1% (0.57% C) would require the addition of roughly 12.5 metric tons of C per hectare. Continuous no-till cropping systems can sequester C at a rate of 500 kg or more per hectare/year (Lal et al., 1998). At this rate, this 1% increase can be achieved in roughly 25 years, depending upon the weather and other pertinent variables. Even though no-till farming has been demonstrated to save farmers money in the long run, a risk premium would still be necessary to be paid to farmers to achieve the necessary rate of adoption.

Kurkalova and his colleagues at Iowa State University estimate that an annual payment of $11.12 per hectare ($4.50 per acre) in 2002 dollars would achieve a 90% rate of adoption of no-till. Applying this figure for the per hectare payment necessary to achieve adoption of no-till cropping in the Red River Basin, for this 25-year period, the cost would amount to $130 per hectare, discounted for the future value of money and assuming a 7% discount rate.

* A study by the Minnesota Department of Natural Resources estimated the impact of crop residue management on downstream flooding in the Red River Basin from the adoption of no-till and other soil conservation practices that increase the amount of soil organic matter would result in a 10% reduction in peak flood flows and volumes, if employed throughout a subwatershed (Solstad, 1998).

† Having the capacity to store sufficient floodwater to have prevented a flood of a certain volume is not synonymous with having the potential to prevent the flood damages associated with an event of a certain volume. Neither rain nor snow are likely to fall evenly across the land within a watershed. Thus, a detailed time-to-travel analysis of floodwater flows within would be required to determine how much water would have to be stored at any specific location within the watershed. Thus, although on average a given percent increase, in absolute terms, would be sufficient to meet the storage requirements, some lands would need greater than the average percent increase, whereas others would require less.

TABLE 11.2
Cost of Increased Soil Organic Matter (SOM) for Floodwater Storage

Hectares of Flood Storage	Percent of Red River Basin Storage Needs	Cubic Meters of Flood Storage	Total Cost[a]
Additional 1% Soil Organic Matter (over a 25-year period)			
3,000,000	7%	112,500,000	$390,000,000
5,000,000	12%	187,500,000	$650,000,000
Additional 2% Organic Matter (over a 50-year period)			
3,000,000	14%	225,000,000	$459,000,000
5,000,000	23%	375,000,000	$765,000,000

[a] Differences in the 25-year versus 50-year cost to achieve equivalent volumes of water storage are explained by the time-value of money.

As indicated in Table 11.2, meeting 7% of the flood storage need to prevent a flood as occurred in 1997 would cost $390 million and require a 1% increase in soil organic matter on 3 million hectares. Adding 2% organic matter over a period of 50 years on 5 million hectares would cost $765 million and potentially meet 23% of the temporary water storage requirement.

HURRICANE KATRINA EXAMPLE OF THE ROLE OF SOIL C IN MITIGATING FLOOD SURGES

A substantial fraction of the estimated $200+ billion cost (CRS, 2005) of Hurricane Katrina to New Orleans and the neighboring coastal development could have been avoided had there been investments in the form of C in the restoration of coastal wetlands (Kousky and Zeckhauser, 2006). Over one million acres of coastal wetlands have been lost since the 1930s (Coalition 2000). Three to twenty-six foot storm surges (NWS, 2006; USACE, 1963) ranged along the Gulf Coast and overwhelmed the dyke and levee system that protected New Orleans. A mile of restored coastal wetland could have reduced the storm surge by as much as 3.1 inches (10 centimeters) and at least one inch (2.6 cm) (Louisiana Coastal, 2000). At a rough cost of restoration of $10,000 per acre ($24,700 per hectare) to restore (Gulf of Mexico Alliance, 2005), a million acres of wetlands or roughly 1,563 square miles (4000 square km) would amount to $10 billion. Assuming 100 miles (161 km) of coastline to protect, an almost 16 mile (26km) wide wetland buffer could provide as little as four and as much as sixteen feet (1.2 to 4.9 m) of surge protection.

The one-time cost is significant, but the financial returns continue to accrue long into the future. Costanza, Farber, and Maxwell (1989) estimate the short-term annual benefits of coastal wetlands in the range of over $200 million to almost $900 million. Four thousand square kilometers of coastal wetlands also sequester roughly 100,000 metric tons of C per year (Chmura et al., 2003), thus offsetting greenhouse gas emissions of fossil fuels and reducing the cost of mitigating climate change.

Reducing the ecological impact of hurricanes has significance for energy and global warming policy as well. A single storm can convert 10% of the total annual U.S. C sequestration to dead and downed forest biomass that also becomes an economic cost to eliminate (Uriarte, 2005). The nation can spend billions of dollars each year on voluntary C sequestration projects. Yet, the greenhouse gas offset benefits can be undone in an instant by a major hurricane coming ashore where the coastal protection provided by wetlands have been impaired or lost.

Slowing water down and holding it longer provides not just flood protection but also environmental benefits. Excessive runoff, as occurs during infrequent, intense storm events, exaggerates stream flow that can degrade stream banks and cause the loss of sediment. When precipitation or irrigation exceeds soil infiltration, the runoff water can contain nitrogen (N) and phosphorus (P) from the pool of residual nutrients in soils, thus contributing to the contamination of surface water (Keeney, 1983) and degrade water quality (NRC, 1993). Runoff from snowmelt and spring rains can also contain high concentrations of N due to the leaching of plant residues (White and Williamson, 1973). Sediment and nutrients are major contaminants of water quality and source of impairment of rivers, estuaries, and lakes (Manale, 2000; US EPA, 2006). Sediments in waterways cause many billions of dollars in damages each year (Clark, Haverkamp, and Chapman, 1985).

Reducing runoff by increasing SOM leads to less erosion and loss of nutrients. By literally slowing down the movement of water from the land, soil C reduces the potential of water to erode soil and deliver sediment and nutrients to streams and rivers. Increasing SOM from one to 3% can reduce soil erosion by 20–33% through greater water infiltration and a more stable soil (Funderberg, 2001).

Managing agricultural landscapes to maintain and restore soil C provides multiple benefits to both the environment and to society, including temporary water storage for flood mitigation. Linking landscape-scale management for soil C and flood mitigation can serve as an entry point for a multitude of other conservation amenities for which the tangible social or private benefits are not so measurable. In many areas of the country, particularly agricultural watersheds prone to repeated flooding, the cost of practices that contribute to restoration of soil C can, in large measure, be compensated through the savings from reduced flood damage and other outlays associated with excessive runoff. Wildlife benefits, nutrient recycling, and improved quality of the water released from agricultural land are all additional ecosystem services potentially enabled through landscape-scale conservation of soils, creating greater social return.

REFERENCES

Blann, K., Webb, T., Keeney, D., and Light, S., Alternatives to crisis: an adaptive management model for the Red River Basin of the U.S. and Canada, in *The Role of Biodiversity Conservation in the Transition to Rural Sustainability*, NATO Science Series: Science and Technology Policy, 41, Light, S.S., Series Ed., 2004.

Chmura, G., Anisfeld, S., Cahoon, D., and Lynch, J.C., Global carbon sequestration in tidal, saline wetland soils, *Global Biogeochemical Cycles*, American Geophysical Union, Washington, D.C., 2003.

Clark, E.H., Haverkamp, H.J.A., and Chapman, W., *Eroding Soils: The Off-Farm Impacts*, The Conservation Foundation, Washington, D.C., 1985.

Coalition to Restore Coastal Louisiana (Louisiana Coastal), *No Time to Lose: Facing the Future of Louisiana and the Crisis of Coastal Land Loss*, 2000. Available at http://www.crcl.org/no_time_to_lose.pdf. Accessed on August 12, 2006.

Congressional Research Service (CRS), Hurricane Katrina: Insurance Losses and National Capacity for Financing Disaster Risk, CRS Report for Congress, prepared by Rawle O. King, 2005. Available at: www.crs.gov. Accessed on August 14, 2006.

Costanza, R., Farber, S.C., and Marshall, J., Valuation and management of wetland ecosystems, *Ecological Econ.*, 1, 335–361, 1989.

Funderberg, E., *What Does Organic Matter Do in Soil?* The Samuel Roberts Noble Foundation, 2001. Available at: www.noble.org/Ag/Soils/OrganicMatter/Index.htm. Accessed on February 10, 2006.

Gulf of Mexico Alliance, *White Paper: Restoration of Coastal Wetlands/Estuarine Ecosystems*, Version 8, 2005. Available at: www.dep.state.fl.edu/gulf/coastalwetlands_louisiana.pdf. Accessed on August 22, 2006.

Huntington, T.G., Available Water Capacity and Soil Organic Matter, *Encyclopedia of Soil Science*, Lal, R., Ed., 1-5, doi: 10.1081/E-ESS120018496, Marcel Dekker, New York, 2003.

Interagency Floodplain Management Review Committee (IFMRC), *Sharing the Challenge: Floodplain Management into the 21st Century*, The President's Floodplain Management Task Force Assembled for the 1993 Midwest Flood, Washington, D.C., 1994.

International Joint Commission: Canada and the United States, *Living with the Red*, published by the International Red River Basin Board of the International Joint Commission, 2000. Available at: www.ijc.org/phhph/publications/html/living.html. Accessed on August 16, 2006.

Keeney, D., Transformation and Transport of Nitrogen, in *Agricultural Management and Water Quality*, Schaller, F., and Bailey, G.W., Eds., Iowa State University Press, Ames, Iowa, 1983.

Kousky, C. and Zeckhauser, R., JARring actions that fuel the floods, in *On Risk and Disaster: Lessons from Hurricane Katrina*, Daniels, R.J., Kettl, D.F., and Kunreuter, H., Eds., University of Pennsylvania Press, Philadelphia, 2006.

Kurkalova, L., Kling, C., and Zhao, J., *Green Subsidies in Agriculture: Estimating the Adoption Costs of Conservation Tillage from Observed Behavior, Working Paper 01-WP-286*, April, Center for Agricultural and Rural Development, Iowa State University, 2003. Available at: www.card.iastate.edu. Accessed on August 15, 2006.

Lal, R., Kimble, J.M., Follett, R.F., and Cole, C.V., *The Potential of U.S. Cropland to Sequester Carbon and Mitigate the Greenhouse Effect*, Ann Arbor Press, Chelsea, MI, 1998.

Leopold, L. and Maddock, T. Jr., *The Flood Control Controversy: Big Dams, Little Dams, and Land Management*, The Ronald Press Company, New York, 1954.

Manale, A., Flood and water quality management through targeted, temporary restoration of landscape functions: paying upland farmers to control runoff, *J. Soil.Water Conserv.*, 55(3), 2000.

Manale, A., Waffles are not just for breakfast anymore: the economics of mitigating flood risks in the Red River basin, *J. Soil Water Conserv.*, 2006.

Mitsch, W.J. and Gosselink, J.G., *Wetlands*, 3rd ed., Van Nostrand Reinhold, New York, 2000.

National Research Council (NRC), *Soil and Water Quality: An Agenda for Agriculture*, The National Academies Press, Washington, D.C., 1993.

NRC, *Valuing Ecosystem Services: Toward Better Environmental Decision Making*, The National Academies Press, Washington, D.C., 2005.

National Weather Service (NWS) Forecast Office New Orleans/Baton Rouge, Louisiana, *Post Tropical Cyclone Report for Hurricane Katrina,* 2006. Available at: http://www.srh.noaa.gov/lix/html/psh_. Accessed on January 9, 2006.

Olness, A. and Archer, D., Effect of organic carbon on available water in soil, *Soil Science*, 170, 90–101, 2005.

Patriquin, D., Water, soil and organic matter: a complex relationship, *Canadian Organic Growers*, 2006. Available at: www.cog.ca. Accessed on January 19, 2006.

Rawls, W.J., Pachepsky, Y.A., Ritchie, J.C., Sobecki, T.M., and Bloodworth, H., Effects of soil organic carbon on soil water retention, *Geoderma*, 116, 61–76, 2003.

Saxton, K., *Soil, Plant, Atmosphere, Water, Field, and Plant Hydrology (SPAW)*, 2005. Available at: http://hydrolab.arsusda.gov/SPAW/Index.htm. Accessed on September 6, 2006.

Solstad, J., *Working Paper Number 6: Watershed Modeling of Various Flood Damage Reduction Strategies*, 1998. Available at: www.rrbdin.org/do/communication/listFiles?d-49520-p=6. Accessed on March 3, 2006.

University of Minnesota (UMN) Extension Service, *Organic Matter Management*, 2002. Available at: www.extension.umn.edu/distribution/cropsystems/components/7402_02.html. Accessed on February 10, 2006.

Uriarte, M., Potential ecological effects of increased severity and frequency of tropical storms, 2005. Available at: www.earthinstitute.columbia.edu/crosscuttig/cciseminars/2005fall/documents/Uriarte_UnderstandingKatrinaforWeb.pdf. Accessed on September 11, 2006.

U.S. Army Corps of Engineers (USACE), Overlands surge elevations coastal Louisiana: Morgan City and Vicinity, File No H-2-22758 plate a-4, 1963. Available at: http://dnr.louisiana.gov/crm/background/. Accessed on August 17, 2006.

U.S. Department of Agriculture Natural Resources Conservation Service (USDA-NRCS), *Soil Quality Concepts*, 2006. Available at: www.soils/usda.gov/sqi/concepts/index.html. Updated on August 8, 2006. Accessed on October 10, 2006.

U.S. Environmental Protection Agency (US EPA), 2000 National Water Quality Inventory, 2006. Available at: www.epa.gov/305b/2000report. Updated on March 22, 2006. Accessed on August 10, 2006.

U.S. Global Change Research Program (USGCRP), *Climate Change Impacts on the United States: The Potential Consequences of Climate Variability and Change Overview*, Great Plains, 2001. Available at: www.usgcrp.gov/usgcrp/Library/nationalassessment/overviewgreatplains.htm#Conservation%20of%20Soil%20Organic%20Matter. Updated on August 20, 2004. Accessed on August 29, 2006.

Wang, X., Groenewold, G.H., Steadman, E.N., Bolles, B.A., Peck, W.D., de Silva, L.L., and Dokken, H.M., *A Coupled Hydrologic—Hydraulic Model for Flood Reduction Analysis in the Red River of the North Basin*, Energy and Environmental Research Center, University of North Dakota, Grand Forks, ND, 2003. Available at http://www.undeerc.org/publications/. Accessed on February 10, 2005.

White, E.M. and Williamson, E.J., Plant nutrient concentrations in runoff from fertilized cultivated erosion plots and prairie in eastern South Dakota, *J. Environ. Qual.*, 2, 453–455, 1973. As reported in Keeney (1983).

12 Environmental and Ecological Benefits of Soil Carbon Management: Surface Water Quality

Gary Pierzynski, Dan Devlin, and Jeff Neel

CONTENTS

INTRODUCTION

For purposes of this chapter, soil carbon (C) management refers to practices that maintain or enhance soil organic C concentrations. The intent is to demonstrate the potential benefits of soil C management on surface water quality. Practices that can enhance soil C concentrations include reducing tillage intensity and increasing biomass production through intensified crop rotations, fertilization, and crop selection. These practices increase C loading to the soil through increased biomass production, decrease the rate of C decomposition in the soil, retain C at the soil surface, or some combination of these mechanisms. Many of these practices are already in use, primarily as a means of reducing soil erosion on soils that are prone to erosion, and soil C management would provide another reason to promote their use on a greater proportion of the landscape.

Soil C management differs from some other atmospheric C mitigation strategies in that it is a *no-regrets approach* to the issue. In this instance, a no-regrets approach is one in which benefits are realized whether or not an anthropogenic connection to global climate change occurs. The primary benefits of soil C management, in addition to slowing the increase in atmospheric CO_2 concentrations, are decreased soil erosion, improved surface water quality, and improved soil physical properties. In contrast, practices such as pumping CO_2 into geologic formations, another form of C sequestration, have no ancillary benefits. In fact, resources utilized for such activities would be wasted if there were no anthropogenic connection to global climate change.

SURFACE WATER QUALITY CONCERNS IN THE UNITED STATES

Water is one of our most precious natural resources. We all depend on water for drinking and household use. In addition, we depend upon water for agriculture, industry, and recreation. Concerns exist in the United States about the quality of our surface and groundwater.

A report (305(b) by the U.S. Environmental Protection Agency (EPA)) in 2000 of surface water conditions in the United States indicated that 39% of assessed rivers and streams (miles), 45% of assessed lakes (acres), and 51% of assessed estuaries (square miles) were considered to be polluted.

The leading causes and sources of impairments in rivers and streams, lakes, and estuaries are provided in Table 12.1. Agriculture is listed as a major source of impairments for rivers and streams and lakes, ponds, and reservoirs. Improved land management practices, such as reduced or no-tillage, which can lead to increased soil C, can reduce soil erosion and soil water runoff and lead to lower nutrient and sediment losses to surface waters.

SURFACE RUNOFF LOSSES OF POLLUTANTS

To appreciate how soil C management can improve water quality, it is helpful to have an understanding of the processes involved in movement of potential pollutants from the landscape to surface waters via surface runoff. When the rate of water addition to the soil exceeds the infiltration rate of the soil, surface runoff occurs. Water can be added to a given soil location via precipitation, irrigation, or from surface runoff generated upslope. Interestingly, the layer of soil that interacts with surface runoff, called the mixing zone, is relatively shallow and typically is only

TABLE 12.1
Water Quality Concerns in the United States

Causes/ Sources	Rivers and Streams	Lakes, Ponds, and Reservoirs	Estuaries
Causes	Pathogens (bacteria) Siltation (sedimentation) Habitat alterations	Nutrients Metals (primarily mercury) Siltation (sedimentation)	Metals (primarily mercury) Pesticides Oxygen-depleting substances
Sources	Agriculture Hydrologic modifications Habitat modifications	Agriculture Hydrologic modifications Urban runoff/storm sewers	Municipal point sources Urban runoff/storm sewers Industrial discharges

1–5 cm in depth. Thus, the chemical and physical characteristics of the mixing zone are critical in determining both the amount of surface runoff and the chemical characteristics of that runoff. The physical properties of the soil underlying the mixing zone are also important, primarily because of the influence on the overall rate of infiltration that determines the fraction of incoming water that becomes surface runoff. In addition, the infiltration rate for a given soil location will vary with time. Factors such as the antecedent moisture content when precipitation begins, recent tillage, or the formation of surface crusts, will greatly influence the infiltration rate.

Most of the research on surface runoff has utilized measurements of "edge of field" or "edge of plot" characteristics of surface runoff. These studies generally collect runoff from relatively small areas (1–10,000 m^2) and make relative comparisons in runoff characteristics between various treatments used on the small plots. Such studies are very useful for estimating the potential impact of management practices on surface water quality, but are difficult to extrapolate to the larger watershed scale. One can collect surface runoff from larger areas but then it is difficult to determine how the various landscape characteristics and management practices within that area influence surface runoff characteristics. Models can also be used to try to estimate the influence of landscape characteristics and management practices on surface runoff characteristics at the watershed scale, and this will be discussed in more detail later in this chapter.

For a variety of reasons, the concentrations of nutrients, pesticides, sediment, and bacteria measured in surface runoff collected edge of field are typically much higher than those measured in nearby surface water bodies. These reasons can be placed into two general categories: (1) processes that influence surface runoff composition between the edge of field and when it arrives at the surface water body, and (2) processes that occur within the surface water body. Runoff leaving the edge of field can be diluted or enriched in constituents by runoff from other areas and additional filtering of constituents can occur in vegetative buffer and riparian areas. Decomposition of organic constituents and die-off of microbial constituents generally occurs over time and can further reduce concentrations.

Within the surface water body, additional dilution, enrichment, decomposition, and die-off of constituents can occur and eroded soil can settle out of the water and become part of the sediment load. The discrepancy between concentrations of various constituents measured in surface runoff and within surface water bodies makes it difficult to interpret edge of field pollutant concentrations or losses because the net impact on nearby surface waters is unknown. For this reason, critical concentrations or losses (e.g., unacceptable or acceptable) for edge of field runoff cannot be established with any degree of certainty and are not generally used. Still, comparisons of runoff characteristics from edge of field measurements provide useful information on the relative efficacy of various management practices. Critical concentrations can be established for the surface water body itself and serve to indicate whether improvements in surface runoff characteristics within the watershed are needed.

The edge of field perspective can best be used to illustrate the two primary factors that influence the potential net impact of surface runoff on surface water quality. The factors are runoff volume and runoff composition, and soil C management can influence both. The product of runoff volume and runoff composition, with the appropriate units, determines the loss of a constituent (mass per area) from the landscape. A decrease in runoff volume or in the concentration of a constituent in surface runoff will reduce the loss of that constituent from the landscape. Some management practices will increase one factor and decrease the other, in which case the net effect may be an increase or decrease in the loss of that constituent from the landscape.

The following example illustrates how edge of field measurements are converted to a common way of expressing the loss of a constituent in units of kg/ha. Assume that a 0.5 ha plot generates 62,500 L of runoff from a storm event. This is equivalent to 0.5 inches of runoff, which is a reasonable amount in a large storm event (> 1 inch total precipitation). The runoff has 20-mg/L total phosphorus (P). We have multiplied the total volume of runoff from the area by the total P concentration in that runoff and applied some unit conversions to get the appropriate units for loss of P per unit area (ha = hectare).

$$\frac{62500\ \text{L}}{0.5\ \text{ha}} \times \frac{20\ \text{mg total P}}{\text{L}} \times \frac{1\ \text{kg}}{10^6\ \text{mg}} = \frac{1.25\ \text{kg total P}}{\text{ha}}$$

It can easily be demonstrated that doubling the runoff volume or the total P concentration would double the total P loss. Similarly, if runoff volume is doubled and the total P concentration is cut in half the total P loss would be unchanged.

Figure 12.1 presents a schematic of a hypothetical watershed illustrating the various types of contributing areas and the various factors that can influence contaminant loads and concentrations, and potential contaminant impacts, between the edge of field and surface water bodies. In addition, note the concept of the mixing zone, which is the shallow layer of soil at the surface that interacts with surface runoff.

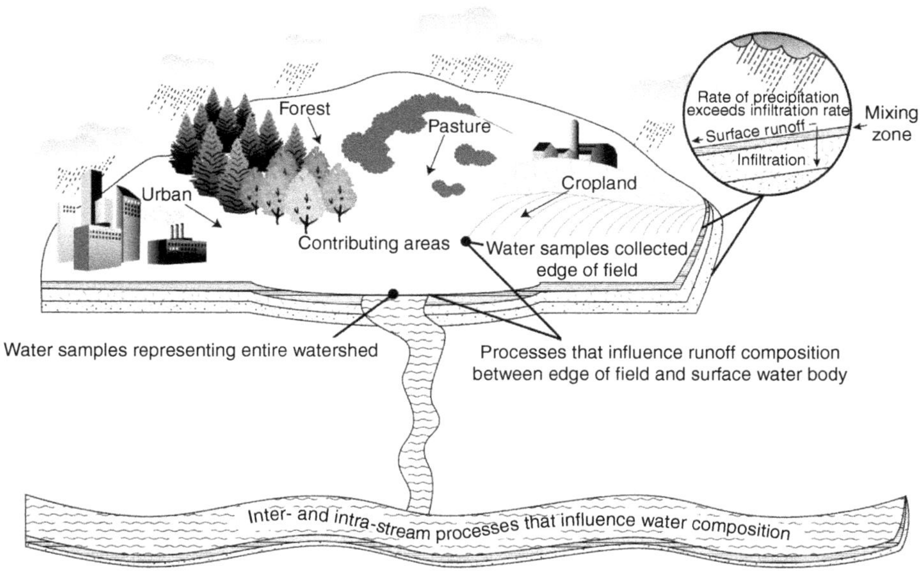

FIGURE 12.1 Hypothetical watershed illustrating the various types of contributing areas and the various factors that can influence contaminant concentrations and loads, as well as potential impacts, between the edge of field and surface water bodies.

RUNOFF VOLUME

Runoff volume is influenced by many factors (Table 12.2). One would assume that humans have little control over climatic or topographic factors; however, in fact, practices such as irrigation, land leveling, or the installation of terraces for erosion control can effectively influence these parameters and therefore runoff volume. Complex interactions exist between all these factors, and it is beyond the scope of this chapter to illustrate and explain them all. One example relevant to this chapter would be tillage intensity. As tillage intensity is reduced, organic C content at the soil surface would increase over time, and this change could reduce bulk density, increase average soil moisture content, increase aggregation, and decrease crusting. Of greatest interest in this chapter are the factors that influence or are influenced by efforts to increase soil organic C content. These factors are in indicated in bold in Table 12.2.

One of the primary practices for increasing soil organic C content is reducing tillage intensity. Tillage practices can range from efforts to produce a seedbed that is entirely free of any crop residues (bare soil) to the complete absence of traditional tillage operations, with the only soil disturbance being the one that is required to plant the crop. The former practice is generally called clean till or conventional tillage, whereas the latter practice is usually described as no-till. The decision to reduce tillage intensity involves the consideration of many factors including the

TABLE 12.2
Climatic, Topographic, and Soil Characteristics that Influence Surface Runoff Volume

Category[a]	Comments
Climatic	
Precipitation/irrigation intensity	Runoff volume increases as intensity increases
Precipitation/irrigation amount	Runoff volume increases as precipitation amount increases during a single storm event
Topographic	
Slope	Runoff volume increases as steepness and length of slope increase
Tillage practices	Runoff volume may increase, decrease, or be unchanged as tillage intensity decreases
Soil characteristics	
Soil chemical properties	
Organic C content	Runoff volume decreases as organic C content increases
Salinity/sodicity	High sodium levels can increase runoff volume
Soil physical properties	
Soil texture	Runoff volume decreases as the sand content increases
Soil moisture content	Runoff volume increases as the antecedent moisture content increases
Crusting	Surface crusting increases runoff volume
Aggregation	Aggregation decreases runoff volume
Bulk density	Runoff volume increases as bulk density increases

[a] Categories in bold influence or are influenced by soil C content or C additions to the soil.

crops to be grown, soil characteristics, climate, pest control, and economic factors (e.g., labor and fuel costs). The discussion in this chapter will be limited to the potential water quality benefits of reducing tillage intensity.

Runoff volume can increase, decrease, or remain unchanged by changing from clean till practices to no-till on cropland. Evaluation of the influence of tillage on runoff volume must be done carefully. If tillage occurred recently, clean till areas will generally have lower runoff volume than no-till because freshly tilled soil has a very high infiltration rate. As the growing season progresses this relationship may reverse because clean-till areas can become more compact and form surface crusts, thus increasing runoff volume relative to no-till. An annual assessment of runoff volume would be needed to determine the net effect of tillage on runoff volume. However, for contaminants such as pesticides, nutrients, and bacteria, runoff volume shortly after the time of application is critical because the concentrations of these constituents can decrease rapidly with time after application regardless of runoff volume relationships over the remainder of the year.

Increasing the organic C content of a given soil will generally reduce runoff volume. The positive effect of soil organic C on soil physical properties and the high water retention capacity of the organic C increases the infiltration rate and therefore decreases runoff volume. Data in Figure 12.2 indicate that cattle manure applications decrease runoff volume immediately after application and 6 months

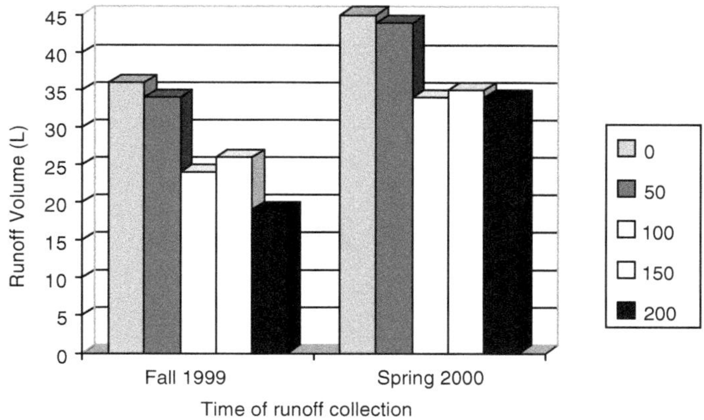

FIGURE 12.2 Runoff volume immediately after and 6 months after incorporated applications of cattle manure. The units of application are Mg/ha. (From Ethridge, K., Influence of Land Application of Cattle Manure on Phosphorus Losses in Runoff, M.S. thesis, Kansas State University, Manhattan, KS, 2002. With permission.)

after application. The manure was incorporated into the soil with a disk. Treatments not receiving manure were also disked so that tillage was uniform across the study. The decrease in volume was attributed to the increase in soil organic C induced by the manure applications.

Dry soil has a higher infiltration rate than wet soil. Therefore, antecedent moisture content has an influence on runoff volume for each runoff event. Dry soils will produce less runoff than wet soils for a given precipitation event. Soil organic C allows soils to retain more water and, on average, soils with higher organic C contents may have higher antecedent moisture contents as compared with similar soils with low organic C contents.

Increasing soil organic C will increase the likelihood that the soil will have favorable soil structure. Thus, aggregation will increase with increasing soil organic C content and the tendency to form surface crusts will decrease. Both factors will work to reduce runoff volume. High soil organic C content also reduces the soil bulk density and makes the soil more resistant to compaction. Again, both factors work to reduce runoff volume.

RUNOFF COMPOSITION

The typical analysis of a water sample involves the determination of the sediment concentration, total and soluble forms of nutrients, pesticide concentrations, and the concentrations of fecal coliform and *E. coli* bacteria. Sediment concentrations are determined by simply filtering a known volume of water and weighing the sediment captured on the filter after drying. Total nutrients are determined after an unfiltered water sample is digested in strong acids, whereas soluble nutrients represent the nutrients that are present in a filtered water sample. Pesticides are generally extracted from filtered water samples using various solid phase absorbents. For bacteria,

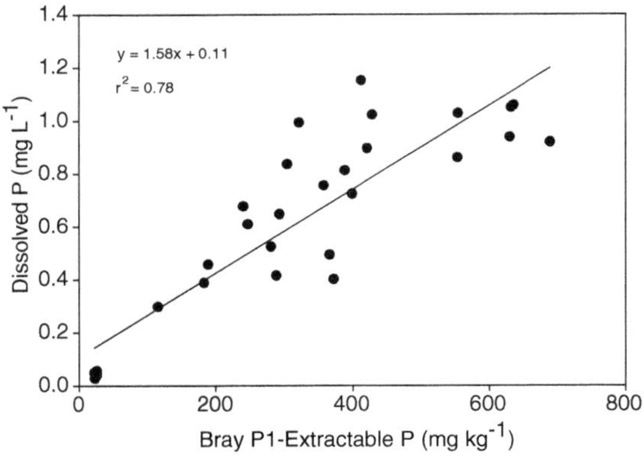

FIGURE 12.3 The relationship between Bray P1-extractable P in manure amended soil and dissolved P in runoff. (From Ethridge, K., *Influence of Land Application of Cattle Manure on Phosphorus Losses in Runoff*, M.S. thesis, Kansas State University, Manhattan, KS, 2002. With permission.)

various dilutions of unfiltered water are plated on growth media and the number of colony forming units determined by visual inspection.

The primary factors that influence runoff composition are the chemical and physical properties of the mixing zone including the concentrations or presence of potential water contaminants such as nutrients, pesticides, and bacteria. Logic would suggest that there may be a relationship between soil properties in general, and properties of the mixing zone in particular, and runoff generated from such areas. This relationship has best been established with P where soil extractable P concentrations are often highly correlated with soluble P in runoff (Figure 12.3). Phosphorus extraction procedures that are commonly used to assess soil fertility status are adequate for predicting soluble P in runoff in some circumstances, so the development of separate tests has not been necessary. The other issue that arises in this situation is the depth of soil sampling. In routine testing for soil fertility, a common sampling depth is 15 cm. This depth is much greater than the mixing zone yet most studies suggest that soil samples taken from 15 cm are as well or nearly as well correlated with runoff composition as those taken from the mixing zone (approximately 5 cm), regardless of the tillage system. This is particularly true when tillage operations mix nutrients in the surface soil horizon. Again, this is fortunate because land managers can gain considerable information from soil samples collected in the traditional fashion.

The concept of relating soil characteristics, particularly for P, to runoff composition has led to the use of threshold values for soil extractable P as a means of estimating the risk of P loss from the landscape. The primary limitation to this approach is that it, in theory, requires an edge of field interpretation of runoff composition because one must match soil composition to a critical P concentration in runoff. The limitation in establishing critical runoff concentrations was discussed earlier. However, this limitation can be overcome somewhat by using reasonable safety factors and accepting that critical soil concentrations do not necessarily relate directly to a critical runoff composition, but instead to a P concentration that is likely to be sufficiently low.

As suggested previously, tillage practices can affect the distribution of nutrients and C in the surface soil horizon. In general, as tillage intensity decreases the stratification of nutrients and C will increase with the resulting distribution having the highest concentrations at the soil surface where the nutrients can readily interact with runoff water. Indeed, many studies have reported that no-till areas will have higher runoff concentrations, and possibly losses, of soluble forms of nutrients as compared with more intensive tillage systems, particularly when nutrients are broadcast on the soil surface without incorporation. However, no-till areas will generally have much lower runoff concentrations and losses of total nutrients because of reduced sediment and the associated nutrients.

When manures or other by-products are surface broadcast on the soil without subsequent incorporation, runoff water interacts directly with these materials as part of the mixing zone and, therefore, the properties of the materials themselves can greatly influence runoff composition. Under this circumstance, analysis of the manure (e.g., water extractable P) may be more useful for predicting potential contaminant losses in surface runoff than an analysis of the soil.

The potentially strong relationship between the composition of the mixing zone soil and the concentrations of contaminants in runoff suggests that incorporating or injecting fertilizers and manures or other by-products upon application may reduce runoff concentrations of nutrients and bacteria because the materials are no longer in the mixing zone. In a conventional tillage system with little crop residue at the soil surface, this is true. It is also true for pastures and no-till systems but the soil disturbance may increase the potential for sediment loss.

The one physical property of the mixing zone that can influence runoff composition is the presence of plant residues or living plant material at the soil surface. These act to absorb the kinetic energy of falling raindrops and reduce the potential for the detachment of soil particles that begins the soil erosion process. Consequently, sediment losses decrease as the amount of plant residue or vegetative cover at the soil surface increases.

CARBON MANAGEMENT EFFECTS ON SURFACE WATER QUALITY — EDGE OF FIELD

Long-term changes in soil physical properties can be expected when using reduced or no-tillage compared with conventional tillage. Generally, increases in soil organic C, surface soil porosity, soil macropores, and soil surface infiltration rates occur as fields are managed so that tillage is reduced or eliminated. Soil C can be increased or conserved in reduced or no-tillage systems. This occurs as: (1) reducing tillage reduces soil organic matter oxidation and leads to increases or conservation of soil C, and (2) reducing tillage reduces soil erosion.

Reduced or no-tillage can result in greater formation and stability of macropores in the soil. Macropores are small channels in the soil caused by earthworms, soil cracking, or root growth. Tillage destroys macropores. Increased macropores allow increased water infiltration reducing the rate and amount of water running off the field to surface water.

Some studies have found that no-till and other conservation tillage systems reduce water runoff volume compared with other more intensive tillage systems (Blevins et al., 1990; Seta et al., 1993). Other studies comparing runoff volume and tillage find greater runoff volume with no-till than with other tillage systems (Lindstrom and Onstad, 1984; Gaynor and Fidlay, 1995). These conflicting results may be due to differences in soils, rainfall amounts and intensities, slope, and antecedent moisture conditions. Contaminants can leave the field in both water and the soil particles portion of runoff. In general, reducing tillage reduces soil erosion, but pesticide and nutrient runoff may or may not be reduced.

SUSPENDED SOLIDS

Suspended solids (soil particles) increase the cloudiness of water, reduce light penetration, which adversely affects photosynthesis, and can lead to reduced levels of dissolved oxygen and food supplies for aquatic organisms. This may cause shifts in the populations and species of aquatic organisms present in a water body, and can increase the cost of treatment for drinking water. Nutrients and pesticides may also be attached to soil particles running off fields. The nutrients, especially P, attached to soil particles may lead to algal blooms in lakes. Many lakes are being filled by sediments, which reduce the effective water storage capacity.

Numerous studies have demonstrated that soil erosion and sediment yields are decreased when moving from conventional tillage to a reduced tillage or no-till system (Andraski, Mueller, and Daniel et al., 1985; Chichester and Richardson, 1992; Gaynor and Fidlay, 1995). For example, Baker (1987) found that no-till reduced erosion by 50 to 90% when compared with conventional tillage.

PESTICIDES

Pesticides may be moved by surface runoff into streams and lakes. Pesticides may also leach into groundwater. Changing to a no-till system can influence the amount of herbicide runoff to surface water.

The chemical characteristics of a pesticide determine whether that pesticide is more likely to move off a field in the sediment or the water portion of runoff. One of the major chemical characteristics is adsorption. Adsorption is a term that describes a chemical's tendency to bind or stick to soil particles, primarily clay or organic matter. Some pesticides are strongly adsorbed, whereas others, such as atrazine, are weakly adsorbed. Weakly adsorbed pesticides are more likely to leave the field in the water and not with eroding soil particles, whereas strongly adsorbed pesticides are more likely to be lost from the field with eroding soil particles than with the water.

Reduced and no-tillage will reduce the loss of pesticides that are attached to soil particles by reducing soil erosion. However, no-till or reduced tillage may have limited value in reducing the movement of weakly adsorbed pesticides from the field. The potential for movement of weakly adsorbed pesticides is generally directly related to the amount of water runoff from the field, and to the placement of the pesticide. Weakly adsorbed pesticide runoff can only be reduced by management practices that reduce runoff volume as well as sediment loss (Wauchope, 1978). Studies examining the effect of different tillage practices on herbicide loss in surface runoff have had mixed results. Some studies found greater losses of herbicides in surface runoff from conservation tillage systems than from conventional tillage systems, due to greater water runoff volumes (Baker and Johnson, 1979; Baker, Laflen, and Johnson et al., 1978; Isensee and Sadeghi, 1993; Gaynor, MacTavish, and Fidlay et al., 1995; Rector et al., 2003). However, other studies found reduced pesticides losses in runoff for conservation tillage practices due to reduced sediment and runoff volume (Felsot et al., 1990; Glen and Angle, 1987; Hall, Mumma, and Watts et al., 1991; Witt and Sander, 1990). Gaynor, MacTavish, and Fidlay et al. (1995) found no differences in herbicide runoff losses due to tillage system when considering both surface runoff and tile drainage.

NUTRIENTS

Movement of P and nitrogen (N) into surface water and leaching of N to groundwater are serious environmental concerns. Changing tillage practices affects the amount of nutrient losses.

Phosphorus can move into surface waters with soil particles either during erosion or as soluble P in runoff water. Generally, most P leaving fields is associated with or bound to soil particles (Siemens and Oschwald, 1976). Much less is soluble P. Therefore, soil erosion has a direct influence on total P losses in runoff. Studies have reported that total P losses decrease with no-till practices, compared with conventional tillage practices (Cox and Hendricks, 2000; Andraski, Mueller, and Daniel et al., 1985, Sharpley and Smith, 1994).

Soluble P losses are generally found to be higher from reduced till or no-till fields (Barisas et al., 1978; Kimmell et al., 2001). This was attributed to leaching of nutrients from plant residue and decreased fertilizer incorporation. However, others have reported soluble P losses to be reduced from reduced tillage or no-till fields (Andraski, Mueller, and Daniel et al, 1985; Chichester and Richardson, 1992; Sharpley and Smith, 1994).

Nitrogen, in the nitrate form, readily moves with soil water and can move downward as water moves through the soil profile. Eventually, the water and nitrate may leach far enough to enter groundwater or tile lines and be an environmental concern. In the past, there was some concern that no-till may result in greater nitrate leaching because of an increase in macropores under no-till. However, research with various soil types comparing tillage systems and nitrate leaching has not found higher nitrate leaching with no-till. Kanwar and Baker (1993) found fewer nitrates for no-till compared with plowed soils to a 1.5-m depth, and Eisenhauer et al. (1993) found similar results for deeper depths. Kelly and Blevins (1993) also reported less nitrate movement downward in no-till than in plowed soils. Karlen et al. (1994) demonstrated higher nitrate concentrations in the upper 30-cm of no-till soils, but lower nitrate concentrations below 30 cm. However, others reported no consistent effects of tillage on nitrate leaching (Kanwar and Colvin, et al., 1995; Lamb et al., 1998).

COST AND EFFECTIVENESS OF BEST MANAGEMENT PRACTICES

The effectiveness of no-till and reduced (conservation) tillage practices in reducing N, P, atrazine herbicide, and suspended solids runoff to surface waters has been estimated considering farm data, professional estimates, and the results of numerous field, laboratory, and computer modeling studies on the effect of crop management practices (Table 12.3). The typical costs and effectiveness of these practices are presented for situations of conventional tillage and moving to no-till and reduced (conservation) tillage. In all cases, the reported water quality benefits are from the adoption of a single new best management practice (BMP). Unfortunately, the effects are not cumulative if multiple BMPs are used.

A reported BMP cost is the expected reduction in producer profits associated with adopting that BMP on a single hectare. The data on reduction of runoff by adopting a BMP are relative to an assumed base practice for corn and grain sorghum fields where atrazine herbicide is applied preemergence (i.e., herbicide broadcast, surface applied following crop planting, but before crop emergence); where P and N fertilizer is broadcast applied before planting the crop and unincorporated; conventional tillage (i.e., less than 30% residue cover following planting); and greater than 1% slope on upland clay or clay loam soil. For wheat and other crops, the comparison benchmark is P and N fertilizer broadcast applied, unincorporated, with conventional tillage, and greater than 1% slope on upland or clay loam soil.

TABLE 12.3
The Estimated Typical Cost and Runoff Reducing Effectiveness of Improved Crop Rotations and Reduced and No-Till Practices for Reducing Atrazine Herbicide, P, N, and Suspended Solids in Runoff from Crop Fields

Best Management Practice	Cost/Acre ($)	Atrazine Herbicide	Nutrients: Soluble P	Nutrients: Total P	Nutrients: N	Suspended Solids
		(% Reduction in Runoff by Adopting BMP)				
Crop rotations	0	30	25	25	25	25
Reduced/conservation tillage (> 30% residue cover following planting)	0	20	0	35	15	30
No-till farming	0	0	0	40	25	75

Source: Adapted from Kansas State University Extension Publication MF-2572, Manhattan, KS.

WATERSHED SCALE EFFECTS OF CARBON MANAGEMENT ON SURFACE WATER QUALITY

The environmental effects of land management, including soil C management, are integrated at the watershed scale and are expressed in the quality of water in streams and other water bodies. A shift in the emphasis of nonpoint source (NPS) pollution research from edge of field to watershed scales has followed this expanded awareness. U.S. farm policy has undergone a similar evolution by expanding its focus from soil conservation to include water quality, water conservation, habitat and ecosystem protection, C sequestration, and air quality as high-priority concerns. The multiple ecological and environmental benefits of sound land management practices beyond production considerations are being articulated in some of our newest national conservation programs. Implementation of conservation programs at watershed scales and targeting of resources to address problematic areas within watersheds have been recognized as strategies for improving the cost-effectiveness of program delivery and for integrating environmental benefits to the soil, water, and biological resource base. Implementation of practices that emphasize sound management of soil C and vegetative protection of soil is key to the long-term sustainability of our food systems and overall protection of surface water quality in our nation's watersheds.

WATERSHEDS AND MANAGEMENT OF THE AGROLANDSCAPE

A watershed is the area of land that drains or seeps into a marsh, stream, river, lake, or groundwater and represents a clearly defined physical unit. A watershed contains the elements of an open system, with recognizable inputs, throughputs, and outputs of matter and energy. Ultimately, spatially diffuse inputs, such as precipitation and eroded materials, are transferred via a matrix of hillslopes and channels into a single outlet at the mouth of a watershed, where runoff and its concomitant loads can be assessed and quantified. The transfer of water and its load from land surface to rivers or lakes and eventually to the ocean demonstrate a tendency toward increased concentration and organization downstream. The implication is that downstream consequences have upstream causes. This is a fundamental reason why management of land and water resources has shifted its focus to watersheds as its basic unit.

In the United States, agriculture plays a dominant role in a majority of our land use decisions. Croplands, including hayfields and pastures, cover between 430 and 500 million acres or nearly a quarter of the land area (Heinz Center, 2002). Croplands are intermingled among a much broader backdrop that includes grasslands, shrublands, field borders, windbreaks, farmsteads, woodlots, ponds, lakes, streams, wetlands, feedlots, small villages, suburban areas, roads and other built-up areas. This type of heterogeneous agrolandscape generally represents a dominant condition in the United States. The composition of agricultural watersheds typically exists in dynamic flux as factors such as management systems, development interests, urban sprawl, farm programs, and crop prices interact to determine if and how property owners manage the landscape. Because the agrolandscape is a highly managed system covering a great deal of the land area, and because its by-products have been implicated as significantly impacting the water quality of our nation's watersheds, agricultural management practices that reduce impacts are a key area of applied research, especially as the concept of a sustainable agriculture becomes a political and social goal. From an environmental standpoint, the development of more sustainable agricultural management practices will undoubtedly have to include improvements in soil and water quality (i.e., environmental benefits) as key components, and this will likely require study at the watershed level if improvements in water quality are to be made.

Water quality in agrolandscapes is largely determined by the crop production systems and animal operations present in the watershed. Proximity of agricultural operations to streams and other water bodies is also important, as is the presence of buffering vegetation, and practices and structures to reduce runoff losses from agricultural fields. The transport of nutrients and pesticides from land areas into streams and water bodies is a result of soil weathering, erosion and runoff processes, which are all sensitive to soil C management and the presence of vegetative cover. For the most part, soil erosion is considered the dominant driver of NPS pollution in watersheds, with sediment also being the most visible pollutant. Although erosion is a natural process, soil loss from agricultural fields is at least an order of magnitude greater than background levels and can be two or more orders of magnitude higher than that occurring under natural vegetative conditions. Important reasons for this are the lack of soil cover afforded by many agricultural operations, disturbance occurring as a result of tillage practices, and the lack of continuous, deeply rooted perennial vegetation to maintain soil structure. Approximately 90% of the nutrient (N and P) losses to receiving water bodies have been reported to result from erosive soil losses (Alberts, Schuman, and Burwell, 1978; McDowell, 1989).

Pesticides attached to soil particles may also be delivered to water bodies because of erosive processes, but the magnitude of the losses is dependent on chemical properties of the pesticide (e.g., half-life, adsorption, and degradation). Soluble losses of both pesticides and nutrients to surface water occur via interactions between precipitation and upper soil layers to produce runoff and as a result of percolation through the soil profile with subsequent return to the surface via lateral flow. Both erosive and soluble losses of nutrients and pesticides from the agrolandscape contribute to the overall water quality in a watershed.

Excessive loading of sediment, nutrients, and pesticides to streams and water bodies can adversely affect water quality and ecosystem health by reducing light penetration and levels of dissolved oxygen, causing eutrophication, producing toxic conditions harmful to aquatic life and causing shifts in communities of aquatic organisms and rendering water unfit for human consumption. A primary question then is how to reduce loading to streams and water bodies resulting from agricultural practices while also maintaining some balanced level of production for human food and fiber needs.

Controlling erosion and reducing runoff within the agrolandscape are logical approaches to reducing soil degradation and, in turn, reducing nutrient and pesticide loading to streams and water bodies. Besides removing fertile topsoil outright, large rainstorms and windstorms selectively remove the reactive clay-sized fraction, materials high in organic matter and nutrients. This process depletes the surface soil of plant-available nutrients and organic matter, increases soil bulk density, and decreases soil porosity and capacity for water intake. For farmers, the majority believe that the most cost-effective means for controlling erosion is to manage organic matter at the soil surface (Moldenhauer and Mielke, 1995), which for the most part means managing crop residues and returning degraded and highly erodible lands back to vegetative cover, usually native grasses and forbs. Practices other than managing crop residues and planting back to native vegetation or some other grass species, but which focus on C management as a key include reducing or eliminating tillage, planting cover crops, and installing field buffer strips to filter out sediment and nutrients which have been lost from farm fields. Surface cover greatly reduces soil erosion, but the amount of cover required to attain erosion rates in the range of natural ecosystems depends on the site and other conservation practices included in a total soil and water conservation plan. However, which practices are most effective and efficient at reducing erosive losses? Which practices are the most ecological, sustainable, and economically viable? Moreover, can these practices be combined in ways that compound their positive effects on the watershed system and overall soil and water quality while also maintaining an adequate level of production?

FIELD-SCALE AND EXPERIMENTAL WATERSHED RESEARCH

One way to explore these questions is to do field-scale plot research to evaluate practices and combinations of practices over a range of soils and cropping systems to evaluate effects on crop yields and soil quality in the field, and to assess the water quality running off the field (edge of field studies). Testing hypotheses in agronomy, hydrology, or watershed management often involves the use of such plot studies (Brooks et al., 2003), and is generally the most common approach to a great deal of agronomic and environmental research in which small, homogeneous plots are isolated under controlled conditions to elucidate cause and effect information. Results from plot studies can often indicate a need for larger-scale (i.e., watershed-level) experiments. Process studies on plots can help to obtain a better understanding of cause and effect relationships, can serve as a basis for subsequent interpretations of results from watershed experiments, and help to identify key relationships that can be developed into mathematical algorithms for computer simulation models (Brooks et al., 2003).

The homogenous nature of plot studies, however, often limits the applicability of this method for studying the system-level responses of a heterogeneous watershed to the effects of land management practices. Watershed-level experiments are often used to evaluate and determine responses of an experimental watershed to management practices, as they provide a more realistic picture of responses than do plot studies. Such experiments have been employed worldwide to determine the effects of vegetative management practices on the magnitude, timing and quality of water regimes, and have greatly contributed to the understanding of the hydrologic cycle in various regions and the effects of watershed management on them (Brooks et al., 2003). Indeed, this approach has proven indispensable in the pursuit of a more complete understanding of watershed management effects (Bosch and Hewlett, 1982). This method does not come without its criticisms, however, because numerous people have criticized the use of experimental watersheds.

Perhaps the most serious criticism of using experimental watersheds is that they are not representative of river basins and, consequently, transferring results from a small, watershed microcosm to a larger, heterogeneous river basin is difficult and may be scale-dependent. Furthermore, integration of factors, relationships, and processes occurring at the watershed scale likely conceals information necessary to understand and interpret how individual components of the watershed system are responding to experimental changes in land management practices at field and farm scales. Because this approach requires an extensive investment in cost, time, and resources as well, the strategic and judicious use of the experimental watershed method to evaluate the effects of land management practices on soil and water quality is often necessary due to funding limitations and may be best utilized as a regional research strategy.

WATERSHED MODELS

Another approach commonly employed today is to apply a computer simulation model. Computer simulation models are representations of actual systems that allow one to explain and oftentimes predict watershed responses to management practices and ultimately gain a better understanding of the impacts of such practices on the system. Based largely on a systems approach, these models are a composite of mathematical and geospatial relationships that attempt to quantitatively describe the natural processes and relationships occurring in a watershed, either empirically or theoretically. As the complexity of the system of study increases, more detail and complexity are required in the formulation of the model. Often, the mathematical relationships comprising the model are determined based on results that are discovered during field-scale and watershed experimentation, and that are later incorporated into a model for application at watershed and river basin scales. Because our understanding of watershed systems and hydrologic processes is not complete and is evolving based on the state of scientific research, the development, integration, and testing of models to explain such systems and processes is ongoing. With the increased processing speeds and data storage capabilities of personal computers and the progress in development of geographic information system (GIS) data layers and software, the use of integrated GIS-based models has become common for understanding and predicting the behaviors of natural processes and better defining the complex relationships among processes at work in systems such as watersheds.

Essentially, models allow our minds to scale processes and relationships occurring in the real world to a level that can be more easily comprehended and evaluated. Such models may also be used as tools for developing successful planning strategies and defining future research needs. Because computer simulation models offer flexibility as an analytical tool and provide a savings of time and costs compared with traditional research methods, scientists, land managers, regulators, and policy makers are using them to investigate the effects of land management on water quality and to inform regional research needs. However, it should be noted that models are simplifications of real world phenomena. They should be calibrated based on measured data, and validated and replicated in many different settings to evaluate how well the model is representing the real world. Because of this and ongoing validation and improvements, watershed models and are still viewed with skepticism.

Though watershed models have many uses, an important application today is to use these models as tools for predicting sediment, nutrient, and pesticide losses occurring within the agrolandscape, and how such losses influence loading to streams and water bodies. Developments in remote sensing technology, GIS, and watershed models have helped open the door to the use of watershed models to evaluate the effectiveness of agricultural management practices, alone or in combination, for their effects on water quality. These types of applications may also allow for better targeting of resources to problematic areas in watersheds. Two watershed models that have gained federal agency support and widespread use in government, academic, and private sectors are the Annualized Agricultural Non-Point Source (AnnAGNPS) (Bingner and Theurer, 2003) and the Soil and Water Assessment Tool (SWAT) (Arnold et al., 1998) models. Both of these models are currently being extensively evaluated as part of the U.S. Department of Agriculture (USDA) Conservation Effects Assessment Project (CEAP) and are supported by the USDA.

Both the AnnAGNPS and SWAT watershed models were developed by the USDA, and are based on over 30 years of research by the Agricultural Research Service on soil erosion and hydrologic processes, as well as over 20 years of precursor model development to simulate erosive processes, plant growth, hydrology, and chemical fate and transport. AnnAGNPS and SWAT are both quasi-physically based, distributed, continuous time, long-term, simulation models integrated with GIS. These models are based on similar mathematical routines to describe continuous processes such as movement of water and constituents to and within streams. Notable differences exist between the models, but discussion of these points is beyond the scope of this chapter. Suffice it to say that both watershed models are capable of predicting and assessing the impact of management practices on water, sediment, and nutrient movement to surface water bodies at watershed scales. Practices that can be modeled with AnnAGNPS and SWAT include soil C management practices such as reduced tillage, no tillage, field buffer plantings, conversion of cropland to grass or native vegetation, terrace installation, and contour farming. Both models are generally considered watershed-scale models, but AnnAGNPS has often been used in field-scale applications and SWAT in several basin-scale studies.

An example of an AnnAGNPS application used to evaluate management practices can be illustrated in a study by Yuan, Bingner, and Rebich (2001) at the Mississippi Delta Management Systems Evaluation Area in the Deep Hollow Lake watershed of Leflore County Mississippi. In this study, AnnAGNPS was used to evaluate the effectiveness of agricultural management practices at reducing sediment losses to streams (Figure 12.4). Results indicated that sediment yield doubled if winter cover crops were not planted and the soil remained fallow following fall tillage operations. Changing all crops to no-tillage cotton reduced sediment yield, and changing all crops to no-tillage soybeans increased the sediment yield, due to the influence of cropping and management practices on erosion rates (i.e., RUSLE

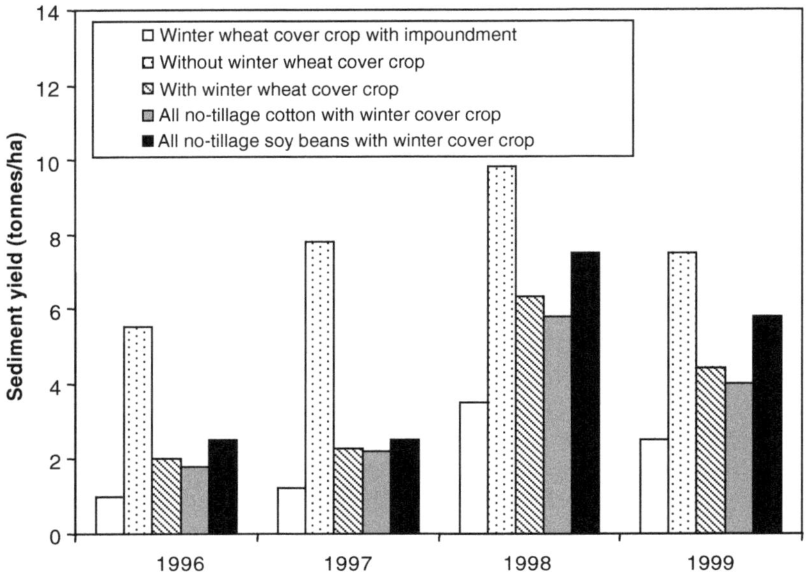

FIGURE 12.4 Annual sediment yield predicted by AnnAGNPS watershed model based on five alternative management scenarios for the Deep Hollow Lake watershed of Leflore County, Mississippi. (From Yuan, Y., Bingner, R.L., and Rebich, R.A., Evaluation of AnnAGNPS on Mississippi Delta MSEA Watersheds, *Trans. ASAE*, 44, 1183–1190, 2001. With permission.)

C-factor). By adding an impoundment near the mainstream channel, sediment yield to the stream was reduced by 50%.

In a similar application, SWAT was used to evaluate the effects of management practices and conservation structures on sediment, nutrient and pesticide yields from an agrolandscape in the Little Blue River watershed of south central Nebraska and north-central Kansas (Neel, Devlin, and McVay, 2005). A number of practices were evaluated alone or in combination and included different tillage systems, different residue levels, and the addition of conservation practices (Figures 12.5–12.8). The effects of tillage and cover were significant factors in determining runoff losses of sediment, total N, total P, and atrazine; conventional tillage systems had the highest losses, followed by mulch tillage, no tillage, and finally conversion to native vegetation (i.e., mixed-grass prairie). No tillage systems had higher runoff losses of soluble P compared with conventional tillage due to enrichment of upper soil layers with organic matter. Farming on the contour and installation of terraces reduced sediment losses, which, in turn, reduced runoff losses of organic N, organic P, and sediment P. Contours and terraces had minimal to no effect on runoff of nitrate-N and generally increased runoff of soluble P. Contours and terraces had minimal to no effect on runoff of atrazine due to its moderate solubility in water and relatively low adsorption coefficient. Field buffers significantly reduced runoff losses of all constituents (i.e., sediment, total N, nitrate-N, organic N, total P, sediment P, organic P, soluble P, and atrazine). Several combinations of BMPs in no-tillage and conventional tillage systems were considered moderately to highly effective at reducing

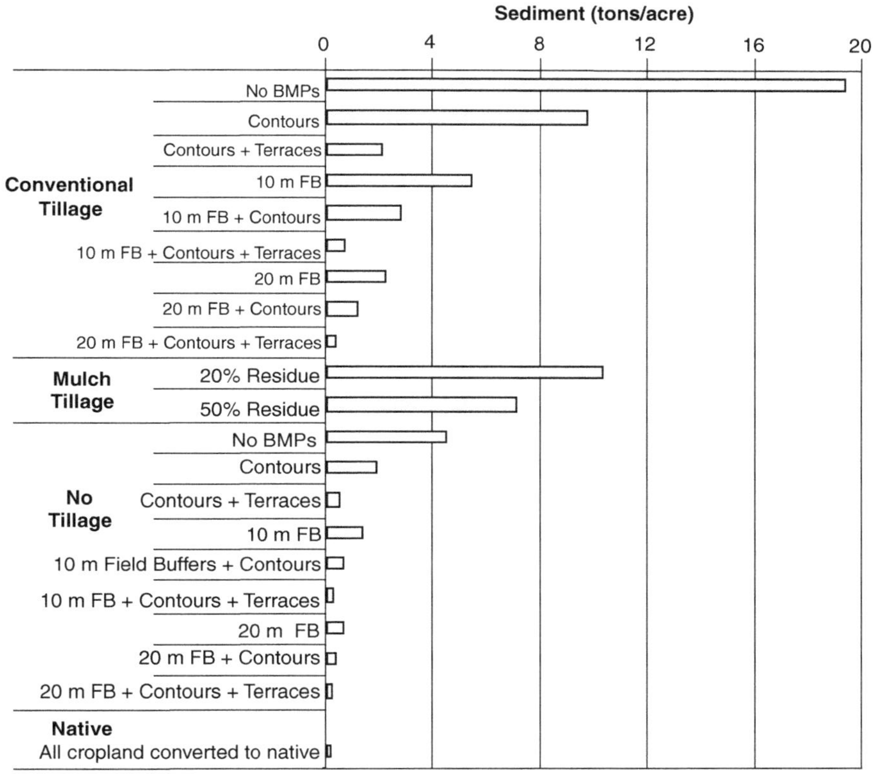

FIGURE 12.5 Annual sediment yield predicted by SWAT watershed model based on numerous management scenarios for the Lower Little Blue River watershed of south-central Nebraska and north-central Kansas. FB indicates simulated vegetative field buffer. (From Neel, J.C., Devlin, D.L., and McVay, K.A., Assessment of best management practices in a Kansas-Nebraska watershed using the SWAT model, in *Proc. USDA-CSREES National Water Qual. Conf.*, La Jolla, CA, February 7–9, 2005, USDA-CSREES, Washington, D.C., 2005. With permission.)

runoff losses. Native mixed-grass prairie had the lowest runoff losses for all constituents, but minimal levels of constituents were still present in runoff. Some combinations of BMPs in no-till systems approached the level of losses predicted under native conditions.

Both the AnnAGNPS and SWAT modeling examples provide an indication of the utility of watershed models for evaluating the effects of soil C management practices on water quality at watershed scales. From these examples, the potential impacts of reducing tillage, managing crop residues, and planting cover crops on water quality is demonstrated. Currently, however, many of the watershed-scale models being used to evaluate management practices, including AnnAGNPS and SWAT, simulate the effects of tillage and cover management using empirical relationships described by the revised universal soil loss equation (RUSLE) or modified universal soil loss equation (MUSLE) equations, particularly the C-factor. This is

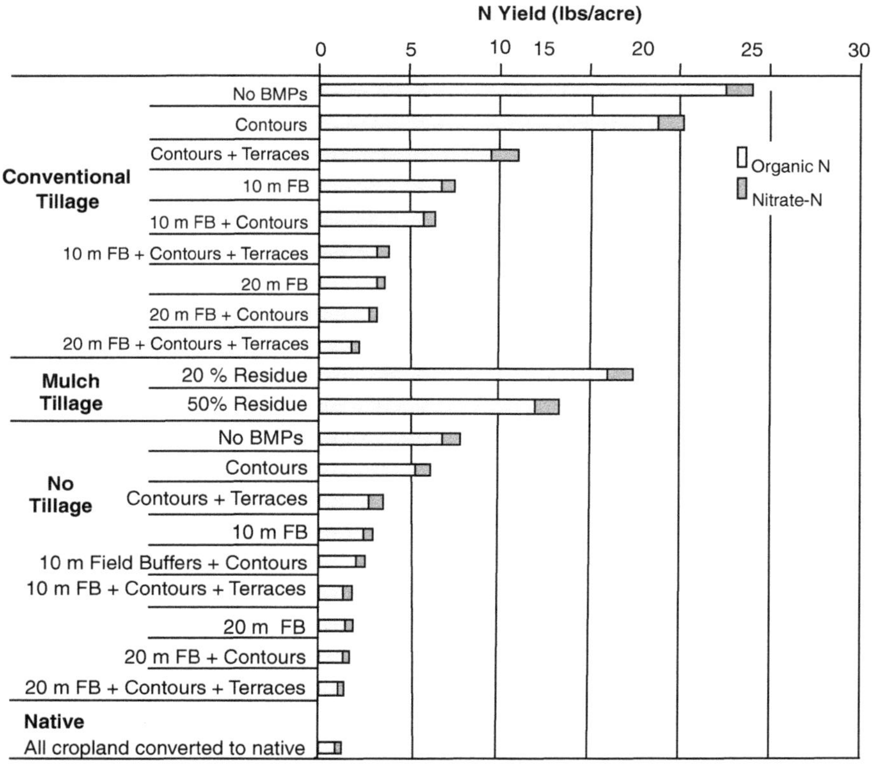

FIGURE 12.6 Annual N yield predicted by SWAT watershed model based on numerous management scenarios for the Lower Little Blue River watershed of south-central Nebraska and north-central Kansas. FB = simulated vegetative field buffer. (From Neel, J.C., Devlin, D.L., and McVay, K.A., Assessment of best management practices in a Kansas-Nebraska watershed using the SWAT model, in *Proc. USDA-CSREES National Water Qual. Conf.*, La Jolla, CA, February 7–9, 2005, USDA-CSREES, Washington, D.C., 2005. With permission.)

likely a limitation of the current state of watershed models, but which might be greatly enhanced by incorporation of algorithms from a plant-soil nutrient cycling model, such as the Century model (Parton et al., 1987, 1993, and 1994; Vitousek et al., 1994), into AnnAGNPS or SWAT to simulate C and nutrient dynamics for different types of ecosystems. Such an occurrence would create a robust, analytical tool for investigating the complex relationships dominating soil organic matter dynamics and overall soil quality in eroding landscapes as well as linking these effects to water quality at watershed scales.

Linking soil quality models to water quality models at watershed scales will likely represent an important research focus in natural resource and conservation sciences in the upcoming years. As more detailed data inputs for models become available to users, as computing power increases, and as GIS technology proliferates, likely advances in continuous-time, physically based watershed simulation models will keep pace.

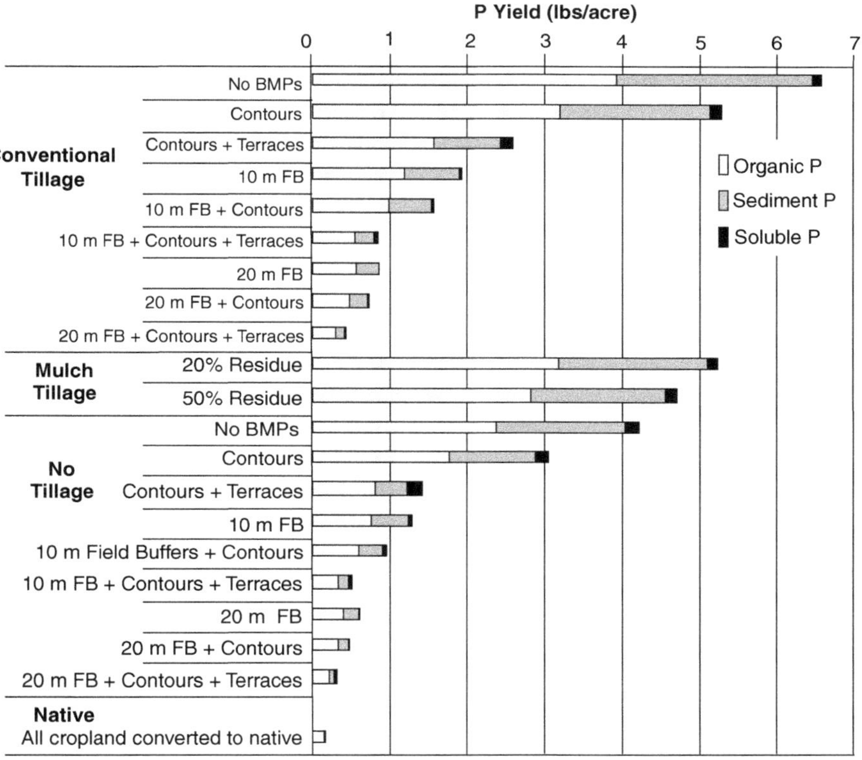

FIGURE 12.7 Annual P yield predicted by SWAT watershed model based on numerous management scenarios for the Lower Little Blue River watershed of south-central Nebraska and north-central Kansas. FB = simulated vegetative field buffer. (From Neel, J.C., Devlin, D.L., and McVay, K.A., Assessment of best management practices in a Kansas-Nebraska watershed using the SWAT model, in *Proc. USDA-CSREES National Water Qual. Conf.*, La Jolla, CA, February 7–9, 2005, USDA-CSREES, Washington, D.C., 2005. With permission.)

Integrating watershed models with ecosystem and landscape ecology metrics that better describe the temporal and spatial movements of energy, matter and nutrients into, through, and out of interrelated hydrologic systems may represent the future of watershed science. Managing agrolandscapes in an environmental, ecological, and economically profitable manner will likely require this type of interdisciplinary and comprehensive understanding, especially if agriculture, nature, and society are to reach a compromise capable of rendering a sustainable solution for our future.

CONCLUSIONS

Efforts to increase soil C concentrations represent a no regret strategy for mitigating the potential effects of increased atmospheric CO_2 concentrations on the climate. In addition to helping reduce the rate of increase in atmospheric CO_2 concentration, increasing soil C concentrations can act to significantly improve surface water

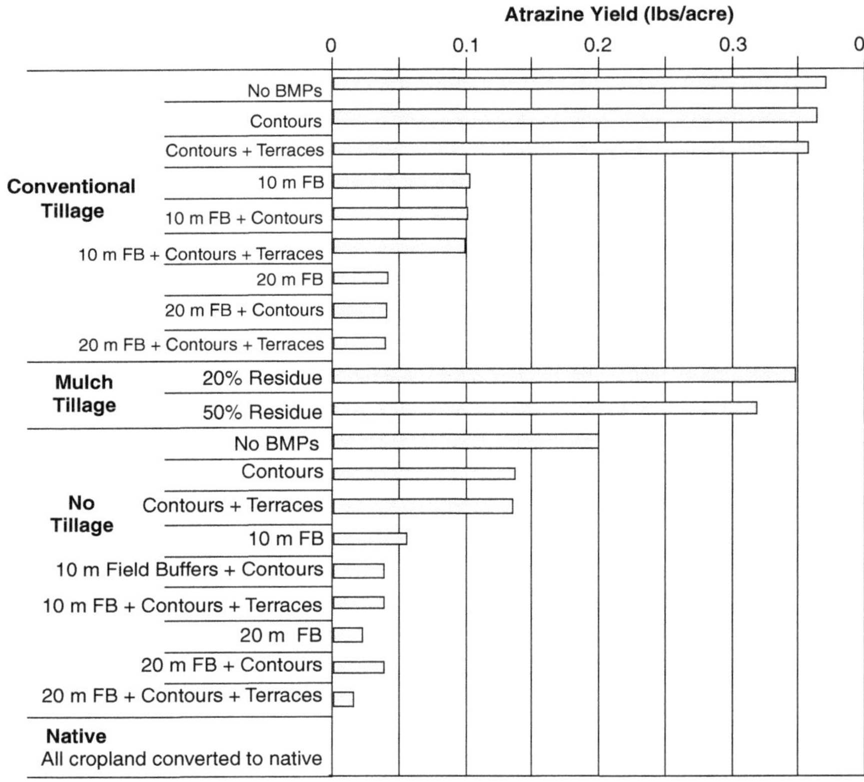

FIGURE 12.8 Annual atrazine yield predicted by SWAT watershed model based on numerous management scenarios for the Lower Little Blue River watershed of south-central Nebraska and north-central Kansas. FB = simulated vegetative field buffer. (From Neel, J.C., Devlin, D.L., and McVay, K.A., Assessment of best management practices in a Kansas-Nebraska watershed using the SWAT model, in *Proc. USDA-CSREES National Water Qual. Conf.*, La Jolla, CA, February 7–9, 2005, USDA-CSREES, Washington, D.C., 2005. With permission.)

quality, primarily by influencing surface runoff characteristics. The most common contaminants found in surface waters in the United States are sediments, nutrients, bacteria, and herbicides. Improved soil C management can help reduce the impact of land management practices on these contaminants.

The net impact of surface runoff on surface water bodies is a product of runoff volume and runoff composition. Improved soil C management can reduce runoff volume and can reduce the concentrations of sediments, nutrients, and herbicides such that the total amount lost from the landscape is decreased. Most of the research in this area has utilized an edge of field approach where the relative differences in runoff volume and composition induced by various practices are measured. It is difficult to extrapolate this to the watershed scale as numerous processes may affect the characteristics of surface runoff between the edge of field and the surface water body that may be impacted. In addition, multiple land management practices within a watershed make it difficult to estimate the effect of BMPs on a watershed scale. For a given land use practice, the effects of BMPs are not additive. Watershed modeling can overcome some of these difficulties by attempting to simultaneously consider all land use and best management practices on surface water quality within a watershed. The models can be calibrated against stream monitoring data and then used to predict the impact of best management practices on water quality over a large area. Overall, results from watershed models are still viewed with some skepticism, but they have clearly progressed to the point of being able to be used for planning purposes and to reduce the need for direct monitoring of water quality.

REFERENCES

Alberts, E.E., Schuman, G.E., and Burwell, R.E., Seasonal runoff losses of nitrogen and phosphorus from Missouri Valley Loess watersheds, *J. Environ. Qual.*, 7, 203–208, 1978.

Andraski, B.J., Mueller, D.H., and Daniel, T.C., Phosphorus losses in runoff as affected by tillage, *Soil Sci. Soc. Am. J.* 49, 1523–1527, 1985.

Arnold, J.G., Srinivasan, R., Muttiah, R.S., and Williams, J.R., Large area hydrologic modeling and assessment. Part 1: model development, *J. Am. Water Resour. Assoc.*, 34(1), 73–89, 1998.

Baker, J.L., Hydrologic effects of conservation tillage and their importance relative to water quality, in *Effects of Conservation Tillage on Water Quality: Nitrate and Pesticides*, Logan, T.J., Ed., Lewis Publishers, Chelsea, MI, 1987.

Baker, J.L., Laflen, J.M., and Johnson, H.P., Effect of tillage systems on runoff losses of pesticides, a rainfall simulation study, *Trans. ASAE*, 21, 886–892, 1978.

Baker, J.L. and Johnson, H.P., The effect of tillage system on pesticides in runoff from small watersheds, *Trans. ASAE*, 22, 554–559, 1979.

Barisas, S.G., Baker, J.L., Johnson, H. P., and Laflen, J.M., Effect of tillage systems on runoff losses of nutrients a rainfall simulation study, *Trans. American Society of Agriculture Engineers (ASAE)*, 21, 893–897, 1978.

Bingner, R.L. and Theurer, F.D., AnnAGNPS Technical Processes, U.S. Department of Agriculture, Agricultural Research Service (USDA-ARS), National Sedimentation Laboratory, Oxford, MS, 2003.

Blevins, R.L., Frye, W.W., Baldwin, P.L., and Robertson, S.D., Tillage effects on sediment and soluble nutrient losses from a Maury silt loam soil, *J. Environ. Qual.*, 19, 683–686, 1990.

Bosch, J.M. and Hewlett, J.D., A review of catchment experiments to determine the effect of vegetation changes on water yield and evapotranspiration, *J. Hydrol.*, 55, 3–23, 1982.

Brooks, K.N., Folliott, P.F., Gregersen, H.M., and DeBano, L.F., Hydrologic methods, in *Hydrology and the Management of Watersheds*, 3rd ed., Iowa State Press, Ames, IA, 2003, 435–460.

Chichester, F.W. and Richardson, C.W., Sediment and nutrient loss from clay soils as affected by tillage, *J. Environ. Qual.*, 21, 587–590, 1992.

Cox, F.R. and Hendricks, S.E., Soil test phosphorus and clay load effects on runoff water quality, *J. Environ. Qual.*, 24, 734–741, 2000.

Eisenhauer, D.E., Norton, M.J., Martin, D.L., and Spalding, R.F., Impact of tillage and crop rotation on water movement with furrow irrigation, in *Proc. Agric. Res. to Protect Water Quality*, Minneapolis, MN, February 21–24, 1993, Soil and Water Conservation Society, Ankeny, IA, 1993, 533–536.

Ethridge, K., *Influence of Land Application of Cattle Manure on Phosphorus Losses in Runoff*, M.S. thesis, Kansas State University, Manhattan, KS, 2002.

Felsot, A.S., Mitchell, J.K., and Kenimer, A.L. et al., Assessment of management practices for reducing pesticide runoff from sloping cropland in Illinois, *J. Environ. Qual.,* 19, 539–545, 1990.

Gaynor, J.D., MacTavish, D.C., and Fidlay, W.I., Atrazine and metolachlor loss in surface and subsurface runoff from three tillage treatments in corn, *J. Environ. Qual.*, 24, 246–256, 1995.

Gaynor, J.D. and Fidlay, W.I., Soil and phosphorus loss from conservation and conventional tillage in corn production, *J. Environ. Qual.*, 24, 734–741, 1995.

Glen, S. and Angle, J.S., Atrazine and simazine in runoff from conventional and no-till corn watersheds, *Agric. Ecosyst. Environ.*, 18, 273–280, 1987.

Hall, J.K., Mumma, R.O., and Watts, D.W., Leaching and runoff losses of herbicides in a tilled and untilled fields, *Agric. Ecosyst. Environ.*, 37, 303–314, 1991.

Heinz Center, *The State of the Nation's Ecosystems: Measuring the Lands, Waters, and the Living Resources of the United States*, The H. John Heinz III Center for Science, Economics and the Environment, Cambridge University Press, Cambridge, U.K., 85–110, 2002.

Isensee, A.R. and Sadeghi, A.M., Impact of tillage practice on runoff and pesticide transport, *J. Soil Water Conserv.*, 48, 523–527, 1993.

Kanwar, R.S. and Baker, J.I., Tillage and chemical management effects on groundwater quality, in *Proc. Agric. Res. to Protect Water Quality*, Minneapolis, MN, February 21–24, 1993, Soil and Water Conservation Society, Ankey, IA, 455–459, 1993.

Kanwar, R.S. and Colvin, T.S., Tillage and crop rotation effects on drainage water quality, in *Clean Water, Clean Environment, 21st Century: Team Agriculture, Working to Protect Water Resources*, Vol. III, Conf. Proc., Kansas City, MO, March 5–8, 1995, ASAE, St. Joseph, MI, 163–166, 1995.

Karlen, D.I., Wollenhaupt, N.C., Erbach, D.C., Berry, E.C., Swan, J.B., Nash, N.S., and Jordahl, J.I., Long-term tillage effects on soil quality, *Soil Tillage Res.*, 32, 313–327, 1994.

Kelly, B.P. and Blevins, D.W., Hydrology and nitrogen distribution in claypan soil and glacial till near Centralia, MO-May 1991 to May 1992, *Proc. Agric. Res. to Protect Water Quality*, Minneapolis, MN, 21–24 Feb. 1993, Soil and Water Conservation Society, Ankeny, IA, 545–547, 1993.

Kimmell, R.J., Pierzynski, G.M., Janssen, K.A., and Barnes, P.L., Effects of tillage and phosphorus placement on phosphorus runoff losses in a grain sorghum-soybean rotation, *J. Environ. Qual.,* 30, 1324–1330, 2001.

Lamb, J.A., Dowdy, R.H., Anderson, J.L., and Allmeras, R.R., Water quality in an irrigated sandy soil: ridge tillage in rotated corn and soybean compared with full-width tillage in continuous corn, *Soil Tillage Res.,* 48, 167–177, 1998.

Lindstrom, M.J. and Onstad, C.A., Influences of tillage systems on soil physical parameters and infiltration after planting, *J. Soil Water Conserv.*, 39, 149–152, 1984.

McDowell, L.L., Nitrogen and phosphorus yields in runoff from silty soils in the Mississippi delta, *Agric. Ecosyst. Environ.*, 25, 119, 1989.

Moldenhauer, W.C. and Mielke, L.N., Crop residue management to reduce erosion and improve soil quality. *Conserv. Res. Report No. 42*, USDA-ARS, Washington, D.C.,1995.

Neel, J.C., Devlin, D.L., and McVay, K.A., Assessment of best management practices in a Kansas-Nebraska watershed using the SWAT model, in *Proc. USDA-CSREES National Water Qual. Conf.*, La Jolla, CA, February 7–9, 2005, U.S. Department of Agriculture, Cooperative State Research, Education, and Extension Service (USDA-CSREES), Washington, D.C., 2005.

Parton, W.J., Schimel, D.S., Cole, C.V., and Ojima, D.S., Analysis of factors controlling soil organic matter levels in Great Plains grasslands, *Soil Sci. Soc. Am. J.,* 51, 1173–1179, 1987.

Parton, W.J., Scurlock, J.M.O., Ojima, D.S., Gilmanov, T.G., Scholes, R.J., Schimel, D.S., Kirchner, T., Menaut, J.C., Seastedt, T., Garcia Moya, E., Kamnalrut, A., and Kinyamario, J.I., Observations and modeling of biomass and soil organic matter dynamics for the grassland biome worldwide, *Global Biogeochemical Cycles*, 7, 785–809, 1993.

Parton, W.J., Ojima, D.S., Cole, C.V., and Schimel, D.S., A general model for soil organic matter dynamics: sensitivity to litter chemistry, texture and management, in *Quantitative Modeling of Soil Forming Processes, Soil Science Society of America (SSSA) Special Publication No. 39,* SSSA, Madison, WI, 1994, 147–167.

Rector, R.J., Regehr, D.L., Barnes, P.L., and Louglin, T.M., Atrazine, S-metolachlor, and isoxaflutole loss in runoff as affected by rainfall and management, *Weed Science*, 51, 810–816, 2003.

Renard, K.G., Foster, G.R., Weesies, G.A., McCool, D.K., and Yoder, D.C., Predicting soil erosion by water, A guide to conservation planning with the Revised Universal Soil Loss Equation (RUSLE), *USDA Agriculture Handbook No. 703*, USDA, Washington, D.C., 1997.

Seta, A.K., Blevins, R.L., Frye, W.W., and Barfield, B.J., Reducing soil erosion and agricultural chemical losses with conservation tillage, *J. Environ. Qual.,* 22, 661–665, 1993.

Sharpley, A.N. and Smith, S.J., Wheat tillage and water quality in the Southern Plains, *Soil Tillage Res.*, 30, 33–48, 1994.

Siemens and Oschwald, Erosion for corn tillage systems, *Trans. ASAE,* 19, 69–72, 1976.

Vitousek, P.M., Turner, D.R., Parton, W.J., and Sanford, R.L., Litter decomposition on the Mauna Loa environmental matrix, Hawaii: Patterns, mechanisms, and models, *Ecology*, 75, 418–429, 1994.

Wauchope, R.D., The pesticide content of surface water draining from agricultural fields — a review, *J. Environ. Qual.*, 7, 459–472, 1978.

Witt, W.W., and Sander, K.W., Movement of triazine herbicides in conventional, reduced tillage, and no-tillage corn production, *Soil Sci. News Views*, 11(5), 1990.

Yuan, Y., Bingner, R.L., and Rebich, R.A., Evaluation of AnnAGNPS on Mississippi Delta MSEA Watersheds, *Trans. ASAE*, 44, 1183–1190, 2001.

13 Urban Lands and Carbon Management

Steven I. Apfelbaum

CONTENTS

THE CHALLENGES

Urban areas represent one of the single largest sources of carbon emissions. Substantial headway is currently being made in addressing some of the key carbon emission sources:

- Improvement of stormwater best management practices (BMPs)
- Changing development strategies to more consolidated areas to free up and improve open space lands where carbon can be managed a market shift to green buildings and hybrid and carbon neutral fuels for vehicle use
- Some shifts to mass transportation
- Clean streams and lakes programs and ecological restoration projects in urban areas, which engage the public in viewing urban open lands as treasures that need management and stewardship, just like gardens

Wetlands (i.e., natural areas such as prairies and forests) and urban environments can be managed to foster greenhouse gas (GHG) management to mitigate climate change. Currently, however, no coherent call to action to reduce the urban growth and development impact on our waterways and water quality, through soil carbon management, has been proposed nationally or even regionally.

This chapter presents many of the factors that influence carbon loss from urban areas and discusses the consequences of this loss in terms of reduced ecosystem services. The chapter also discusses possible solutions to mitigate this loss and use terrestrial ecosystems to sequester additional carbon.

For purposes of this chapter, urban landscape is defined broadly to include urban core areas, and exurban fringe lands that are entering a development cycle. It also includes first and second ring suburbs and urban core areas that are becoming zones of disinvestment and decline.

This chapter has purposely not remained focused on urban carbon management in soils. We have integrated overviews of the significant magnitude of other carbon uses and releases from urban areas, that have a primary or secondary impact on global soil carbon management, and identified some key nonsoil carbon issues that do effect carbon emissions and soil carbon management opportunities in urban areas.

Urban landscapes are net consumers of energy, food, water, gasoline, and other resources and commodities. They are also net consumers of land, through development, conversion to parking lots, sewage treatment facilities, and the range of other land uses. Urban landscapes are also one of the net producers of waste: municipal garbage, sanitary wastes, medical wastes, air quality constituents from fuel combustion, automobile exhausts, and erosion and sedimentation from land development activities, as well as a range of other sources.

Virtually each land use and resource use and waste stream created from urban areas represents a broken cycle that releases wastes, including carbon and other GHGs into the atmosphere. The challenge in urban landscapes is recognizing and reconnecting the linkages between resource use and waste, and leveraging efficiencies and strategies to reduce the level of resource use and waste.

Many programs from many governmental agencies target pieces of the puzzle, but no coherent single or even a few strategies are available to bring greater efficiency to the programming strategies through integration and this is one of the challenges to managing carbon (and other GHGs) in urban landscapes.

BACKGROUND

Urban areas occupy and supplant landscapes that historically sequestered carbon with buildings, impervious land cover uses, and disproportionately high carbon emission producing systems. Automobiles, home heating, and industrial operations and power plants concentrate the carbon emissions in urban areas. At the same time increased erosion and sedimentation, stream and other water body deterioration from this erosion and sedimentation, and discharge of warmed waters (e.g., rainwater washing over pavements, from industrial operations, and so forth) all accelerate and contribute to disproportionately increased emissions of carbon and other GHGs from urban areas. On a per-acre basis, excluding coal-fired and oil-burning power plants and such natural emitters as volcanoes, urban areas by far exceed all other human sources of carbon emissions and contribute significantly to soil carbon losses.

Sprawling subdivisions remove productive lands and soils in the Midwestern United States and elsewhere. They are replaced with global climate deteriorating land uses and landscapes. More, and accelerated, stormwater runoff from these lands contributes significantly to erosion of uplands and riparian corridors.

The U.S. population growth rates and the number of new home sales are expected to increase significantly over the next 20 years (Palmer et al., 2004). Acting now in urban areas to reduce runoff, increase the quality of water from runoff, reduce erosion and sedimentation in our nations waterways are crucial initiatives. Likewise, it is important to rethink the way land development is done and use development as an economic engine to finance conservation investments and restoration and stewardship by restoring conservation lands. These changes are important and might be expected to result from managing soil carbon in urban areas. Specifically, the integration of stormwater management programs that use native vegetation for landscaping that does not require irrigation, fertilizer and minimum seasonal mowing, can substantially reduce carbon emissions. The absence of coherent single regulatory agency and programs that administer urban area development and land use makes it extremely difficult to consistently bring important changes to the urban landscapes. Most land use controls are locally administered and therein lays a significant problem. Most local regulations focus on just that, but an interesting movement is afoot in the United States, with regions, counties, and even municipalities now passing resolutions, ordinances, and making policy that would suggest that it is becoming increasingly recognized that global issues must become local issues, and that local policy and decisions have a very important role in packaging global solutions. Reducing global carbon emissions will require a circumspect and large-scale perspective that is larger than local ordinances and controls.

Many models for alternative types of urban land developments exist that have been extremely practical and developers both save money and make higher profit (see Urban Land Institute). Inspired private sector members have largely pushed the creative side of the market around the United States. Some municipalities, counties, and cities have engaged in "smart growth principles" to varying degrees.

INTRODUCTION

Technical and laypersons read newspapers daily that are replete with recurring themes as well as the increasing costs of living as a result of these themes:

- Water pollution
- Decline and loss of open space
- Stream channel erosion and increasing frequency and risk of flooding
- Developed communities that don't work for pedestrians, and every trip requires use of an automobile because of poor interconnectivity in the fundamental community designs
- Mass transportation systems operating in the red
- Increasing energy costs and miles traveled in internal combustion vehicles per mile
- Landfill spaces filling up with garbage

Think tanks, foundations, agencies, associations, and others have been striving to find and integrate solutions for some of these problems, whereas associations (e.g., National Home Builders and National Green Building Alliance) have focused in on specific elements of the larger problem.

Expanding transportation systems and needs of urban societies appear to be poised for explosive growth. The number of automobiles, traveled miles, airline miles, and passenger use are all projected to increase. Only recently have the contribution of airline soot and its interaction with ozone and other higher atmosphere functions been examined (Wei et al., 2001). New types of engine technologies, include cleaner diesel engines hold promise for reducing some carbon emissions.

The environmental impacts to water resources that results from soil erosion and stream sedimentation, stream and river channel deterioration, increased runoff from increased imperviousness, increased air and water pollution from highways and combustion byproducts from urbanization have long been identified (Apfelbaum, 1993; Apfelbaum et al., 1995, 1996). Yet, solutions and strategies to address the source of the problems have customarily focused in addressing industrial point source polluters. In Britain, some of the first measurements of the soil carbon loss rates through erosion into rivers have been made (Hope et al., 1997). Estimates have also been made in the United States (Lal et al., 1999). Only in the past decade or so have more wide ranging approaches to understanding and addressing solutions made its way to the forefront (US EPA). Clean streams programs that have engaged public volunteer groups (e.g., stream teams and lake monitoring teams) have begun to identify problems at a more resolved level and watershed approaches to working with solutions and awareness are well underway in some parts of the United States.

Erosion of soils has been a major problem with urban land development. Soil erosion results in depleted soil carbon stocks, and once the eroded soils settle into aquatic environments, particularly in eutrophied and seasonally anaerobic below-water settings, methane, and other more potent GHGs are released to the atmosphere. Such development is a major carbon emission source. Various erosion and sedimentation control ordinances including the National Pollution Discharge Elimination System (NPDES) permitting requirements have been developed because earth moving disrupts soils, compacts and reduces stormwater infiltration in heavier soils, and documented impacted include a subsequent increased erosion, sediment, and runoff (US EPA). These impacts and changes in soil carbon, nitrogen and phosphorus and resulting water pollution have been examined in many areas of the United States (US EPA), and elsewhere, including China (Genxu, 2004). The process of stream degradation from this urbanization, as a root cause, has been detailed (Dunne and Leopold, 1978).

Conservation development strategies have been used to curtail many of the problems with conventional development. In Illinois, the benefits to reduced erosion, stormwater runoff, and improved water quality from predevelopment agricultural conditions (Figure 13.1) have been documented (Apfelbaum, 1993; Apfelbaum et al., 1995, 1996) and the use of conservation development design can result in significant cost reduction and environmental benefits (AES unpublished data). Conservation development strategies typically use and disturb less land and many protect and restore greater than 50% of each parcel. Many use alternative stormwater management systems such as Applied Ecological Services' Stormwater

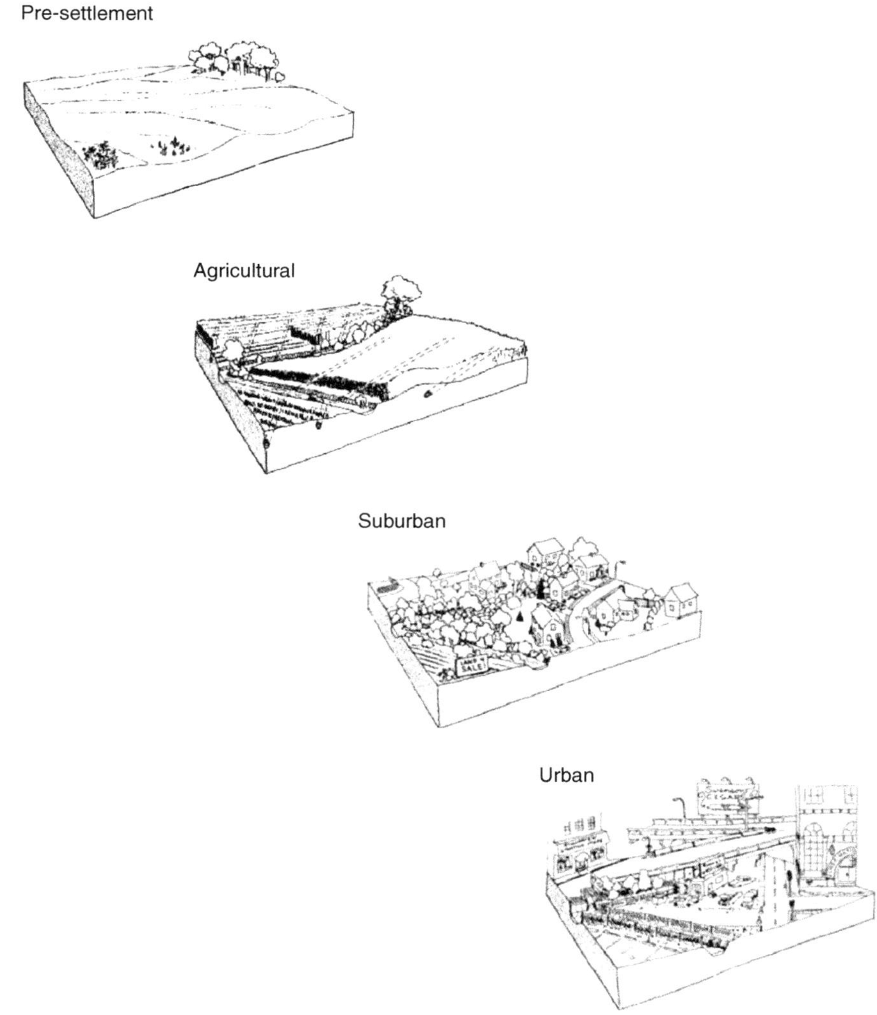

FIGURE 13.1 The process of urbanization changes watersheds and has associated increased soil loss, runoff, and emissions from other urban carbon sources. (From Apfelbaum, S.I., Applied Ecological Services, Inc., unpublished M.S. thesis.)

Treatment Train™ (Figure 13.2), which can completely eliminate costly stormwater sewers, thereby reducing or alleviating typical water quality and downstream erosion problems. In addition, many use native landscaping features that can be less costly, lower maintenance, and that become assets in a development. Including the added costs for restoration of the open space in conservation developments, we have demonstrated substantially reduced overall development costs by consolidating the developed areas and substantially higher market premiums as lots face open space and residents have access to trails. (Figure 13.3, Figure 13.4, and

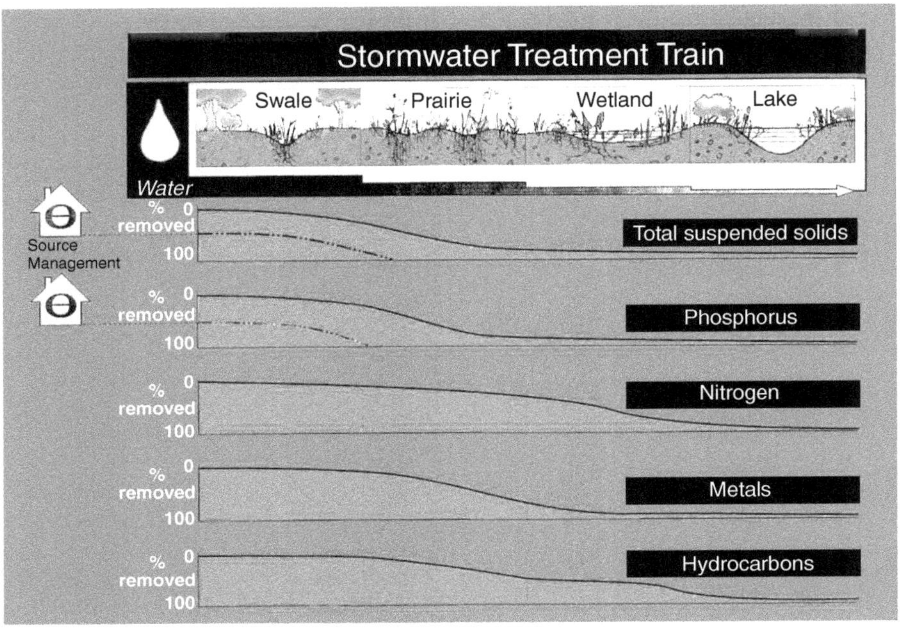

FIGURE 13.2 Instead of stormwater sewers, in conservation developments restored native landscapes are designed to reduce runoff rates and volumes, and increase the quality of water leaving the land. These Stormwater Treatment Trains™ have proven very effective at producing lower cost solutions for stormwater management. Soil carbon management can benefit from reduced erosion, and open space restoration and stabilization can result from use of these restored systems. (From Apfelbaum, S.I. et al., The prairie crossing project: attaining water quality and stormwater management goals in a conservation development using ecological restoration to meet clean water act goals, National Symposium on Using Ecological Restoration to Meet Clean Water Act Goals, Chicago, IL, March 14–16, 1995, US EPA, Washington, D.C., 1995, 33–38, and Apfelbaum, S.I., et al., On conservation developments and their cumulative benefits, in *Proceedings of National Symposium, Assessing the Cumulative Impacts of Watershed Development on Aquatic Ecosystems and Water Quality*, U.S. Environmental Protection Agency (US EPA), Chicago, IL, 1996, 181–188. With permission.)

Applied Ecological Services, Inc. (AES) unpublished data). Importantly, the included open space, especially where it is restored can attain carbon sequestrating rates and soil carbon stock rebuilding benefits that are higher than the remnant open space or manicured park lawnscapes found in conventional land developments and higher than for the former row cropped agricultural land uses that often preceded the development.

Costs and opportunities for the building sector to reduce carbon emissions have been explored (Koomey et al., 1998). Similar investigations for the steel industry, paper and pulp producers, and other materials demonstrate promising technological changes (Ruth et al., 2004; Gielen, 2003).

Redevelopment projects including blighted urban core areas can bring back green areas that can provide a multitude of benefits including carbon management.

Many cities founded on river, ocean, and lakefronts historically built industrial complexes that needed good water sources and access. With the decline of these industrial operations, many cities are redeveloping waterfront parks and greenways that offer a complete range of environmental, recreational, aesthetic, and other benefits, including carbon management opportunities. Many of these greenway projects and regreening around neighborhoods and buildings can create significant carbon benefits. For example, in Chicago, Illinois the planting of trees for landscaping could shade building and provide a net utilization of electricity for cooling. Reduced use for heating is also a benefit where trees provide a windbreak in winter (Jo and McPherson, 2001). Similar findings have been made in other U.S. cities (Akbari, 2002).

Mass transportation systems and personal vehicle fuel options are becoming available in some urban areas from the use of virtually carbon neutral fuels (e.g., biodiesel, ethanol) or electricity from alternative energy systems (e.g., wind, geothermal, landfill methane, and beneficial reuse of landfill gases for electrical generation) may change the carbon emissions in future cities. Studies have explored various mixes of electrical generation strategies to reduce or manage carbon emissions (Whittington, 2002). The potential benefits from conversion from diesel to cleaner fuels have been identified. However, new strategies for leveraging the conversion to clearer and carbon neutral fuels are a strongly pursued prospect in only a few U.S. cities. Hybrid technologies are on the rise and appear to have near-term benefits in reducing carbon emissions, but without important national policy and a more diverse economic stimulus, the benefits may never be realized as more private vehicles are being used annually. A serious and focused national need exists for addressing this prevalent and increasing source of GHG emissions. Policy discussions are under way (e.g., Fleming and Webber, 2004) in the United States, China, Britain, and in other nations.

Climate change is predicted to be associated with significantly more unpredictable weather patterns. Erosion, greater instability in stream channels, more destruction of forests, higher cost for urban landscape mangement, and greater variability in the productivity of plant communities are some predicted changes. McNulty (2002) documents that expected increased carbon emissions from hurricane damage on U.S. forests. Lawn maintenance (mowing, and disposal of clippings) contributes to a large carbon emission from urban environments (Jo and McPherson, 1995). Urban treescapes and especially landscapes dominated by exotic trees provide substantial carbon storage (Freedman, Love, and O'Neil, 1996). Soil carbon turn over rates in disturbed environments can be much greater than in mature ecosystems (Trumbore, 1997) suggesting that urban soils may be prone to faster turnover and higher release rates. Increased decomposition and carbon emissions from soil organic matter with soil warming in forested and other ecosystem types has been a documented concern in urban and wildlands (Melillo et al., 2002). However, basic relationships between plant diversity, net primary production, and carbon dynamics in soils need to be better understood (Catovsky, Bradford, and Hector, 2002).

Coastal shelve environments, near urban areas have been vulnerable to nutrient loading and water pollution. Declining diversity and abundance of phytoplankton and sea grasses suggest urban areas can contribute to significant changes in carbon

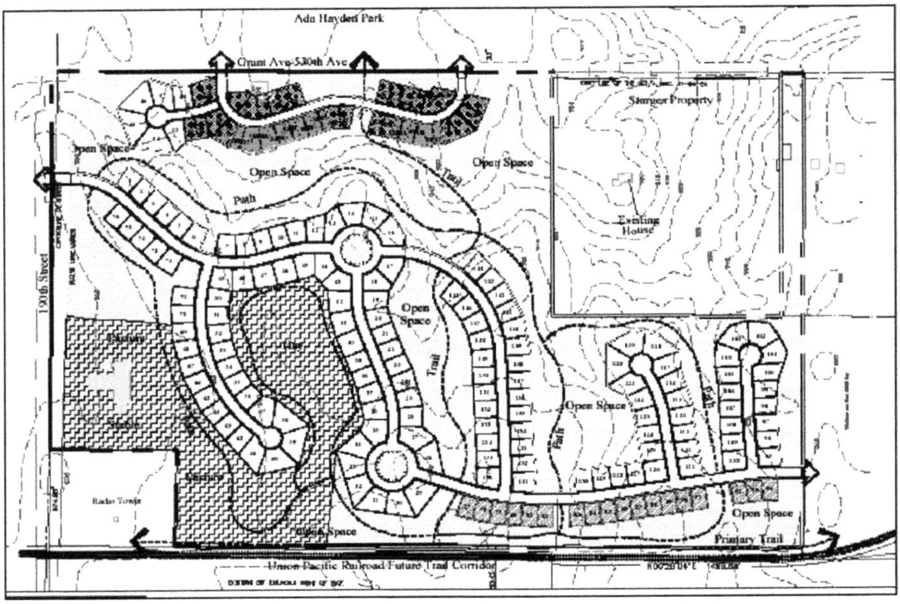

(a)

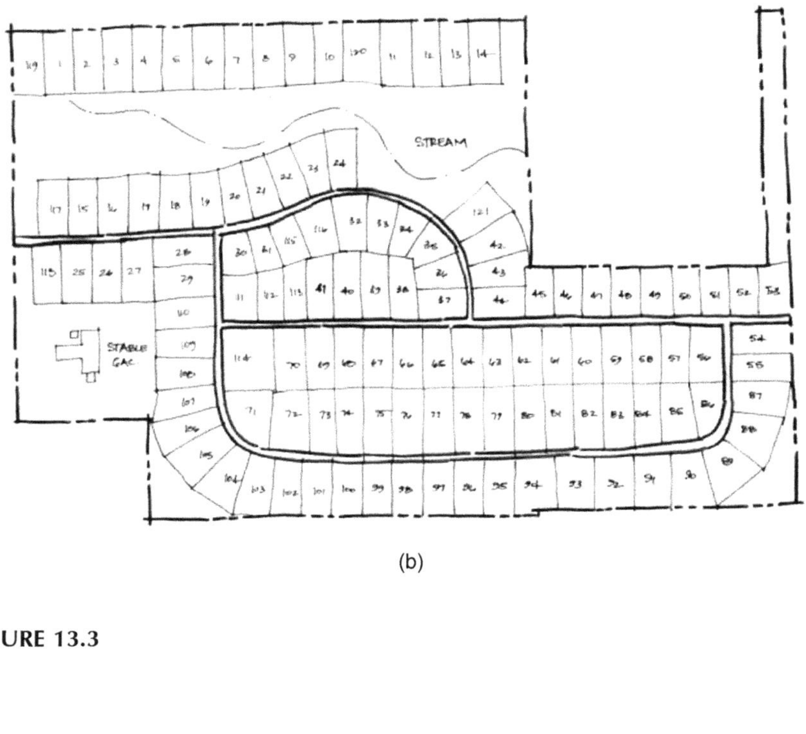

(b)

FIGURE 13.3

	Conservation Development	Traditional Development
Total Acreage	***180.8 AC***	***180.8 AC***
Open Space	***111.2 AC***	***33.4 AC***
Total Lots	***20***	***21***
Roadways (est. cost)	$1,415,040	$1,800,960
Storm Sewers (est. cost)	$27,152	$840,000
Curb/Gutter (est. cost)	$339,400	$600,000
Restoration (est. cost)	$877,500	$334,00
TOTAL	**$1,659,292**	**$3,574,960**
	~53% Savings	

FIGURE 13.4 Conservation development can cost less and can generate more profitable returns than conventional development. This figure compares the development costs in Figure 13.3. (From Applied Ecological Services, Inc.™, unpublished paper. With permission.)

emissions from these coastal waters (Thom et al., 2003). Nutrient loading from aerosol contaminants, such as from nitrogen deposition, is a key concern as greater available nitrogen levels in soils can accelerate the decomposition of soil carbon levels (Waldrop et al., 2004). Urban areas and agricultural lands generate a disproportionate quantity of aerosol nitrogen loading into the atmosphere (Figure 13.5).

Carbon consumption directly relates to the economic status in urban areas. Wood from urban and rural forests and peat materials are used extensively in some parts of the United States and other areas of the world for fuel for heating and cooking. Studies of carbon consumption have found strong correlations with income classes in India and elsewhere (Parikh, Panda, and Murthy, 1997). Higher-income populations had the highest carbon consumption, using more carbon for transportation and purchasing more carbon intensive resources (e.g., steel, cement, electrical power).

FIGURE 13.3 (See figure, facing page.) Conservation development (a) can protect and restore > 50% of a parcel compared with conventional development (b). After restoration to prairies, wetlands, and forests, the open space typically has reduced soil erosion, increased infiltration, and reduced stormwater runoff of predevelopment row crop agricultural lands. (From Apfelbaum, S.I., The role of landscapes in stormwater management, National Conference on Urban Runoff Management: Enhancing Urban Watershed Management at the Local, County, and State Levels, Chicago, IL, March 1993, US EPA, Washington, D.C., 1993, 165–169; Apfelbaum, S.I. et al., The prairie crossing project: attaining water quality and stormwater management goals in a conservation development using ecological restoration to meet clean water act goals, National Symposium on Using Ecological Restoration to Meet Clean Water Act Goals, Chicago, IL, March 14–16, 1995, US EPA, Washington, D.C., 1995, 33–38; and Apfelbaum, S.I. et al., On conservation developments and their cumulative benefits, in *Proceedings of National Symposium, Assessing the Cumulative Impacts of Watershed Development on Aquatic Ecosystems and Water Quality*, U.S. Environmental Protection Agency (US EPA), Chicago, IL, 1996, 181–188. With permission.)

(a)

(b)

FIGURE 13.5 Vegetative biofilter wetlands (a) provide improved aesthetic and functional stormwater management systems compared with National Urban Runoff Program (NURP) pond systems (b). They also can denitrify excessive nitrogen loads while these materials pass through NURP ponds. (From Apfelbaum, S.I. et al., The prairie crossing project: attaining water quality and stormwater management goals in a conservation development using ecological restoration to meet clean water act goals, National Symposium on Using Ecological Restoration to Meet Clean Water Act Goals, Chicago, IL, March 14–16, 1995, US EPA, Washington, D.C., 1995, 33–38, and Apfelbaum, S.I. et al., On conservation developments and their cumulative benefits, in *Proceedings of National Symposium, Assessing the Cumulative Impacts of Watershed Development on Aquatic Ecosystems and Water Quality*, U.S. Environmental Protection Agency (US EPA), Chicago, IL, 1996, 181–188. With permission.)

The urban rich were found to generate about 15 times more carbon emissions than the rural poor (Parikh, Panda, and Murthy, 1997). This suggests that differences in income can translate to vastly different carbon use and emissions. Other studies have warned of increasing socioeconomic disparities over the next 50–100 years with the support and maintenance of an extended human population of 8–11 billion people (Palmer et al., 2004). These future population and urban expansion projections suggest that without policy, economic, and cultural changes urban areas are likely to continue to contribute increased carbon emissions by the following vectors:

1. Soil disruption
2. Displacement of open lands
3. Increased imperviousness, erosion, deposition
4. Resource use inefficiency
5. Waste stream reuse inefficiency and lack of integration
6. Increasing gasoline/diesel consumption; soot from airplanes and interaction with ozone layer
7. Lack of consistent, coherent, durable policies and programs that integrate solutions to these highly related problems
8. Socioeconomics
9. Soil warming and aerosol nitrogen loading and resulting soil organic matter decomposition
10. Ecosystem disruptions (e.g., soil disruption, increased severe weather effects such as blowdown and erosion)

THE SOLUTIONS

Program Incentives for Urban Carbon Management

1. Encourage conservation developments wherein development activities are massed closer together, which retains open space systems in what would otherwise be developed into a conventional development scenario. Use ecological design principles by managing stormwater in open space systems to eliminate storm sewers. Storm sewer elimination can save substantial money for the developer and long-term maintenance by the responsible public agency (e.g., municipality, county, etc.). Keeping the water within the soil system instead of in subterranean pipes restores local hydrology, feeds depleting groundwater supplies, and maintains soils systems against deterioration typically associated with dewatering soils, particularly those with moderate and higher levels of soil carbon and organic matter.
2. Encourage metropolitan areas of greater than 25,000 persons to perform a natural resource inventory of their metropolitan area and to then incorporate the findings in their master plans. The goal should be to provide each community with this important data and mapping to enable them to leverage the protection of their important natural resources areas so that the soil carbon continues to be effectively managed. Such planning can

also contribute to stormwater management functions, providing habitat, air quality cleansing functions, and others. These natural resource areas could be protected through permit and entitlement negotiations with developers. Once the distribution and importance of the resources are mapped and known, the communities should prepare conservation plans to indicate the protection areas, restoration and land management plans to improve the soil building and soil protection capacity, and to develop strategies for funding the stewardship of these set-aside areas.

3. Encourage metropolitan regional economic incentives to restore the degraded lands and open lands included in developments. If a network of all the protected parcels in each development were viewed and designed regionally to produce stormwater management needs, carbon management could be conducted through ecological restoration and soils management programs in the approximate 50% (or greater) open space typically found in each conservation development. The sheer quantity of retained open lands would have other important regional values including recreation, habitat, and park uses. Such open lands would be low maintenance, thus reducing petrochemical uses and carbon emissions associated with the lawn and landscape management that is typical for conventional urban parks.
4. Encourage the development of a federal initiative that combines the Department of Energy's "EnergyStar" program, which is geared toward increasing the energy efficiency of homes, with a new program that maps natural resources and prepares a conservation plan for each region that would reduce carbon emissions. Such an initiative should be followed by the formulation of needed policies by an economics and policy team to move forward at a regional scale and to encourage regional energy conservation. The Kansas City Metropolitan area is involved in this process currently and it may serve as a useful model for other parts of the United States. In Kansas City, covering parts of two states, and nine counties, natural areas were mapped over the region, to identify restoration needs, important conservation priorities, stormwater buffering needs along drainageways and streams, and numerous other elements (Mid-America Regional Council).
5. Encourage the use of native and low-maintenance landscaping in urban areas that need no watering, herbicides, pesticides, or mowing. Use of these landscaping and planting strategies in some areas of parks, right of ways, vacant lands, some areas of school and municipal building properties (e.g., public works buildings, fire department buildings, and sanitary facility grounds) can serve as regional demonstrations for private landowner understandings and perhaps conversion of some of their yard acreage to similar vegetation.
6. Create opportunities, through national incentives, for reducing carbon emissions (and other air pollutants) from gasoline and diesel fuel consuming private, fleet, and government vehicles. This can be accomplished through incentives for the replacement of older vehicles with hybrid technologies and carbon neutral fuels. Fuel efficiency increases provide

yet another way to reduce per capita and per mile consumption and emissions of carbon.

7. Host forums on sustainability and economic competitiveness of our urban areas. As traffic gridlock increases, more roads and increased capacity on our road systems for more automobiles does not appear to provide a viable solution to address long-term problems. A diverse range of solutions should be contemplated and encouraged through regulatory change, incentives, monetizing the full costs in the short- and long-term to human health, costs to repair and restore rivers and maintain potable water supplies, and the increasing costs in maintaining the highway and roadway infrastructure under the increasing traffic volumes.

CONCLUSIONS

Despite the wonderful strides that have been made to date, there remains a significant need for a coherent, consistent, and integrated series of solutions to help stem the carbon emission from urban areas. The soil carbon management strategies suggested here are a vital, economically beneficial part of a necessary larger policy construct. Coherent policies with economic, education, new technology, and demonstration programs could be the focus in the future.

Urban area carbon management strategies represent a treasure trove where creative ideas and solutions can be used to simultaneously address the constantly changing needs for infrastructure, open space, rethinking, and for recasting the organization, structure, and relationships present. From energy conservation programs to the use of conservation development design principles and land use strategies, as well as alternative fuel transportation systems, significant benefits can be derived for global carbon management in urban areas.

REFERENCES

Akbari, H., Shade trees reduce building energy use and CO_2 emissions from power plants, *Environ. Pollut.*, 116(suppl. 1), S119–S126, 2002.

Apfelbaum, S.I., Sands, M., Price, T., Eppich, J.D., Margolin, P., and Hoffman, D., On conservation developments and their cumulative benefits, in *Proceedings of National Symposium, Assessing the Cumulative Impacts of Watershed Development on Aquatic Ecosystems and Water Quality*, U.S. Environmental Protection Agency (US EPA), Chicago, IL, 1996, 181–188.

Apfelbaum, S.I., Eppich, J.D., Price, T.H., and Sands, M., The prairie crossing project: attaining water quality and stormwater management goals in a conservation development using ecological restoration to meet clean water act goals, National Symposium on Using Ecological Restoration to Meet Clean Water Act Goals, Chicago, IL, March 14–16, 1995, US EPA, Washington, D.C., 1995, 33–38.

Apfelbaum, S.I., The role of landscapes in stormwater management, National Conference on Urban Runoff Management: Enhancing Urban Watershed Management at the Local, County, and State Levels, Chicago, IL, March 1993, US EPA, Washington, D.C., 1993, 165–169.

Applied Ecological Services, Inc. (AES), unpublished data, studies comparing conventional development and conservation development ecological, hydrological, water quality, and biodiversity impacts and economics in Southeast Wisconsin, for ZonaRosa development in Kansas City, MO, and for Wild Meadows in Medina, MN. Funded by Great Lakes Protection Fund Grant, Evanston, IL, ZonaRosa development, and Metropolitan Council, respectively, 1996–present.

Catovsky, S., Bradford, M.A., and Hector, A., Biodiversity and ecosystem productivity: implications for carbon storage, *Oikos*, 97(3), 443–448, 2002.

Dunne, T. and Leopold, L.B., *Water in Environmental Planning,* W.H. Freeman and Company, San Francisco, 1979.

Fleming, P.D. and Webber, P.H., Characterization of diesel particles: effects of fuel reformulation, exhaust after treatment, and engine operation on particle carbon composition and volatility, *Energy Policy*, 32(6), 761–771, 2004.

Freedman, B., Love, S., and O'Neil, B., Tree species composition, structure, and carbon storage in stands of urban forest of varying character in Halifax, Nova Scotia, *Canadian Field-Nat.*, 110(4), 675–682, 1996.

Genxu, W, Haiyan, M, Ju, Q, Juan, C., Impact of land use changes on soil carbon, nitrogen and phosphorus and water pollution in an arid region of northwest China, *Soil Use Manage.*, 20(1), 32–39, 2004.

Gielen, D., CO_2 removal in the iron and steel industry, *Energy Conserv. Manage.*, 44(7), 1027–1037, 2003.

Hope, D., Billett, M.F., Milne, R., and Brown, T.A.W., Exports of organic carbon in British rivers, *Hydrology Process.*, 11(3), 325–344, 1997.

Jo, H.-K. and McPherson, E., Feb 2001. Indirect carbon reduction by residential vegetation and planting strategies in Chicago, USA, *J. Environmental Manage.*, 61(2), 165–177, 2001.

Jo, H.-K. McPherson, E.G., Carbon storage and flux in urban residential greenspace. *J. Environmental Manage.*, 45(2), 109–133, 1995.

Koomey, J.G., Martin, N.C., Brown, M., Price, L.K., and Levine, M.D., Costs of reducing carbon emissions: U.S. building sector scenarios, *Energy Policy*, 26(5), 433–440, 1998.

Lal, R., Kimble, J.M., Follett, R.F., and Cole, C.V., *The Potential of U.S. Cropland to Sequester Carbon and Mitigate the Greenhouse Effect*, Ann Arbor Press, Chelsea, MI, 1999.

McNulty, S.G., Hurricane impacts on U.S. forest carbon sequestration, *Environmental Pollut.*, 116(suppl. 1), S17–S24, 2002.

Melillo, J.M., Steudler, P.A., Aber, J.D., Newkirk, K., Lux, H., Bowles, F.P., Catricala, C., Magill, A., Ahrens, T., and Morrisseau, S., Soil warming and carbon-cycle feedbacks to the climate system science, *Science*, 298(5601), 2173–2176, 2002.

Palmer, M., Bernhardt, E., Chornesky, E., Collins, S., Dobson, A., Duke, C., Gold, B. et al., Ecology for a crowded planet, *Science*, 304(5675), 1251–1252, 2004.

Parikh, J.K., Panda, M.K., and Murthy, N.S., Consumption patterns by income groups and carbon-dioxide implications for India: 1990–2010, *Int. J. Global Energy Issues*, 9(4-6), 237–255, 1997.

Ruth, M., Davidsdottir, B., and Amato, A., Climate change policies and capital vintage effects: the cases of U.S. pulp and paper, iron and steel, and ethylene, *J. Environ. Manage.*, 70(3), 235–252, 2004.

Thom, R.M., Borde, A.B., Williams, G.D., Woodruff, D.L., and Southard, J.A., Climate change and seagrasses: climate-linked dynamics, carbon limitation and carbon sequestration, *Gulf Mexico Sci.*, 21(1), 134, 2003.

Trumbore, S.E., Potential responses of soil organic carbon to global environmental change, *Proc. Natl. Acad. Sci.*, 94(16), 8284–8291, 1997.

U.S. Environmental Protection Agency (US EPA), See Total Maximum Daily Load Program (TMDL) descriptions.

US EPA, See National Pollution Discharge Elimination System (NPDES) Program descriptions.

Waldrop, M.P., Zak, D.R., Sinsabaugh, R.L., Gallo, M., and Lauber, C., Nitrogen deposition modifies soil carbon storage through changes in microbial enzymatic activity, *Ecol. Appl.*, 14(4), 1172–1177, 2004.

Wei, C-F., Larson, S.M., Patten, K.O., and Wuebbles, D.J., Characterization of diesel particles: effects of fuel reformulation, exhaust after treatment, and engine operation on particle carbon composition and volatility: modeling of ozone reactions on aircraft-related soot in the upper troposphere and lower stratosphere, *Atmos. Environ.*, 35(35), 6167–6180, 2001.

Whittington, H.W., Electricity generation: options for reduction in carbon emissions, *Philos. Trans., Ser. A, Math Phys. Eng. Sci.*, 360(1797), 1653–1668, 2002.

BIBLIOGRAPHY

Giese, L.A., Aust, W.M., Kolka, R.K., Trettin, C.C., Biomass and carbon pools of disturbed riparian forests, *For. Ecol. Manage.,* 180(1–3), 493–508, 2003.

Giese, L.A., Aust, W.M., Trettin, CC, and Kolka, R.K., Spatial and temporal patterns of carbon storage and species richness in three South Carolina coastal plain riparian forests, *Ecol. Eng.*, 15(suppl. 1), S157–S170, 2000.

Gitz, V. and Ciais, P., Amplifying effects of land-use change on future atmospheric CO_2 levels, *Global Biogeochemistry Cycles*, 17(1), 2003.

Groffman, P.M., Boulware, N.J., Zipperer, W.C., Pouyat, R.V., Band, L.E., and Colosimo, M.F., Soil nitrogen cycle processes in urban riparian zones, *Environmental Science Technol.*, 36(21), 4547–4552, 2002.

Hardner, J.J., Frumhoff, P.C., and Goetze, D.C., 2000. Prospects for mitigating carbon, conserving biodiversity, and promoting socioeconomic development objectives through the clean development mechanism, *Mitigation Adaptation Strategies Global Change*, 5(1), 61–80, 2000.

Hayhoe, K.A.S., Kheshgi, H.S., Jain, A.K., and Wuebbles, D.J., Trade-offs in fossil fuel use: the effects of CO_2, CH_4 and SO_2 aerosol emissions on climate, *World Resour. Rev.*, 10(3), 321–333, 1998.

Hoffmann, C.C., Rysgaard, S., and Berg, P., Denitrification rates predicted by nitrogen-15 labeled nitrate microcosm studies, in situ measurements, and modeling, *J. Environmental Qual.*, 29(6), 2020–2028, 2000.

Hood, E.W., Williams, M.W., and Caine, N., landscape controls on organic and inorganic nitrogen leaching across an alpine/subalpine ecotone, Green Lakes Valley, Colorado Front Range, *Ecosystems*, 6(1), 31–45, 2003.

Huston, M.A. and Marland, G., Carbon management and biodiversity, 1 CO_2 removal in the iron and steel industry, *J. Environmental Manage.*, 67(1), 77–86, 2003.

Irons, J.G., III and Oswood, M.W., Organic matter dynamics in 3 subarctic streams of interior Alaska, USA, *J. North Am. Benthological Society*, 16(1), 23–28, 1997.

Klara, S.M., Srivastava, R.D., and McIlvried, H.G., Integrated collaborative technology development program for CO_2 sequestration in geologic formations — United States Department of Energy R&D, *Energy Conserv. Manage.*, 44(17), 2699–2712, 2003.

Lovett, G.M.,Weathers, K.C., and Arthur, M.A., Control of nitrogen loss from forested watersheds by soil carbon: nitrogen ratio and tree species composition, *Ecosystems*, 5(7), 712–718, 2002.

Maddison, D., A cost-benefit analysis of slowing climate change, *Energy Policy*, 23(4–5), 337–346, 1995.

Read, P., Food, fuel, fibre and faces to feed: simulation studies of land use change for sustainable development in the 21st century, *Ecological Econ.*, 23(2), 81–93, 1997.

Ross, D.J., Tate, K.R., Scott, N.A., and Feltham, C.W., Land-use change: effects on soil carbon, nitrogen and phosphorus pools and fluxes in three adjacent ecosystems, *Soil Biol. Biochem.*, 31(6), 803–813, 1999.

Smith, P., Powlson, D.S., Smith, J.U., Falloon, P., and Coleman, K., Human population and natural resource management. Meeting Europe's climate change commitments: quantitative estimates of the potential for carbon mitigation by agriculture, *Global Change Biol. J.*, 6(5), 525–539, 2000.

Staver, K.W. and Brinsfield, R.B., Using cereal grain winter cover crops to reduce groundwater nitrate contamination in the mid-Atlantic coastal plain, *J. Soil Water Conserv.*, 53(3), 230–240, 1998.

Taylor, G.E., Jr., Schaller, K.B., Geddes, J.D., Gustin, M.S., Lorson, G.B., and Miller, G.C., Microbial ecology, toxicology and chemical fate of methyl isothiocyanate in riparian soils from the upper Sacramento River, *Environmental Toxicol. Chem.*, 15(10), 1694–1701, 1996.

Whelan, M.J. and Gandolfi, C., Modelling of spatial controls on denitrification at the landscape scale, *Hydrology Process.*, 16(7), 1437–1450, 2002.

14 Prairies, Savannas, and Forests and Global Carbon Management – The Challenges

Steven I. Apfelbaum

CONTENTS

THE CHALLENGES

Extensive loss of native grasslands, wetlands, savannas and forests to agricultural, urban, and other land uses has reduced their productivity resulting in less carbon removal and storage from the atmosphere. The removal of the deep and dense plant root systems and conversion to sparse, short-lived, and shallower rooted agricultural crops has contributed to significant erosion and sedimentation, nutrient loss, flooding, impaired water quality, and increased threats to human health and safety.

Similarly, overgrazing and extensive, intensive logging potentially leaves soils without vegetative cover and reduces their productivity. These factors directly reduce the extraction of carbon from the atmosphere as effectively and efficiently by photosynthesizing vegetation as historically occurred. As the land condition and habitat quality declines, so too does wildlife. Typically, such degradation results in the "weedier" species of plants and animals becoming more prevalent on land and within the nation's waterways. These conditions favor invasive plant and animal species that are tolerant of degradation and are symptomatic of the poor carbon and soil management occurring on the landscapes.

Several existing programs supported by the U.S. Department of Agriculture (USDA) are providing incentives to slow the rate of conversion and in some cases return previously developed areas back to their native condition.

BACKGROUND

The grasslands, collectively referred to as prairies, were one of the largest historic ecological systems in the United States. Spanning from the Front Range of the Rocky Mountains and into Ohio, and well north of the Canadian border and the U.S. Great Plains, and south to Texas and Arizona, these grasslands varied in type, plants, and wildlife species, but one thing that was common was their innate diversity and ability to protect the soil. Even though they may have provided ground cover, productivity in some parts of this ecosystem could be low (i.e., in the shortgrass prairies). However, collectively, these native grasses and wildflowers used huge quantities of carbon from the atmosphere annually and created the deep black topsoil's of the breadbasket states (e.g., Illinois, Iowa, Indiana, Nebraska, Kansas, and parts of the Dakotas) and slightly shallower and brown soils of Michigan, Ohio, parts of Missouri, and central Minnesota. In addition, in droughty lands (e.g., eastern Colorado, Arizona, New Mexico, Idaho, and parts of Utah) grasslands created the important productive soils.

The prairie occupied the rolling exposed uplands and seasonal wetlands, as well as other relatively level lands, for example, in Illinois. Much of the original tallgrass prairie has been converted to row crop agriculture in the eastern areas. And, much of the prairie has been actively drained with ditching and drain tile systems to keep the soil predictably dry enough to support farming activities. This additional drainage, just like the drainage of millions of acres of wetlands, has contributed to problems such as flooding, water quality impairments, degraded water sources, reductions in wildlife and fisheries, and erosion and sedimentation in the nation's waterways, wetlands, lakes, and estuaries.

Over the prairie landscape, protected draws and steep elevated slopes grew many of the same prairie grasses and wildflowers, but also trees, such as bur oak. These stalwart oaks and the "oak savanna" community type remains in grazed landscapes, but is also a prized location for urban development near cities (Haney and Apfelbaum, 1990, 1991) (Figure 14.1).

In the most protected locations, such as along rivers, deeper forests grew. The forests, savannas, and grasslands in combination provided a very effective buffer against direct stormwater water and erosion and sedimentation into the nation's waterways. Forests also grow in upper Midwest, Northeastern, Eastern, and Southern landscapes, as well as in the Western mountain areas of the United States.

Large areas of former floodplain and river bottom forests have been converted to development and deforested for agricultural uses. We are learning about the highly vulnerable conditions of some of our nation's waterways because of the losses of these ecological systems. The Missouri, Mississippi, and many other rivers now harbor declining populations of once common, now decreed endangered fishes and other wildlife. The National Research Council's review and proposals for restoring the Missouri river and its floodplains is a monument to the huge scale of the disruption and imperativeness for promptly making changes to recover and restore carbon management, hydrological functions, and habitat conditions in large acreages of land along our nation's waterways.

Depletion of tens of millions of acres of our nation's floodplain forests and declining access of floodwaters to the historic floodplains, such as along the Missouri

FIGURE 14.1 As healthy Oak Savanna systems degrade, biological diversity collapses, soils systems can erode, and increased stormwater runoff occurs. (From Haney, A. and Apfelbaum, S.I., Structure and dynamics of Midwest oak woods, in *Proceedings of the Oak Woods Management Conference*, Bradley University, Peoria, IL, Eastern Illinois University, Charleston, IL, 1991, 81–89. With permission.)

and Mississippi rivers, has had significant impacts on the nation's carbon management capacity and impacts on the global atmospheric carbon.

Extensive forested and savanna areas bordered the grasslands in all directions. The deciduous forests of the eastern United States occurred in eastern Indiana and Ohio where the prairie transitioned to these forested lands. Many forest types occurred further southeast including the Allegheny, Cumberland Plateau, Smoky Mountains, Coastal Plain, and Piedmont forests that ranged from pine lands through deep deciduous and mixed stands of hardwoods and conifers. Northern hardwood forests bordered the prairies to the north, such as in Minnesota, Wisconsin, and Canada, and this forest type stretched through New England and slightly further north into Canada.

The western forests fingered into the prairie ecosystem along river and protected ridges and draws from the strongholds in the mountains and foothills of the western mountain ranges: the Rockies, the Black Hills, among others. These ranges, and those in California Nevada, Idaho, Washington state, and Oregon, among others, were also heavily forested. Extensive coastal forests bordered the Pacific Ocean producing some of the most extensive and productive coniferous forested systems in the world.

Savannas were typically at the transitions between grassland and forest, sometimes representing a transition zone of sparsely treed lands hundreds of miles wide, whereas in other locations, such as where the prairie met the western foothills of the Rockies, changes were abrupt.

Historically, the transitions between savanna, forest, and prairie were maintained by wildfire, soil and moisture and other meteorological conditions, insect infestations, and grazing animals including bison and elk. The natural patterns and transitions and their influence on soil building and carbon stocks are largely disrupted or now sometimes gone. Agriculture, development, and altered hydrology, altered wildfire regimes and replacement of wild grazing animals with cattle have blurred the past patterns on the land.

INTRODUCTION

Suggestions for better managing grasslands and forested systems for carbon and other outcomes begins with an understanding of the science of carbon sequestration in these systems. The first realization one has is that not all forests, grasslands, and savanna's perform the same. Great variability in these ecological systems is present on earth, and in the United States. The following attempts to provide an overview of the known range of information and variability. For a more comprehensive review and understanding of the derived principles, the reader is encouraged to review the books by Follett, Kimble, and Lal (Follett, Kimble, and Lal, 2001; Kimble, Follett, and Lal, 2002; Kimble et al., 2003; Lal et al., 1995a, 1995b, 1995c; Lal et al., 1997, 1998, 2000a, 2000b, 2000c, 2001).

Managing private and public forests and grasslands can be used to enhance carbon sequestration and soil management and other multiple-environmental services, products and functions through policy refinements, incentives for private landowners, and even environmental service payments to participating landowners who demonstrate ecological performance benefits on their land. Most important, the use of ecological restoration strategies to attain the functions of healthy ecosystems from the often highly modified conditions represents one of the most important future changes to more effectively manage carbon.

Public forests represent around 37% of all forests in the conterminous United States. They represent a substantial area of potential carbon sequestration and storage. Changes in forest management from 1953–2004 have resulted in a projected increase in stored carbon from 16.3–19.5 Gigatonnes. The proportion of stored carbon in public forestlands compared with all other forests increased from 35–37%. Depending on the future scenario, the national forests can greatly influence U.S. carbon sequestration (Smith and Heath, 2004). National grasslands and restored/reforested prairies planted in private lands that are included in Conservation Reserve Programs and Wildlife Habitat Incentives Programs, among others, also have the potential to provide substantial carbon sequestration and storage in the future (Lal et al., 1999).

Plantations of highly productive trees, such as poplar, have been studied to determine if they could be used as a biomass crop and to increase the carbon sequestration on the land (Coleman et al., 2004). This investigation found plantations of poplar and planted switchgrass fields to have more predictable and similarly elevated levels of soil carbon compared with adjacent agricultural soils, which had more variable levels of soil carbon. They also noted other benefits such as reduced soil erosion, runoff abatement, and wildlife habitat benefits from growing highly productive forest and prairie grass plantations.

Changing forest rotations by lengthening the time between harvests offers significant opportunity for carbon sequestration and storage (Kaipainen et al., 2004). Their study found a shift to increased carbon storage could result in the standing forest depending on the forest type but that some of the carbon storage gains may be offset by a concurrent slight reduction in soil carbon levels. Other studies have examined reduced logging impact as a strategy to increase carbon retention in forested landscapes and the costs of reduced impact logging strategies. Healey, Price, and Tay (2000) found that with a 44% reduction in the logged area, there was a 22% reduction in timber yield per logged hectare and an increase of 18% carbon retention when comparing conventional logging to reduced impact logging. Reduced impact logging was found to damage wildlife, soils, and water quality less than conventional logging. Under various discounting scenarios, the reduced impact logging appeared to be the most cost effective as a carbon management strategy when the carbon storage could be sold to offset timber yields.

Numerous studies have examined reforesting lands that were previously forest but that had been converted to row-crop agricultural or pastured uses. These investigations generally found that for up to 10 years the reforested lands had similar soil carbon levels to pastures of the same age and to standing crops with similar carbon levels. After reforested lands achieved 20–30 years of age, there was generally an increase in soil carbon stocks. This study also identified the importance and influence of site selection on performance. Poor soil areas neither grew good pasture nor good reforested stands and consequently, carbon stocks remained lower and more variable than on more favorable soil areas (de Koning, Veldkamp, and Lopez-Ulloa, 2003).

Land use changes are viewed as important determinants for the ultimate carbon management benefits that may accrue from reforestation, forest management, grassland management, management of peat lands, or other ecological systems (Gitz and Ciasis, 2003; Giese et al., 2003; Hood et al., 2003; Groffman et al., 2002; Lovett, Weathers, and Arthur, 2002; Ross et al., 1999). These authors suggest that changes from indirect land-use changes (e.g., aerosol nutrient deposition) and direct impacts (e.g., silviculture) must be understood to make accurate predictions of carbon management benefits in any ecological setting.

In Texas, dry grasslands were heavily grazed during the 1800s through the mid-1900s, then many were less intensively grazed because depletion of soil carbon and increased erosion left the land at lower levels of grazing productivity. Invading woody vegetation (e.g., mesquite among others) is now predominant on these historic grasslands. Several studies have examined soil carbon and standing carbon stocks through this progression of change and made projections of the future carbon sequestration and stocking projections. These studies have suggested that 3× to as high as 24× higher carbon stocking rates will be associated with the conversion from historic grasslands through this savanna to forest within a time period of around 100 years (Hibbard et al., 2003). Similar patterns of woody vegetation invasion into former savannas and prairies have documented canopy closure and shade suppression of understory vegetation and subsequent erosion of topsoil in Midwestern U.S. savannas (Haney and Apfelbaum, 1990, 1991).

Agricultural policy and program compliance and the future potential use of these programs for encouraging carbon management in the United States has been examined by Claassen et al. (2004). This study found that approximately 25% of the reduction in soil erosion from agricultural lands, including pastured grasslands, highly erodible croplands, and conserved wetlands, has resulted from compliance with agricultural programs. They also suggested that compliance incentives have likely deterred conversion of noncropped, highly erodible lands and wetlands to croplands and that a compliance approach could be used effectively to address nutrient and carbon runoff from crop production lands.

Many programs of carbon management have focused on reforestation of lands. Studies of the longevity of carbon stocks in forested landscapes have documented that age-dependent relationships exist where forests between 150–200 years of age have achieved the highest levels of carbon stocks and storage following which time carbon stocks become more dynamic, meaning more variable (Sun et al., 2004).

Threats to forests resources (and thus potentially to aboveground carbon stocks), particularly fire, blowdown, and insect infestations have been extensively studied (Amiro et al., 2001; Sohngen and Haynes, 1997; McNulty, 2002; Read, 1997; Trumbore, 1997). The anticipated increase in damages from meteorological changes, such as increased hurricane damage to forests, has been viewed to have a strong influence over presumed carbon stocks. Increased forest fires associated with drought expectations represent a significant impact on standing crop carbon stores, particularly in forests and less so in prairies, wetlands, and savannas where carbon storage is primarily belowground in the soil systems. Studies suggest these impacts need to be factored into carbon management plans globally. Because of the benefits of belowground storage in grasslands and wetlands, mixing reforestation with the more protected carbon stocks produced belowground needs further scientific understanding.

THE SOLUTIONS

1. Encourage short rotational grazing on national forest and grassland grazing allotments which encourages higher rates of carbon removal and efficiency, reduces the soil damages from intensive overgrazing, and reduces soil erosion, the rate and total amount of water running off the land and degrading our nation's waterways.
2. Continue the Conservation Reserve Program (CRP) and Grassland Reserve Program by expanding the acreages in these programs to even include more productive as well as marginally productive agricultural lands, and by extending benefits to indefinitely maintain the federal investment that resulted in setting aside approximately 45 million acres of CRP lands. Explore a new federal perpetual conservation easement tax benefit program for landowners willing to ensure CRP lands will not be plowed and farmed again.
3. Encourage the restoration of native grasslands, savannas, and forest by removal or disabling of the drainage ditching and tiles that currently dewater millions of acres of historic prairie, savanna, and forested lands. Expand the Wetland Reserve Program (WRP) to include these transitional

ecological systems that historically supported the functions millions of healthy acres of wetlands.

4. Require and encourage former deforested and drained wetlands in floodplain environments, which are now converted to agricultural cropping uses, to convert to different soil management strategies (e.g., use of no-till soil preparation and management) or different cropping strategies (e.g., reforest and grow forest crops, game and nongame wildlife as a crop and source of income), and financially reward farmers to produce atmospheric carbon removal benefits.
5. Encourage sound forestry practices that do not degrade soils and their productivity. Use reforestation practices that can heal harvested lands quickly so that productive atmospheric carbon removal vegetation systems are restored.

SUMMARY/CONCLUSIONS

The most extensive ecosystems in the conterminous United States, the grasslands, forests, savannas, and freshwater wetland systems have been highly modified and their functions diminished. Grasslands and wetlands can contribute some of the highest carbon stocks present in the temperate zone ecosystems in the United States. The lands formerly occupied by enormous areas of these ecosystems for over 100 years have been converted to the economic engine of the U.S. agro-industrial operations. Decisive policy changes and economic incentives along with integrating ecosystem restoration strategies into agricultural operations and in marginal lands can be used to again rebuild soils, reestablish carbon stocks, and create environmental service benefits such as reduced flooding, improved groundwater supplies through restoring recharge, and a range of biodiversity and water quality benefits. Because of the extensiveness of these ecosystems, significant carbon benefits can be derived from policy and restoration programs that can operate at large scales.

REFERENCES

Amiro, B.D., Stocks, B.J., Alexander, M.E., Flannigan, M.D., and Wotton, B.M., Fire, climate change, carbon and fuel management in the Canadian boreal forest, *Int. J. Wildland Fire*, 10(3), 405–413, 2001.

Claassen, R., Breneman, V., Bucholtz, S., Cattaneo, A., Johansson, R., and Morehart, M., Environmental compliance in U.S. agricultural policy: Past performance and future potential, *Agric. Econ. Rep.*, 832, 52 pp., 2004.

Coleman, M.D., Isebrands, J., Tolsted, D.N., and Tolbert, V.R., Comparing soil carbon of short rotation poplar plantations with agricultural crops and woodlots in North Central United States, *Environ. Manage.*, 33(S1), S299–S308, 2004.

de Koning, G.H.J., Veldkamp, E., and Lopez-Ulloa, M., Quantification of carbon sequestration in soils following pasture to forest conversion in northwestern Ecuador, *Global Biogeochem. Cycles*, 17(4), 2003.

Follett, R.F., Kimble, J.M., and Lal, R., Eds., *The Potential of U.S. Grazing Lands to Sequester Carbon and Mitigate the Greenhouse Effect*, Lewis Publishers, Boca Raton, FL, 2001.

Giese, L.A., Aust, W.M., Kolka, R.K., Trettin, C.C., Biomass and carbon pools of disturbed riparian forests, *For. Ecol. Manage.,* 180(1–3), 493–508, 2003.

Gitz, V. and Ciais, P., Amplifying effects of land-use change on future atmospheric CO_2 levels, *Global Biogeochemistry Cycles*, 17(1), 2003.

Groffman, P.M., Boulware, N.J., Zipperer, W.C., Pouyat, R.V., Band, L.E., and Colosimo, M.F., Soil nitrogen cycle processes in urban riparian zones, *Environmental Science Technol.*, 36(21), 4547–4552, 2002.

Haney, A. and Apfelbaum, S.I., Structure and dynamics of Midwest oak woods, in *Proceedings of the Oak Woods Management Conference*, Bradley University, Peoria, IL, Eastern Illinois University, Charleston, IL, 1991, 81–89.

Haney, A. and Apfelbaum, S.I., Structure and dynamics of Midwest oak savannas, in *Management of Dynamic Ecosystems*, Sweeny, J., Ed., North Central Section of the Wildlife Society, West LaFayette, IN, 1990, 19–30.

Healey, J.R., Price, C., and Tay, J., The cost of carbon retention by reduced impact logging, *For. Ecol. Manage.*, 139(1–3), 237–255, 2000.

Hibbard, K.A., Schimel, D.S., Archer, S., Ojima, D.S., and Parton, W., Grassland to woodland transitions: Integrating changes in landscape structure and biogeochemistry, *Ecol. Appl.,* 13(4), 911–926, 2003.

Hood, E.W., Williams, M.W., and Caine, N., landscape controls on organic and inorganic nitrogen leaching across an alpine/subalpine ecotone, Green Lakes Valley, Colorado Front Range, *Ecosystems*, 6(1), 31–45, 2003.

Kaipainen, T., Liski, J., Pussinen, A., and Karjalainen, T., Managing carbon sinks by changing rotation length in European forests, *Environ. Sci. Policy*, 7(3), 205–219, 2004.

Kimble, J., Follett, R., and Lal, R., Eds., *Agricultural Practices and Polices for Carbon Sequestration in Soil*, Lewis Publishers, Boca Raton, FL, 2002.

Kimble, J.M., Heath, L., Birdsey, R., and Lal, R., Eds., *The Potential of U.S. Forest Lands to Sequester Carbon and Mitigate the Greenhouse Effect*, Lewis Publishers, Boca Raton, FL, 2003.

Lal, R., Kimble, J.M., Follett, R.F., and Cole, C.V., *The Potential of U.S. Cropland to Sequester Carbon and Mitigate the Greenhouse Effect*, Ann Arbor Press, Chelsea, MI, 1999.

Lal, R., Kimble, J., and Levine, E., Eds., *Soil Processes and Greenhouse Gas Emissions*, U.S. Department of Agriculture, Soil Conservation Service, National Soil Science Center (USDA-SCS-NSSC), Lincoln, NE, 1995a.

Lal, R., Kimble, J., Levine, E. and Stewart, B.A., Eds., *Soil Management and Greenhouse Effect, Advances in Soil Science*, CRC Press, Inc., Boca Raton, FL, 1995b.

Lal, R., Kimble, J., Levine, E., and Stewart, B.A., Eds., *Soil and Global Change,* Advances in Soil Science, CRC Press, Inc., Boca Raton, FL, 1995c.

Lal, R., Kimble, J., and Follett, R., Eds., *Soil Properties and Their Management for Carbon Sequestration,* USDA-NRCS-NSSC, Lincoln, NE, 1997.

Lal, R., Kimble, J.M., Eswaran, H., and Stewart, B.A., Eds., *Global Climate Change and Pedogenic Carbonates*, CRC Press, Inc., Boca Raton, FL, 2000a.

Lal, R., Kimble, J.M., and Stewart, B.A., Eds., *Global Climate Change and Tropical Ecosystems,* CRC Press, Inc., Boca Raton, FL, 2000b.

Lal, R., Kimble, J.M., and Stewart, B.A., Eds., *Global Climate Change and Cold Regions Ecosystems,* Lewis Publishers, Boca Raton, FL, 2000c.

Lovett, G.M., Weathers, K.C., and Arthur, M.A., Control of nitrogen loss from forested watersheds by soil carbon: Nitrogen ratio and tree species composition, *Ecosystems*, 5(7), 712–718, 2002.

McNulty, S.G., Hurricane impacts on U.S. forest carbon sequestration, *Environ. Pollut.*, 116(suppl. 1), S17–S24, 2002.
Read, P., Food, fuel, fibre and faces to feed: simulation studies of land use change for sustainable development in the 21st century, *Ecological Econ.*, 23(2), 81–93, 1997.
Ross, D.J., Tate, K.R., Scott, N.A., and Feltham, C.W., Land-use change: effects on soil carbon, nitrogen and phosphorus pools and fluxes in three adjacent ecosystems, *Soil Biol. Biochem.*, 31(6), 803–813, 1999.
Smith, J.E. and Heath, L.S., Carbon stocks and projections on public forestlands in the United States, 1952–2040, *Environ. Manage.*, 33(4), 433–442, 2004.
Sohngen, B.L. and Haynes, R.W., The potential for increasing carbon storage in United States unreserved timberlands by reducing forest fire frequency: An economic and ecological analysis, *Climatic Change*, 35(2), 179–197, 1997.
Sun, O.J., Campbell, J., Law, B.E., and Wolf, V., Dynamics of carbon stocks in soils and detritus across chronosequences of different forest types in the Pacific Northwest, USA, *Global Change Biol.*, 10(9), 1470–1481, 2004.
Trumbore, S.E., Potential responses of soil organic carbon to global environmental change, *Proc. Natl. Acad. Sci.*, 94(16), 8284–8291, 1997.

BIBLIOGRAPHY

Asner, G.P., Archer, S. Hughes, R.F., Ansley, R.J., and Wessman, C.A., Net changes in regional woody vegetation cover and carbon storage in Texas Drylands, 1937–1999, *Global Change Biol.*, 9(3), 316–335, 2003.
Burkart, M.R. and James, D.E., Agricultural-nitrogen contributions to hypoxia in the Gulf of Mexico, *J. Environ. Qual.*, 28(3), 850–859, 1999.
Burkart, M.R., Simpkins, W.W., Squillace, P.J., and Helmke, M., Tributary stream infiltration as a source of herbicides in an alluvial aquifer, *J. Environ. Qual.*, 28(1), 69–74, 1999.
Campbell, B.D., Smith, D.M.S., and McKeon, G.M., Elevated CO_2 and water supply interactions in grasslands: A pastures and rangelands management perspective, *Global Change Biol.*, 3(3), 177–187, 1997.
Casella, E. and Soussana, J.-F., Long-term effects of CO_2 enrichment and temperature increase on the carbon balance of a temperate grass sward, *J. Exp. Bot.*, 48(311), 1309–1321, 1997.
Demissie, M. and Khan, A., Influence of wetlands on streamflow in Illinois, National Conference on Hydraulic Engineering, Nashville, TN, 949–954, 1991
Englin, J. and Callaway, J.M., Environmental impacts of sequestering carbon through forestation, *Climatic Change*, 31(1), 67–78, 1995.
Fearnside, P.M., Global warming and tropical land-use change: Greenhouse gas emissions from biomass burning, decomposition and soils in forest conversion, shifting cultivation and secondary vegetation, *Climatic Change*, 46(1-2), 115–158, 2000.
Grace, J., Presidential Address: Understanding and managing the global carbon cycle, *J. Ecol.*, 92(2), 189–202, 2004.
Gruenzweig, J.M., Sparrow, S.D., Yakir, D., and Chapin, S.F., Impact of agricultural land-use change on carbon storage in boreal Alaska, *Global Change Biol.*, 10(4), 452–472, 2004.
Hall, G.M.J., Mitigating an organization's future net carbon emissions by native forest restoration, *Ecol. Appl.*, 11(6), 1622–1633, 2001.

Hardner, J.J., Frumhoff, P.C., and Goetze, D.C., Prospects for mitigating carbon, conserving biodiversity, and promoting socioeconomic development objectives through the clean development mechanism, *Mitigation Adaptation Strategies Global Change*, 5(1), 61–80, 2000.

Hatfield, J.L., Jaynes, D.B., Burkart, M.R., Cambardella, C.A., Moorman, T.B., Prueger, J.H., and Smith, M.A., Water quality in Walnut Creek Watershed: Setting and farming practices, *J. Environ. Qual.,* 28(1), 11–24, 1999.

Hope, D., Billett, M.F., Milne, R., and Brown, T.A.W., Exports of organic carbon in British rivers, *Hydrol. Proc.*, 11(3), 325–344, 1997.

Huang, Ching-Hsun, and Kronrad, G.D., The cost of sequestering carbon in forests of the southern United States, *World Resour. Rev.*, 14(3), 385–397, 2002.

Hungate, B.A., Holland, E.A., Jackson, R.B., Chapin, F.S. III, Mooney, H.A., and Field, C.B., The fate of carbon in grasslands under carbon dioxide enrichment, *Nature*, 388(6642), 576–579, 1997.

Jackson, R.B., Canadell, J., Ehleringer, J.R., Mooney, H.A., Sala, O.E., and Schulze, E.D., A global analysis of root distributions for terrestrial biomes, *Oecologia*, 108(3), 389–411, 1996.

Janssens, I.A., Freibauer, A., Ciais, P., Smith, P., Nabuurs, G.J., Folberth, G., Schlamadinger, B., Hutjes, R.W.A., Ceulemans, R., Schulze, E.D., Valentini, R., and Dolman, A.J., Europe's terrestrial biosphere absorbs 7 to 12% of European anthropogenic CO_2 emissions, *Science*, 300(5625), 1538–1542, 2003.

Jaynes, D.B., Hatfield, J.L., and Meek, D.W., Water quality in Walnut Creek watershed: Herbicides and nitrate in surface waters, *J. Environ. Qual.*, 28(1), 45–59, 1999.

Joyce, L.A., Mills, J.R., Heath, L.S., McGuire, A.D., Haynes, R.W., and Birdsey, R.A., Forest sector impacts from changes in forest productivity undermate change, *J. Biogeography,* 22(4-5), 703–713, 1995.

Ju, W. and Chen, J.M., Distribution of soil carbon stocks in Canada's forests and wetlands: Simulated based on drainage class, topography and remotely sensed vegetation parameters, *Hydro. Proc.*, 19(1), 77–94, 2005.

Keane, R.E., Hardy, C.C., Ryan, K.C., and Finney, M.A., Simulating effects of fire on gaseous emissions and atmospheric carbon fluxes from coniferous forest landscapes, *World Resource Rev.*, 9(2), 177–205, 1997.

Lal, R., Kimble, J., Follett, R., and Stewart, B.A., Eds., *Soil Processes and C Cycles, Advances in Soil Science*, CRC Press, Inc., Boca Raton, FL, 1998.

Lal, R. Kimble, J., Follett, R., and Stewart, B.A., Eds., *Management of Carbon Sequestration, Advances in Soil Science*, CRC Press, Inc., Boca Raton, FL, 1998.

Lal, R., Kimble, J.M., Follett, R.F., and Stewart, B.A., Eds., *Assessment Methods for Soil Carbon*, CRC Press, Boca Raton, FL, 2001.

Law, B.E., Sun, O.J., Campbell, J, Van Tuyl, S., and Thornton, P.E. Apr. 2003. Changes in carbon storage and fluxes in a chronosequence of Ponderosa Pine, *Global Change Biol.*, 9(4), 510–524, 2003.

Liski, J., Pussinen, A., Pingoud, K., Maekipaeae, R., and Karjalainen, T., Which rotation length is favourable to carbon sequestration? *Can. J. For. Res./Rev. Can. Rech. For.,* 31(11), 2004–2013, 2001.

Maddison, D., A cost-benefit analysis of slowing climate change, *Energy Policy*, 23(4–5), 337–346, 1995.

McKane, R.B., Rastetter, E.B., Shaver, G.R., Nadelhoffer, K.J., Giblin, A.E., Laundre, J.A., and Chapin, F.S. III, Reconstruction and analysis of historical changes in carbon storage in arctic tundra, *Ecology*, 78(4), 1188–1198, 1997.

McKane, R.B., Rastetter, E.B., Shaver, G.R., Nadelhoffer, K.J., Giblin, A.E., Laundre, J.A., and Chapin, F.S., III, Climatic effects on tundra carbon storage inferred from experimental data and a model, *Ecology*, 78(4), 1170–1187, 1997.

Melillo, J.M., Steudler, P.A., Aber, J.D., Newkirk, K., Lux, H., Bowles, F.P., Catricala, C., Magill, A., Ahrens, T., and Morrisseau, S., Soil warming and carbon-cycle feedbacks to the climate system, *Science*, 298(5601), 2173–2176, 2002.

Mortenson, M.C., Schuman, G.E., and Ingram, L.J., Carbon sequestration in rangelands interseeded with yellow-flowering alfalfa (*Medicago sativa ssp. falcata*), *Environ. Manage.*, 33(S1), S475–S481, 2004.

National Research Council (NRC), The Missouri River Ecosystem: Exploring the Prospects for Recovery, Committee on Missouri River Ecosystem Science, 188 pp., 2002.

NRC, Review of the U.S. Army Corps of Engineers Restructured Upper Mississippi River-Illinois Waterway Feasibility Study, Committee to Review the Corps of Engineers Restructured Upper Mississippi River-Illinois Waterway Feasibility Study, NRC, 66 pp., 2004.

Ogle, S.M., Conant, R.T., and Paustian, K., Deriving grassland management factors for a carbon accounting method developed by the Intergovernmental Panel on Climate Change, *Environ. Manage.* 33(4), 474–484, 2004.

Schimel, D.S., et al., Continental scale variability in ecosystem processes: Models, data, and the role of disturbance, *Ecol. Monogr.*, 67(2), 251–271, 1997.

Schuman, G.E., Janzen, H.H., and Herrick, J.E., Soil carbon dynamics and potential carbon sequestration by rangelands, *Environ. Pollut.*, 116(3), 391–396, 2002.

Scott, N.A., Rodrigues, C.A., Hughes, H., Lee, J.T., Davidson, E.A., Dail, D.B., Malerba, P., and Hollinger, D.Y., Jul 2004. Changes in carbon storage and net carbon exchange one year after an initial shelterwood harvest at Howland Forest, ME, *Environ. Manage.*, 33(S1), S9–S22, 2004.

Silver, L.W., Ostertag, R., and Lugo, E.A., The potential for carbon sequestration through reforestation of abandoned tropical agricultural and pasture lands, *Restoration Ecol.*, 8(4), 394–407, 2000.

Tate, K.R. and Ross, D.J., Elevated CO_2 and moisture effects on soil carbon storage and cycling in temperate grasslands, *Global Change Biol.*, 3(3), 225–235, 1997.

Tomer, M.D. and Burkart, M.R., Long-term effects of nitrogen fertilizer use on ground water nitrate, *J. Environ. Qual.*, 32(6), 2158–2171, 2003.

Zelek, C.A. and Shively, G.E., Measuring the opportunity cost of carbon sequestration in tropical agriculture, *Land Econ.*, 79(3), 342–354, 2003.

Index

D

E

F

G

H

I

J

K

L

M

N

T

U

For Product Safety Concerns and Information please contact our EU
representative GPSR@taylorandfrancis.com
Taylor & Francis Verlag GmbH, Kaufingerstraße 24, 80331 München, Germany

www.ingramcontent.com/pod-product-compliance
Ingram Content Group UK Ltd.
Pitfield, Milton Keynes, MK11 3LW, UK
UKHW020013200926
475702UK00014B/1867
* 9 7 8 0 3 6 7 3 8 9 1 0 9 *